international® AIR POWER REVIEW

AIRtime Publishing
United States of America • United Kingdom

international®
AIR POWER
REVIEW

Published quarterly by AIRtime Publishing Inc.
120 East Avenue, Norwalk, CT 06851
Tel (203) 838-7979 • Fax (203) 838-7344

Created for AIRtime Publishing by Aeromedia Communications Ltd

Softbound Edition ISSN 1473-9917 / ISBN 1-880588-84-6
Hardcover Deluxe Casebound Edition ISBN 1-880588-80-3

Origination by Universal Graphics, Singapore
Printed in Singapore by KHL Printing

International Air Power Review is published quarterly in two editions (Softbound and Deluxe Casebound) and is available by subscription or as single volumes.
Please see details opposite.

Acknowledgments
We wish to thank the following for their kind help with the preparation of this issue:

Eric Heyes and Mike Moore at Lockheed Martin, Craig Luther (AFFTC/HO), Dave Menard and Terry Panopalis for their help with the B-36 article

'Allt om hobby' – publishers of the book *Flygande Tunnan*, Freddy Stenbom, Aeroproduction and Svensk Flyghistorisk Förening (Swedish Aviation Historical Society) for their assistance in compiling the SAAB J 29 article

The editors welcome photographs for possible publication but can accept no responsibility for loss or damage to unsolicited material.

Contact and Ordering Information (hours: 9am-5pm EST, Mon-Fri)
addresses, telephone and fax numbers

International Air Power Review, P.O. Box 5074, Westport, CT 06881, USA
Tel (203) 838-7979 • Fax (203) 838-7344
Toll free within USA and Canada: 1 800 359-3003
Toll free from Australia (13-15 hours ahead) : 0011 800 7573-7573
Toll free from New Zealand (17 hours ahead): 00 800 7573-7573
Toll free from Japan (14 hours ahead): 001 800 7573-7573

International Air Power Review, Postbus 3946, 4800 DX Breda, Holland
Toll free (to our US East Coast office) from the United Kingdom, Belgium, Denmark, France, Germany, Holland, Ireland, Italy, Luxembourg, Norway, Portugal, Sweden and Switzerland (5-6 hours ahead): 00 800 7573-7573
Toll free from Finland (6 hours ahead): 990 800 7573-7573

website
www.airtimepublishing.com

e-mails
airpower@airtimepublishing.com
inquiries@airtimepublishing.com

Subscription & Back Volume Rates

One-year subscription (4 quarterly volumes), inclusive of ship. & hdlg./post. & pack.:

Softbound Edition
USA $59.95, UK £48, Europe EUR 88, Canada Cdn $99, Rest of World US $79 (surface) or US $99 (air)

Deluxe Casebound Edition
USA $79.95, UK £68, Europe EUR 120, Canada Cdn $132, Rest of World US $99 (surface) or US $119 (air)

Two-year subscription (8 quarterly volumes), inclusive of ship. & hdlg./post. & pack.:

Softbound Edition
USA $112, UK £92, Europe EUR 169, Canada Cdn $187, Rest of World US $148 (surface) or US $188 (air)

Deluxe Casebound Edition
USA $149, UK £130, Europe EUR 232, Canada Cdn $246, Rest of World US $187 (surface) or US $227 (air)

Single/back volumes by mail (each):

Softbound Edition
US $16, UK £10.95, Europe EUR 18.50, Cdn $25.50 (plus ship. & hdlg./post. & pack.)

Deluxe Casebound Edition
US $20, UK £13.50, Europe EUR 22, Cdn $31 (plus ship. & hdlg./post. & pack.)

Prices are subject to change without notice. Canadian residents please add GST. Connecticut residents please add sales tax.

Ship. & Hdlg./Post. & Pack. Rates
(for back volume and non-subscription orders)

	USA	*UK*	*Europe*	*Canada*	*ROW (surface)*	*ROW (air)*
1 item	$4.50	£4	EUR 8	Cdn $7.50	US $8	US $16
2 items	$6.50	£6	EUR 11.50	Cdn $11	US $12	US $27
3 items	$8.50	£8	EUR 14.50	Cdn $14	US $16	US $36
4 items	$10	£10	EUR 17.50	Cdn $16.50	US $19	US $46
5 items	$11.50	£12	EUR 20.50	Cdn $19	US $23	US $52
6 or more	$13	£13	EUR 23.50	Cdn $21.50	US $25	US $59

Volume Thirteen
Summer 2004

CONTENTS

Major Features planned for Volume Fourteen

Focus Aircraft: Saab 37 Viggen, **Warplane Classic:** Lockheed P-38 Lightning, **Variant Briefing:** Mikoyan MiG-23/27 Part 1, **Air Power Analysis:** Russia Part 3, **Type Analysis:** Gloster Javelin, **Air Combat:** Football War, **Technical Briefing:** *Charles de Gaulle*, **Photo-feature:** Wessex in Uruguay

PROGRAMME UPDATE

Eurofighter Typhoon

During the first half of 2004 the four Eurofighter partners made significant progress towards service entry when the delivery of production two-seaters allowed training to get under way. On the development front, the most significant event was the first flight of a production single-seater, in the shape of PS001/IPA4, which flew from EADS CASA's Getafe plant on 27 February. It was subsequently painted in full Ejército del Aire colours with serial C.16-20 and Ala 11 wing codes 11-91. As an Instrumented Production Aircraft it will be retained by EADS CASA to complete the trials work that had been undertaken by the first prototype, DA1, which returned to Manching in Germany. DA1 had been at Getafe to cover the shortfall caused by the loss of DA6 in November 2002.

Germany

On 30 April 2004 operational training for the Luftwaffe got under way following the delivery of eight two-seat Typhoons from EADS Manching to Jagdgeschwader 73 'Steinhoff' at Laage in northern Germany. Formerly equipped with MiG-29s – since transferred to Poland – JG 73 will continue as the Luftwaffe's main Typhoon training centre, although it will also have operational commitments. The wing expects to receive its first single-seat aircraft in the autumn following the completion of military certification. Four pilots are assigned to operational evaluation tasks while an initial batch of 10 instructors is being trained.

Germany's Tranche 1 order comprises 16 two-seaters and 28 single-seaters, from the full requirement for 147 single-seaters and 33 two-seaters. In late June funding release for 58 single-seat and 10 two-seat Eurofighters planned for German Tranche 2 procurement was approved by the parliament, allowing Germany to be the first of the four partners to be in a position to sign a contract. Final approval, however, will not be forthcoming

Wearing full EdA colours, this is IPA4, the first production single-seater. It is being used by EADS CASA for ongoing development tests at Getafe.

On display at ILA 2004 was this JG 73 aircraft (GT007), showing the colour scheme adopted for service Luftwaffe aircraft.

PROJECT DEVELOPMENT

Australia

First AAAC Tiger flies

A significant milestone in Australian Army Aviation Corp's procurement of 22 Tiger armed reconnaissance helicopters (ARHs) was achieved on 20 February with the maiden flight by test pilot Jacques Larra of the first production ARH at Eurocopter's Marignane plant. As scheduled, this aircraft started the qualification flight test programme to validate delivery of the first two AAAC ARHs by late 2004. ARH flight-test development started in July 2003, with a modified Franco-German Tiger prototype.

Apart from ARH development, two major Eurocopter contracts in December 2001 provided for licensed assembly by Australian Aerospace of 18 Tigers, following delivery of the first four from Marignane. Two AAAC Tigers are already being produced in Marignane, and three are on the Australian Aerospace assembly line in Brisbane, Queensland. Costing $1.3 billion in all, the RAAF contract also includes 15-year fleet management, through-life maintenance and logistic support from service-entry in 2005.

ARH 1 and 2 will be handed over to the Army on schedule on 15 December 2004, for initial training at Oakey Army Base. The first Australian-assembled aircraft, ARH 5, will begin flight tests in November for scheduled AAAC delivery in mid-2005. Tiger preliminary military certification is planned by March 2005, followed by final clearance a year later. The first AAAC Tiger squadron will become operational with 17 ARHs in the 1st Aviation Regiment at Darwin (North Australia), in June 2007, followed by delivery completion in April 2008.

Chile

Pillán revival planned

Following sales of 130 T-35 Pillán piston-engined basic trainers, including the last four to the Ecuadorean navy in April 2002, ENAER is renewing efforts to resume production, and market the T-35DT Turbo Pillán, for new border surveillance and drug interdiction roles. With a 313.3-kW (420-shp) Rolls-Royce Allison 250-B17D turboprop, the T-35DT has long completed its basic flight development programme. It underwent 90 hours of test-flying from March 2003, however, with a FLIR Systems Ultra 7500 sensor turret beneath the port wing, on one of four stores pylons, and associated Broadcast Microwave Services datalink and cockpit displays.

France

UK carrier co-operation agreed

Long-awaited selection by the Délégation Générale pour l'Armement (DGA) of conventional rather than nuclear propulsion for the French navy's second aircraft-carrier, planned to supplement the nuclear-powered *Charles de Gaulle*, cleared the way for agreement on 8 June for France to co-operate with the UK in the RN's CVF programme. Definition tenders for the new French carrier may be issued this year, for possible contract award in 2005, and construction start in 2007.

As the main partner, with BAE Systems, and design originator in the CVF programme, Thales played a major part in negotiating possible joint development for the French requirement, in which the national DCN shipyard will have a 65 percent share. Thales chairman Denis Ranque claimed that a CVF-class vessel of 50,000-60,000 tonnes, could be built in France for less than Euro2 billion ($2.44 billion), while reducing overall programme costs for the three carriers to below Euro6 billion ($7.32 billion). The RN's two CVFs were originally costed at about £2.7 billion ($4.95 billion) overall, but overruns of up to 33 percent are forecast.

In the absence of STOVL combat aircraft, like the UK's planned Lockheed Martin F-35Bs, a French CVF would require powerful 90-m (295-ft) catapults to launch the Aéronavale's proposed 60 Dassault Rafale M and two-seat N multi-role fighters. The Aéronavale expects to receive its first two-seat Rafale N in 2007.

Pakistan

Sino-Pakistan fighter progress

Powered by a Klimov RD-93 turbofan, and with fly-by-wire controls, the prototype FC-1/JF-17 began flight development on 25 August 2003, followed by a second prototype on 9 April 2004, both from China's Chengdu Wenjiang facility. PAF JF-17 Project Director Air Vice-Marshal Shahid Lateef said that five prototypes were being built, including two for static testing. An initial batch of eight JF-17s for each country, of which Chengdu will assemble 12, would be produced by late 2005, for mid-2006 service entry. Full series production will start a year later in both countries, in which Pakistan's Aeronautical Complex will have a 59 percent share. Pakistan's investment includes full design and development partnership, exemplified in April by the first JF-17 test-flights from 7 April by two PAF pilots, who will continue regular participation in the overall programme.

The PAF has specified an advanced European digital avionics suite, which may include the licence-built Galileo Grifo S7 fire-control radar installed in the prototypes, although AVM Lateef said that the West was reluctant to offer help in this respect. An avionics package had nevertheless been selected, for imminent contract finalisation. "JF-17 weapons will include such short- and medium-range AAMs as the AIM-9P, PL-9, or Magic 2 and the PL-11, Aspide, or AIM-7E", he added, "and the PAF has now selected BVR missiles for integration, having made a recent break-through by acquiring the technology". The JF-17 would also carry H-2 and H-4 missiles, and he claimed that it would be superior to the PAF's existing F-16As.

Last December, the PAF announced acquisition of a new H-4 beyond-visual-range (120-km/65-nm) AAM, developed by the National

In late May 2004 the first Typhoon T.Mk 1 for No. 29 (Reserve) Squadron appeared at BAE Systems Warton. While No. 17(R) Squadron handles operational evaluation, No. 29 has been established for type conversion.

No. 17(R) Squadron's six pilots are engaged in expanding the Typhoon's tactical envelope, a task eased by the 13 May 2004 formal release to service and removal of flight restrictions.

until after the parliamentary summer break. Included in the deal is Euro60 million to cover the development work necessary to intregate the KEPD 350 Taurus missile. Germany has Euro4.6 billion Tranche 3 acquisition plans for 61 single- and seven two-seat Eurofighters.

Italy

The Aeronautica Militare Italiana formally accepted its first Typhoon (IT002) in March and a further two were delivered to the 4° Stormo at Grosseto, whose 9° Gruppo is the initial operating unit and type OCU. The first Italian production aircraft (IT001) went to Cameri in February for training of ground personnel. Italy's Tranche 1 allocation consists of 10 two-seaters and 19 single-seaters from a total requirement of 121 aircraft.

Spain

Typhoon training for the Ejército del Aire was initiated at the EADS CASA facility at Getafe, with an initial eight Ala 11 instructor pilots plus ground personnel receiving instruction from the manufacturer. The Ala 11 detachment transferred, as Escuadrón 113, to its operational base at Morón on 27 June. It had three aircraft, and was due to have six by the end of the year. The squadron will remain the type OCU, although the parent Ala 11 will also have operational commitments. Another Typhoon was delivered to Albacete for ground personnel training.

United Kingdom

Wing Commander David Chan, CO of No. 17 (Reserve) Squadron, the Typhoon's Operational Evaluation Unit, made the first official Eurofighter flight of the new OEU on 11 February, although the RAF has been flying the aircraft since December 2003. No. 17 Squadron is initially based at BAE Systems' Warton factory airfield under the Case White programme.

Training of the first pilots was under way in early 2004, and on 13 May the Typhoon received its formal release to service. A week later, on 20 May, the first aircraft painted in No. 29(R) Squadron colours took to the air. This unit is the type operational conversion unit (OCU). Case White has trained six pilots for the OEU and is training 10 for the OCU, which is due to move to Coningsby in July 2005.

Engineering and Scientific Commission (NESCOM), in close collaboration with Pakistan's missile and air weapons complex. Three successful tests were completed earlier this year of the H-4, reportedly derived from South Africa's Denel/Kentron T-Darter BVRAAM, with passive infra-red guidance, for initial integration with the PAF's 80 or so Mirage III/5s, and later the JF-17. The PAF's BVRAAM missile inventory is also claimed to include the H-2, with a range of up to 60 km (32.4 nm).

Ukraine

Slow An-70 progress

Development is continuing in Kiev by ANTK Antonov of the An-70 short-field heavy-lift prop-fan transport, which earlier this year had completed 80 percent of its joint Ukrainian-Russian state flight-test schedules. Although Russia is continuing its nominal partnership in the An-70 programme, FRAF C-in-C Col Gen. Mikhailov continues to oppose this project on technical grounds, and favours acquiring uprated Ilyushin Il-76MFs. This situation is reflected in funding allocations by Ukraine of 243 million hryvna ($45.5 million) towards An-70 development and production between 2004–2022, compared with a nominal Russian Federation 2004 R&D total of Rbls 38.47 million ($1.3 million).

United States

AIM-9X Sidewinder update

Following 66 development tests, the USAF fired an AIM-9X Sidewinder in an operational environment for the first time during an air-to-air weapons system evaluation program (WSEP) mission at Tyndall AFB, Florida, recently. Flown by an F-15C pilot attending the USAF Weapons School, the mission was managed by the 83rd Fighter Weapons Squadron (FWS), which is responsible for conducting the air-to-air WSEP.

V-22 progress

The V-22 Integrated Test Team began a new round of air-to-air refuelling flights recently. During the initial pair of sorties, which took place in March 2004, flight test crews logged 10 'dry plugs' behind a KC-130F tanker. MV-22B no. 22 (BuNo. 165444), which is equipped with a 3.35-m (11-ft) fixed refuelling probe, supported the tests. A new retractable refuelling probe has been installed on Osprey no. 21 (BuNo. 163443) in preparation for testing. The probe is 2.7-m (9-ft) long when extended but flush with the fuselage when it is retracted.

In related news, Osprey no. 9 (BuNo. 164941), which is modified to CV-22B configuration, flew its first open air range electronic warfare flight recently at NAWS China Lake's Electronic Combat Range in California.

MC²A project update

The USAF has announced that two additional versions of the Multi-Sensor Command and Control Aircraft (MC²A) will follow the E-10A. Designated the E-10B, the second variant will be equipped with an air moving target indicator (MTI), and will replace the E-3B/C Sentry in the AWACS role. A third version, designated the E-10C, is intended as a replacement for the RC-135V/W Rivet Joint electronic intelligence fleet.

Eagle Eye VUAV progress

Bell Helicopter Textron, which has already built a ⅞-scale prototype of its Eagle Eye tiltrotor UAV, has announced construction of a full-scale prototype. The Eagle Eye has been selected as the VTOL UAV (VUAV) portion of the US Coast Guard's Integrated Deepwater System. Bell Helicopter is developing the Eagle Eye as a sub-contractor to Lockheed Martin Naval Electronics & Surveillance Systems and is scheduled to deliver the VUAV in time to meet the Coast Guard's first unit equipped (FUE) date in 2006. First flight of the full-scale prototype will be conducted by 1 November 2004.

Fire Scouts for the army

Northrop Grumman recently received a $115 million sub-contract from Boeing and SAIC associated with the system development and demonstration (SDD) phase for the US Army's Future Combat System (FCS). Under the eight-year SDD phase, the contractor will develop the Class IV

On 18 April 2004 Boeing's X-45A unmanned combat aerial vehicle (UCAV) demonstrator delivered an inert 113-kg (250-lb) precision-guided small diameter bomb. The GPS-guided weapon was delivered from an altitude of 10668 m (35,000 ft) and a speed of 711 km/h (442 mph) at NAWS China Lake, California. The entire flight, including the weapon release sequences, was conducted autonomously. Boeing is currently building a larger X-45C, at its St Louis, Missouri, facility. The first flight of the X-45C is scheduled for mid-2006.

The RAAF's first Boeing 737-700 AEW&C aircraft, equipped with Northrop Grumman's multi-role electronically-scanned array (MESA) L-band dorsal radar, started flight tests at Boeing Field in Seattle on 20 May 2004. Under the $A3.4 billion project, the RAAF's re-formed No. 2 Squadron, based at Williamtown, near Newcastle, NSW, will receive its first two aircraft in November 2006, and two more by 2008. Further purchases are likely.

unmanned aerial system (UAS) architecture, and deliver seven RQ-8B Fire Scout vertical take-off and landing tactical unmanned aerial vehicles (VTUAV). It will also integrate the VTUAV's reconnaissance, surveillance, and target acquisition (RSTA) systems. Similar to the RQ-8A being produced for the Navy, the RQ-8B will feature a new, improved four-bladed rotor system and several performance enhancements that will provide the VTUAV with more than eight hours endurance with a payload of 59-kg (130-lb). Installation of the new main rotor will triple the RQ-8's payload capacity to 272-kg (600-lb) or double its on-station time with a 90.72-kg (200-lb) payload at 204 km (110 nm) radius. The system also enables Fire Scout to carry multiple payloads simultaneously. In addition to the Army VTUAVs, the contractor is already building eight RQ-8As for the Navy under a Lot 2 LRIP contract. It also recently received a $49 million engineering and manufacturing, development (EMD) contract for the Navy RQ-8B Fire Scout UAV that includes two RQ-8B UAVs.

Predator B flies

The first pre-production MQ-9A Predator B UAV is undergoing testing at General Atomics Aeronautical Systems' flight operations facility in El Mirage, California. Powered by a Honeywell TPE-331-10T turboprop engine, the MQ-9A is capable of operations above 15240 m (50,000 ft). It is equipped with a longer span wing and its wider fuselage houses additional fuel, enabling it to fly for over 30 hours while carrying more than 1361 kg (3,000 lb) externally and 363 kg (800 lb) internally. The MQ-9A has a gross take-off weight (GTOW) of 3048 kg (10,000 lb), and features triple redundant avionics and dual mechanical control systems that meet the requirements for flight in the National Air Space (NAS). General Atomics is also developing the Predator B-Extended Range, which will be capable of an increased endurance of up to 49 hours.

C-130J news

Air Mobility Command recently concluded an extensive series of tests of the C-130J at Little Rock AFB, Arkansas, as part of its Procedures Development and Evaluation (PD&E). Completion of the three-week trial moved the C-130J one step closer to full operational capability. Units from the Air National Guard (ANG) and Air Force Reserve Command (AFRC) supported the evaluations. The USAF recently dropped plans to refer to the stretched C-130J aircraft as the CC-130J and will now use the same designation for both the standard length and stretched versions.

Nighthawk delivers JDAM

The F-117 Combined Test Force (CTF) and the 410th Flight Test Squadron (FLTS) recently delivered a joint direct attack munition (JDAM) for the first time. The F-117A released examples of the 907-kg (2,000-lb) GBU-31(V)1/B and GBU-31(V)3/B on a precision impact range at Edwards AFB.

On 28 April 2004 Aermacchi conducted the first taxi tests of the M-346 trainer at its Varese-Venegono facility. The advanced trainer, based on the Yak-130, has both brake-by-wire and steer-by-wire systems.

Joint Strike Fighter update

According to recent reports the F-35 Joint Strike Fighter is between 453 and 680 kg (1,000 and 1,500 lb) overweight and faces a production delay of at least a year. The delay will result in an extension of the development programme. Besides additional costs, the problems could reduce the number of aircraft that the DoD is able to order. As a result, the start of low-rate initial production (LRIP) for the conventional and STOVL variants will be pushed back from 2006 to 2007. The carrier version will still enter initial production as planned in 2008.

The USAF announced plans to purchase an undisclosed number of the STOVL F-35Bs for use in the close air support (CAS) role. The USAF, which will still order the conventional F-35A, will develop training and tactics for the STOVL version in conjunction with the Marines.

In related news, Pratt & Whitney has completed the first full afterburner test run of the initial F135 production configuration test engine at its facility in West Palm Beach, Florida. The JSF Propulsion System Team, which also includes Hamilton Sundstrand and Rolls-Royce, recently delivered the second engine to the test facility and it will begin testing the short take-off and vertical landing (STOVL) system for the F-35B.

USAF tanker revision

The USAF's plan to acquire 100 KC-767A tankers hit another snag recently when the Secretary of Defense ordered two new reviews of the service's tanker plans. The studies are evaluating the current capabilities of the KC-135 fleet and determining whether the KC-135s tankers can continue to meet the DoD's needs or if a replacement is actually required. The USAF had initially planned to acquire 100 KC-767s under a lease-to-buy plan, which was subsequently revised to include the lease of the first 20 and the outright purchases of 80 aircraft. A review of the costs associated with the project and questionable contractual negotiations caused it to be put on hold in December 2003.

As a result of delays in the KC-767 plan, the USAF has revised its schedule for retiring 61 KC-135E tankers. Under the new plan the service will initially retire 12 KC-135Es operated by the ANG and AFRC during 2004. It will subsequently retire 41 aircraft in 2005 and the final eight in 2006. Seven examples will be shifted to back-up inventory status. In its original plan the USAF had intended to remove 37 tankers from the active inventory this year. However, the 2004 Defense Bill prohibited the retirement of more than 12 aircraft. As a result the service increased the planned retirements in 2005. The KC-135Es, which are operated by reserve component units, will be replaced by 48 KC-135Rs transferred from the active component. The first transfers began in February 2004 when the Pennsylvania ANG's 171st Air Refueling Wing (ARW) at Pittsburgh IAP and the Air Force Reserve Command's 927th ARW at Selfridge ANGB, Michigan, received their first KC-135Rs. The same units will respectively send eight and four KC-135Es to AMARC. The USAF had hoped to place the first KC-767As into service in 2006.

New sensors for Longbow

Boeing recently began flight-testing the Modernized Target Acquisition and Designation Sight/Pilot Night Vision Sensor (M-TADS/PNVS) equipment at its Mesa, Arizona, facility. Developed by Lockheed Martin, the new sensors improve performance and reliability by more than 150 percent. Known as the Arrowhead, the upgrade provides the TADS/PNVS with advanced target acquisition/designation and night vision capabilities. Lockheed Martin is currently building 19 Arrowhead kits that will be retrofitted into AH-64Ds in mid-2005. The first operational Arrowhead kits will be fielded by June 2005.

MPATS delayed

The Navy has postponed the expected acquisition of an Undergraduate Military Flight Officer (UMFO) Multi-Place Aircraft Training System (MPATS). More commonly referred to as the T-48TS, the project included at least 16 commercial business-class jets with integrated multi-mode radar; a ground-based training system (GBTS); and logistical support. The aircraft will eventually replace the T-39G/N Sabreliners based at NAS Pensacola, Florida.

Nachshon and Nevatim

The CAEW Nachshon operators' compartment with six stations is in the back of the cabin, as the front section is reserved for the mission equipment supporting the conformal antennas on both sides of the fuselage.

The SIGINT Nachshon fuselage is divided into four sections comprising, from front to rear: cockpit, tactical situation compartment with five stations, operators' compartment with six stations, and the equipment compartment.

The Israel Defence Force/Air Force (IDF/AF) has revealed models of the new Gulfstream G550 Nachshon (Pioneer) SIGnal INTelligence (SIGINT) and Compact Airborne Early Warning (CAEW) aircraft. The SIGINT Nachshon is to be equipped with the IAI Elta EL/I-3001 system, while the CAEW Nachshon is to be equipped with the IAI Elta EL/W-2085 system. The first Nachshon aircraft is scheduled to arrive in Israel by the end of 2004 as a 'green' airframe to be fitted locally with the mission equipment by IAI Elta. First to enter service will be the SIGINT Nachshon, with the CAEW aircraft following. The whole fleet is planned to operate out of Nevatim air base in the Negev.

The Israel Defence Force/Air Force (IDF/AF) launched the transport wing relocation project on 24 June during an official ceremony at Nevatim air base in the Negev. In line with an Israeli Government decision to relocate military installations to the periphery and as part of the ongoing IDF/AF process to strengthen the Negev air bases, the IDF/AF transport aircraft are to be transferred from Lod air base to Nevatim.

Nevatim is currently the home of two F-16 Netz (Sparrowhawk) squadrons and the air base will more than double in capacity when the transport aircraft move in during 2008 and 2009. The project cost is estimated at $350 million and Lod air base will be handed over to the Israeli Treasury by 31 December 2009. Work at Nevatim will start in January 2005 and will cover infrastructure (electricity, fuel, water and communication), construction of two runways, aprons, maintenance centre, terminal, operational and administrative buildings. A new entrance from the south will complement the current entrance from the north and Nevatim will be linked to a railway line.

Relocation will cover the Boeing 707 Re'em (Ram), Lockheed C-130 Karnaf (Rhinoceros) and Gulfstream G550 Nachshon (Pioneer) aircraft. It is also expected that the Navy's IAI 1124N Shahaf (Gull) Maritime Patrol Aircraft (MPA) will move in to Nevatim, though other options are currently explored to base the MPA fleet closer to the Israeli coast rather than in the desert. When the move from Air Base 27 at Lod to the Negev is completed, Nevatim's status will be changed from a Wing to an Air Base.

Shlomo Aloni

Upgrades and Modifications

Israel

Skyhawk service planned

Continued operation of the McDonnell Douglas A-4 Skyhawk, which comprised the backbone of Israel's 1980s attack forces, is planned until at least 2010 by the IDF/AF, which uses 14 or more two-seat TA-4H/J advanced trainer versions both in its flight academy and for lead-in fighter training for its current Boeing F-15 and Lockheed Martin F-16 combat aircraft. Modernisation of the IDF/AF A-4s began in January 2003, when Israel's RADA Electronic Industries was awarded a $2 million contract for installation of an advanced debriefing system. The second step, activated in January in conjunction with Israel Aircraft Industries' (IAI) Lahav division, includes replacing obsolete onboard avionics with modern systems costing $2.8 million, for which a $1.2 million serial production order was imminent.

Italy

Trainer orders and upgrades

Aermacchi is to take back 21 SIAI-Marchetti SF.260AM piston-engined primary trainers from 45 originally delivered to 207º Gruppo/70º Stormo at Rome-Latina in 1976, in part exchange for 30 new upgraded SF.260EAs. A Euro35 million ($43 million) Aermacchi contract received last December also includes establishment of a logistic support spares and maintenance centre for the new SF.260s at Latina. With modernised avionics and cockpit systems, their deliveries are due from early 2005, for initial screening and basic training roles. Aermacchi also received a Euro33 million ($40 million) Logistics Command contract in June for support and upgrading AMI's MB-339A, MB-339CD and SF-260 military trainers, plus flight simulators and ground equipment. Apart from providing spares, overhauls, technical and engineering support, the contract includes mid-life systems and structures updates of 49 MB-339A basic/advanced trainers, for secondary close air support (CAS) and slow moving interceptor (SMI) roles.

Netherlands

Dutch P-3 upgrades continue

While a final decision was still awaited in early 2004 concerning Germany's planned acquisition of eight Dutch Lockheed Orions, costing Euro261 million ($318.8 million), allowing Portugal to buy the remaining five for Euro72 million ($88 million), the first RNN P-3C II.5 to undergo Capability Upkeep Programme (CUP) upgrades flew back to RNLNAS Valkenburg on 23 May from Lockheed Martin's Greenville, South Carolina, plant. Having arrived there on 7 July 2002, its tactical equipment was replaced by a completely new mission suite, for continued operation until 2020.

Lockheed Martin is continuing similar CUP upgrades, each costing some Euro20 million ($24.43 million), of two more Dutch P-3Cs, and is contractually committed to completing the 10th and last aircraft by March 2006. The programme is apparently too far along for cancellation, despite Defence Minister Henk Kamp's parliamentary backing to disband the RNN's Maritime Patrol Group (MARPAT), sell the P-3Cs and close Valkenburg air base by early 2006.

New Zealand

L-3 selected for Orion upgrades

The RNZAF's long-deferred Project Sirius follow-on upgrades for its six Lockheed P-3K Orion maritime patrol and surveillance aircraft was finally revived in March by selection of US L-3 Communications Integrated Systems as preferred contractor. Four companies were invited to respond to requests for tender last year, for the revised $NZ300 million ($189 million) two-phase Project Guardian programme. The NZ Defence Ministry then asked L-3 to submit a best and final offer for presentation later this year. Limited ASW and surface attack capabilities enhancements are envisaged in the revived P-3K radar, digital communication, data-handling, sensor and navigation systems upgrade requirements for No. 5 Squadron's Orions at Whenuapai air base.

As outlined in the NZ Defence Long Term Development Plan (LTDP), the upgrade will follow contract negotiations over several months, to precede another two years for design, development and testing. The first updated P-3Ks will be fully operational between 2006-08. Last November, an RNZAF P-3K reached a milestone 20,000 flight hours during a support mission in the Gulf region, but major Project Kestrel wing and tail structural improvements incorporated in New Zealand's Orion fleet should allow their operation through at least 2016.

Russia

New radar for 'Helix'

Flight testing has begun in Russia of a Kamov Ka-27 'Helix' helicopter that has been equipped with the Kopyo-A [Spear] radar. The Russian Navy hopes to equip the first batch of Ka-27s with the new maritime surveillance radar in 2005. Originally developed for the MiG-21 fighter upgrade programme by Phazotron-NIIR, a version is also planned for the improved Su-25SM.

Slovakia

MiG-29 upgrades planned

Plans were announced in February to upgrade 18 of 20 MiG-29s and three two-seat MiG-29UB combat trainers in Slovakian air and air defence force (LAPSOS) service, with Russian assistance, for a 10-year life extension. Funding shortages have grounded most of these aircraft, but the MiG-29s are now planned to spearhead a force additionally including 10 Aero L-39ZA armed jet-trainers, plus 18 Mil Mi-24D/V attack, 18 Mi-17 transport, and four Mi-2 training helicopters.

Spain

Upgraded F-5s and Hornets

The first three of 17 Spanish air force (EdA) two-seat Northrop SF-5Bs, headed by s/n AE9-008, being upgraded by EADS CASA Military Aircraft with Israel Aircraft Industries' advanced digital avionics as lead-in fighter trainers, were redelivered from Getafe to Talavera air base last December. Seven more SF-5Bs were then also being upgraded, towards 2005 programme completion from a Euro31 million ($38.5 million) two-year production contract.

The Israel Defence Force/Air Force (IDF/AF) has announced that the 'Hornet Squadron' (HS) will operate the AH-64D Sharaf (Serpent). Currently an AH-64A Peten (Python) operator, the HS is scheduled to start Sharaf operations in early 2005. Meanwhile, the first Israeli AH-64D – an upgraded Peten now flying as Sharaf 716 – performed in February a successful first flight following the installation of Israeli systems.

Recently cleared for use by the F-16 is the EDO Corporation BRU-57 double weapons rack, which includes two BRU-46 ejector units with Mil Std 1760 interfaces. The F-16 can now carry four 1,000-lb (454-kg) class GPS-guided weapons, as opposed to two, and is shown here with Wind-Corrected Munitions Dispensers.

The 416th Flight Test Squadron (FLTS) at Edwards AFB completed the first live firing of a Raytheon AIM-9X from an F-16C on 9 April 2004. The first of three unguided separation tests, carried out over the Naval Air Warfare Center at NAWS China Lake, California, will be followed by three guided separation tests against sub-scale targets.

EADS CASA also started work in January at Getafe on a four-year Euro186 million ($230.75 million) mid-life multi-role upgrade of 90 EdA Boeing/MDC EF/A-18A and 12 two-seat EF/A-18B Hornets, on which prototype development started in 2000. EdA's Armament and Experimental Logistics Centre has been involved with EADS CASA in design, development and integration of upgraded software, for a new microprocessor, tactical computer, two MFCDs, Have Quick II secure communications, GPS/INS, EW and night-vision systems.

United Arab Emirates

C-130 upgrades funded

A single Lockheed C-130H-30 and an L-100-30 Hercules operated by the Dubai Air Wing since the 1980s are to be upgraded from a new contract with Canada's CMC Electronics and the Abu Dhabi-located GAMCO group. CMC's CMA-900 digital flight management system will be installed locally to meet new international air traffic requirements. GAMCO has provided depot maintenance for UAEAF C-130s for more than 14 years, and it became an authorised Hercules service centre for 16 Middle Eastern, African and Asian countries in January 2002.

United Kingdom

Harrier upgrades extended

Further upgrades in the RAF's long-running £500 million-plus ($925 million) Harrier GR.Mk 9 sustainment and modernisation programme are in progress from a new £100 million ($185 million) MoD contract, known as Capability C2, received by BAE Systems in January. The systems and software upgrade contract was awarded to BAE Customer Solutions & Support (CS&S) by the UK MoD Defence Logistics Organisation.

Under Capability C2, several new systems will be integrated in the GR.Mk 9, linked by a new onboard computer. These include the RAF's new Paveway IV 227-kg (500-lb) precision guided bomb, for which Raytheon was awarded a £150 million initial contract in December 2003, and infra-red and TV-guided variants of the Raytheon/Hughes AGM-65 Maverick attack missile. Also included is Raytheon's Successor Identification Friend or Foe (SIFF) system.

The full GR.Mk 9 programme began with a 19 April 2002 bridging contract for upgrade engineering and physical modification of the existing UK Harrier GR.Mk 7 fleet with updated integrated digital systems for enhanced smart weapons and operational capabilities. The programme also includes upgrading to GR.Mk 9 standards a dozen two-seat RAF Harrier T.Mk 10 combat trainers as T.Mk 12s.

Harrier fleet upgrades to the new specification is taking place through the HMP3 (Harrier Modification Programme), which included delivery of the first uprated Pegasus Mk 107 turbofans for Harrier GR.Mk 7A/9As in November 2003. Fleet conversions also include fitment of new rear fuselages on specific aircraft, for operational release to service due in 2006.

United States

ICAP III undergoes OPEVAL

The EA-6B ICAP III (Improved Capability III) electronic attack weapon system entered into its operational evaluation (OPEVAL) at NAWS China Lake, California, on 2 April 2004. Although ICAP III has been in low-rate production since mid-2003 it must successfully complete its OPEVAL before full production is authorised. Air Test and Evaluation Squadron VX-9 is conducting the five-month OPEVAL, using two ICAP III test aircraft. The programme successfully completed its Technical Evaluation (TECHEVAL) in February. The first ICAP III EA-6B will be delivered to the fleet in early 2005.

In related news the VAQ-129 'Vikings' recently transferred the last Block 82 EA-6B from NAS Whidbey Island, Washington, to Naval Air Depot Jacksonville, Florida, where it will be updated to the latest Block 89A configuration. The move brought an end to the Block 82 Prowler's service career.

Updated Greyhound flies

The first C-2A carrier onboard delivery (COD) aircraft to undergo the navy's service life extension programme (SLEP) recently made its first flight at NAS North Island, California. As part of the 15-month project, BuNo. 162169 was equipped with a series of structural modifications that will allow the Greyhound to operate for up to 36,000 landings and 15,000 flight hours. The airlifter had originally been designed for a service life that included 15,020 carrier landings and 10,000 flight hours. Over the next seven years the entire fleet, comprising 35 aircraft, will undergo the SLEP, allowing the Greyhound to remain in service until at least 2015. In addition to the structural upgrades, the C-2As are receiving updated avionics and will soon begin testing the eight-bladed Hamilton Sundstrand NP2000 propeller.

The Israel Defence Force/Air Force (IDF/AF) Raytheon B200CT Zufit (Thrush) 3 aircraft has been 'missionised' locally by Elbit Systems' subsidiary Cyclone. The mission equipment has been supplied by Rafael and the IDF/AF is promoting the Zufit 3 as the 'C-47 successor' in the communications relay role, thus closing an operational gap that was created when the C-47 was withdrawn from IDF/AF service in January 2001.

Cockpit upgrades for Hercs

The USAF has announced that the first two C-130 units to be equipped with modernised airlifters will be the AFRC's 908th Airlift Wing at Maxwell AFB, Alabama, and the Idaho ANG's 124th Wing at Boise. The aircraft, which respectively comprise the C-130H-2 and C-130E models, will be equipped with new systems under the Avionics Modernization Program (AMP). Scheduled to begin in late 2007, the programme will bring all 520 C-130E, C-130H-1, H-2 and H-3 to a common avionics configuration. The AMP, which is being developed by Boeing, replaces the aircraft's current analog instrumentation system with six digital displays and the flight management system developed for the Boeing 737 airliner.

The US Navy has awarded Boeing a $3.4 million contract to develop an update programme for the cockpit avionics of 48 C-130T and KC-130T aircraft operated by the Navy and Marine Corps. The project will equip the aircraft with a cockpit configuration similar to that under development as part of the USAF's AMP.

Navy plans update for Talons

Ten T-38As operated by the US Naval Test Pilot School at NAS Patuxent River, Maryland, will be updated to T-38C configuration over the next three years. The aircraft will be individually inputted to the Boeing modification line in Mesa, Arizona.

RC-135 mods

Boeing's Wichita Development and Modification Center in Kansas has received a $17.5 million modification to an existing contract covering the installation of F108 (CFM56) engines and associated modifications on four RC-135 aircraft.

C-5 avionics update

Lockheed Martin has received a $5.9 million contract, associated with the C-5 Avionics Modernization Program (AMP) that covers the installation of the eight Block 1 AMP kits into C-5B aircraft beginning in June 2004.

Saudi Snooper

On 9 January 2004, the Royal Saudi Air Force took delivery of its first upgraded Boeing RE-3A when aircraft serial 1901 was delivered from RAF Mildenhall to Al Kharj Air Base. The aircraft was originally ferried from Saudi Arabia to the USA in December 2001, and has been undergoing upgrade for more than two years. The work was carried out by L-3 Communications Integrated Systems (formerly Raytheon's Aircraft Integrated Systems and before that E-Systems) at Majors Field, Greenville, Texas.

The aircraft was previously a KE-3A tanker (serial 1817) with the Saudi Air Force, and was accepted into service in September 1987 for service with 18 Squadron at King Khalid Airport, Riyadh, and later at Al Kharj. The tanker was subsequently returned to the USA for conversion to become the first of at least two (and possibly three) RE-3As. It is believed that the aircraft left Saudi Arabia in August 1998, and was completed in September 2000. It is not known when the aircraft was delivered to Saudi Arabia, but its period of service in the Kingdom was relatively short considering its departure back to the USA in December 2001.

Two reconnaissance versions of the E-3 are reported to be in service, consisting of the RE-3A, which is the standard Tactical Airborne Surveillance System (TASS), and the Improved Tactical Airborne Surveillance System (ITASS), designated RE-3B. 1901 carried the designation RE-3A on the data block on the port side beneath the pilot window. In antenna fit the RE-3A is somewhat similar in appearance to the RC-135V and W models which serve the US Air Force under the combined name of Rivet Joint. The Saudi jet has similar, but not identical, 'chipmunk' cheeks to those fitted to Rivet Joint. These contain the primary sensors for the collection of electronic intelligence (ELINT) and are referred to as interferometers (and not SLAR – sideways looking airborne radar – as is all too frequently assumed). The cheeks are similar in size to the Rivet Joint versions, but feature a slightly different sensor suite. Beneath the fuselage the RE-3 has numerous small antennas, as well as four large black inverted 'T' Comint antennas. The quantity and location of the antennas was almost identical to when the aircraft returned to the USA in 2001, and therefore it is likely that the upgrade is mostly internal – to enhance the speed, range and capability of the sensor suite.

Bob Archer

Procurement and Deliveries

Algeria

More MiG-29s?

Negotiations were reported earlier this year between the Algerian and Russian governments for AQJJ acquisition of up to 49 upgraded MiG-29M1/M2 advanced combat aircraft. A proposed $1.8 billion contract was expected to include weapons, spares, and support equipment, to supplement 24 'Fulcrum-Cs' delivered in 1999. First export orders for the MiG-AT advanced jet-trainers were also being considered for the AQJJ, together with Ilyushin Il-103 primary trainers.

Deliveries are also expected to start later this year of 10 EADS CASA CN 235M twin-turboprop tactical transports ordered by the Algerian air force (AQJJ) in late 2003. These comprise four equipped for maritime surveillance and patrol roles, and two for government passenger transport, to supplement about 20 Lockheed C-130H/H-30s awaiting systems upgrades.

Australia

Airbus wins tanker contract

Recent UK selection of the AirTanker group, offering customised Airbus A330-200s, in preference to the rival TTSC bid of Boeing KC-767s, for future RAF air-refuelling requirements, was followed in April by RAAF commitments for similar new multi-role tanker/transports (MRTTs), to replace its three venerable Boeing 707-338Cs. Supported by Rockwell Collins Australia and Smiths Aerospace, to help meet Australian Industry Involvement (AII) requirements which include 20-year through-life support, the KC-767 was the only Airbus rival for the RAAF's Project Air 5402 tanker programme.

EADS/Airbus Military Transport Aircraft Division (MTAD) announced RAAF orders for five A330-200 MRTTs, powered by 320.27-kN (72,000-lb) thrust General Electric CF6-80E1 turbofans, from initial $US100 million engine contracts, for delivery from 2006. RAAF service entry for its five A330 MRTTs, equipped with both underwing hose and drogue pods, and a new EADS-developed remotely-controlled rear-fuselage boom system, is planned for 2007.

Two contracts totalling around Euro1.2 billion ($1.46 billion) are now being negotiated with EADS CASA, as prime contractor for supplying the aircraft, and QANTAS Defence Services Pty Ltd (QDS) for 20-year through-life support. As prime sub-contractor to EADS CASA, QDS will also convert the second and remaining aircraft into MRTTs from basic commercial aircraft, as well as customising training systems, and developing and delivering Australian training programmes. QANTAS already flies civil CF6-powered A330-200s, with long-established aircraft support, training, maintenance and overhaul facilities. Brisbane-based EADS subsidiary Australian Aerospace also provides through-life support for RAAF Caribous and AP-3C Orions

Canada

New SAR aircraft

The Canadian government has announced plans to purchase 15 new twin-engined search and rescue (SAR) aircraft at a cost of $950 million over the next four years. The new aircraft will replace Air Command's six remaining CC-115 Buffalos as well as CC-130s that are currently used for SAR duties. The government will initially commit $219 million in 2005, and will purchase the aircraft over the next four years. The aircraft will be stationed at CFB Greenwood, Nova Scotia, Trenton, Ontario, Comox, British Columbia, and Winnipeg, Manitoba. Although the Alenia/Lockheed Martin C-27J appears to be the front-runner to fill the role, the CASA C-295 is also a contender. The Department of National Defence has not yet however begun its tendering process for the new aircraft.

Czech Republic

Gripen MoU signed

Offset contracts for Czech industry and economic aid worth Ckr25.5 billion ($995 million) represented 130 percent of the Ckr19.6 billion ($765 million) MoU agreement signed on 14 June by the Prague and Swedish governments, covering the 10-year lease of 14 Saab/BAE JAS 39C/D Gripen multi-role combat aircraft. The agreement between SAAB, the Swedish Defence Materiel Administration, and the SAF, also covered Gripen modifications for Czech and NATO requirements, plus pilot and technician training now starting in Sweden. The Czech Republic thus becomes the fourth confirmed Gripen customer – after Sweden, South Africa and Hungary, to replace MiG-21s from 2005.

Russian helicopters sought

Repayment of Ckr22 billion ($813.5 million) owed to the Czechs from Russian trading debts, may include about 50 percent from military equipment supplies. Last year, a Czech proposal to include three Antonov An-70 heavy-lift transports, costing $150 million in all, for delivery in 2005-06, was countered by Russia with an offer of three upgraded Ilyushin Il-76MFs instead. These were rejected by the Czech Defence Ministry, which in February switched its requirements to a proposed acquisition of 18 "basically modernised" Kazan-built Mil Mi-17V-5 transport helicopters, and 18 Rostvertol-supplied upgraded Mi-24 attack helicopters, funded from Russian trade debts.

Denmark

C-130J deliveries completed

Deliveries of all three stretched-fuselage Lockheed Martin C-130J-30 Super Hercules to the Royal Danish Air Force (RDAF) took place between 1 March and mid-April, three years after contract signature. Denmark, which is replacing six 1973-vintage C-130Hs, has an option for a fourth C-130J. Deliveries by RDAF crews from Lockheed Martin Marietta, Georgia, coincided with transfer of the RDAF's main air transport base and 721 Squadron's aircraft from Værløse AB, near Copenhagen, to Aalborg AB, in northern Jutland.

As part of force consolidation measures, 721 Squadron began operations on 1 April from Aalborg, after construction of a new C-130J hangar. A joint commercial airfield, Aalborg has accommodated one of three RDAF Lockheed Martin F-16 squadrons (currently No. 726) for over 20 years. 721 Squadron operates four different

The first production NH90 was handed over to the German army on 11 May 2004, during the ILA show at Berlin-Schönefeld.

The Mexican Navy has taken delivery of three E-2C Hawkeyes following their upgrade by Bedek Aviation in Israel. The former IDF/AF aircraft had been withdrawn from service in 1997 and stored pending sale. In Mexican service they will primarily be used as a counter to arms and narcotics smuggling.

aircraft types, exemplified by fly-pasts of two Canadair Challenger 601 government transports, three Saab T-17 trainers and two C-130Hs, as the C-130J arrived, escorted by an F-16A.

Danish C-130J-30s feature a strengthened cargo ramp and improved 463-km/h (250-kt) air-drop capability; enhanced cargo handling; and RDAF-specific items, including a customised EW suite.

Greece

Hellenic Seahawk flies

The first of three new S-70B Seahawk helicopters for Greece's Hellenic Navy made its initial flight at Sikorsky's Stratford, Connecticut, facility on 10 February. Known as the Aegean Hawk, the S-70B-6 is equipped with numerous system enhancements including a glass cockpit featuring four CRTs, a Collins FMS-800 flight management system (FMS) and an improved navigation system. The helicopter is also equipped with the AN/AAS-44 FLIR and is capable of employing laser-guided AGM-114 Hellfire missiles. The Aegean Hawk also has an improved countermeasures capability and electronic support measures (ESM). The initial two aircraft will join the existing fleet of eight S-70B-6 helicopters in June 2004.

Indonesia

Sukhoi procurement continues

Additional procurement is reportedly planned this year of six more KnAAPO-built upgraded Sukhoi Su-27SMKs and two two-seat Su-30MKKs from IAPO. These will supplement two recently-delivered Su-27SMKs and two Su-30MKs, plus two Rostvertol Mil Mi-35V attack helicopters ordered from a $193 million initial Russian contract signed in April 2003. Indonesian requirements reportedly remain for up to 44 more Sukhoi combat aircraft by 2007, to re-equip four fighter squadrons, mainly concentrated at Hasanuddin eastern air base.

Navy lightplanes ordered

A $2.5 million EADS SOCATA contract was announced in February for five 180-hp (134-kW) Textron Lycoming O-360-engined TB10 Tobago GT light primary trainers, and a Frasca fixed-base simulator equipped with FVS-200 visual system, for Indonesian naval aviation (DPAL). This is receiving its first two TB10s in June 2004, and the remaining three in early 2005. They will supplement earlier DPAL procurement of four 160-hp (119-kW) Lycoming-powered TB9 Tampico GTs in 1994.

Jordan

New trainers delivered

Jordan Aerospace Industries (JAI) recently delivered the first Sama CH 2000 trainer to the Royal Jordanian Air Academy. The two-seat, low-wing trainer is powered by an 86.5-kW (116-hp) Lycoming O-235 four-cylinder engine. JAI, which is located at Amman's Queen Alia International Airport, claims to have export orders for at least 25 aircraft from buyers in the Middle East, Africa and Asia.

Blackhawks ordered

Jordan has reportedly signed a letter of acceptance to acquire eight S-70 Blackhawk helicopters from Sikorsky Aircraft at a cost of $200 million. The purchase will likely be financed with US military aid and the aircraft will be delivered by 2009.

Oman

F-16 recce systems

Twelve Block 50 F-16C/Ds to be delivered by Lockheed Martin from 2005 to the Royal Air Force of Oman (RAFO) from a 2002 $825 million FMS contract will be equipped with Advanced Airborne Reconnaissance Systems (AARS) ordered earlier this year from BAE Systems Communication, Navigation, Identification and Reconnaissance (CNIR). Also including a ground processing capability, and integrated logistics support, the new digital reconnaissance system will provide multiple fields-of-view in a single sensor package.

Pakistan

US aids defence expansion

Ending three years of restricted defence spending, Pakistan increased its 2004 military allocations by 7 percent in June to a record Rs194 billion ($3.36 billion), compared with the 2003 arms budget of Rs181 billion ($3.14 billion). The new allocations represent a real-terms increase of 3.1 percent, or 21.7 percent of Pakistan's Rs902 billion total budget, and will contribute towards forecast spending of $12-15 billion for PAF refurbishment and modernisation over the next five years. Defence will be further funded by $300 million from the $701 million US military and economic assistance package expected in November, following Congressional approval.

This year, the US will also write-off some $480 million still owed by Pakistan from an earlier $2 billion loan. It will be additional to the $75 million US aid promised in late 2002, for Pakistan's support for anti-terrorist operations in Afghan border areas, resulting in its recent US elevation to major non-NATO ally (MNNA) status. Further US military equipment is expected to include imminent supply of six refurbished Lockheed C-130Bs from AMARC storage at Davis-Monthan AFB, Ariz., to supplement about 10 remaining PAF C-130B/E/L-100 Hercules operated since the 1960s.

US upgrades are also being sought for the PAF's 20 or so remaining F-16As and 10 two-seat F-16Bs, and clearance has been given to Belgium to offer 25 of its F-16s from storage, for possible purchase by Pakistan. Restoration is planned of the navy's two grounded Lockheed P-3Cs, which may be supplemented by six more from US aid.

Other planned procurement comprises 13 Kazan-built Mi-17 army transport helicopters, costing $50.7 million. Acquisition is also planned of 150 or more PAF FC-1/JF-17 Thunder advanced lightweight multi-role developments of the MiG-21.

Philippines

Hueys for the PAF

The Philippines has purchased 20 refurbished and upgraded UH-1H helicopters at a cost of $12 million. Singapore Technologies Aerospace began delivering the second-hand aircraft at a rate of three per month in March 2004. Although the Philippines AF already has some 60 UH-1Hs in its inventory, less than 30 are believed to be operational. It had previously accepted US offers that will provide 30 UH-1Hs.

Russia

New equipment plans

Construction is nearing completion by the Novosibirsk Chkalov Aviation Production Association (NAPO) of the eighth Sukhoi Su-34 (originally designated Su-27IB, or Su-32 for export) strategic strike aircraft built to date, for extensive development trials. These are continuing at the Zhukovskiy Flight Test Institute (LII) near Moscow, and the Federal Russian air force (FRAF) State Flight Test Centre (GLITs) in Akhtubinsk. During a NAPO visit, FRAF C-in-C General-Colonel Vladimir Mikhailov said that preparations were progressing for full-scale production of the Su-34.

He added that NAPO will build up to 10 new Su-34s by 2005–2006, to replace Tupolev Tu-22M3s and

On 23 June 2004 an Antonov Airlines An-124 arrived at RNAS Yeovilton to transport the first three of 16 AgustaWestland Super Lynx 300 Mk 120s on order for the Royal Air Force of Oman. They are expected to join Nos 3 and 14 Squadrons. The first Omani Lynx flew at Yeovil on 25 October 2003.

Buckeye retirement

In April 2004 the final class of US Navy Student Naval Aviators (SNAs) completed its Intermediate training on the T-2C and the venerable Buckeye was retired from the pilot training role it has fulfilled for the past 40 years. By August 2004 all the Buckeyes operated by VT-9 'Tigers' at NAS Meridian, the final operator of the type in the pilot training role, were due to have been withdrawn. The squadron now operates the T-45C 'glass' cockpit Goshawk.

The T-2Cs at Meridian have all accrued over 11,000 flight hours, so most will go to AMARC at Davis Monthan AFB for permanent storage. Some will find their way to Museums and others may fly on with VT-86 'Sabrehawks', which continues to employ the T-2C in the Naval Flight Officer (NFO) Advanced training role at NAS Pensacola. The only other US operator of the T-2C is the Navy Test Pilot School at Patuxent River, Maryland. Clearly the withdrawal of the T-2C fleet at Meridian will significantly impact the cost of operating the remaining aircraft and could accelerate the full retirement of the type from the Naval inventory.

The T-2C was used by VT-9 for Intermediate Strike Training, which applies a syllabus designed to provide students with the basic skills required for tactical strike and carrier aviation. When the T-2C student class completed its course the Intermediate Strike element of Naval flight training lapsed. With the retirement of the T-2C Buckeye the US Navy will have completed the transition to the T-45 Goshawk as its sole platform for tactical jet pilot training, and all new tactical jet student pilots will in future enter an all T-45 'TS' (Total System) syllabus. The TS course will take the Navy's student pilots direct from Primary Training through to gaining their naval aviator's wings.

The T-2C will be undoubtedly be missed. With its straight wing and wide undercarriage it is immensely stable and robust – ideal attributes for a training aircraft. Additionally, it was unique in being the only jet trainer capable of being routinely spun to provide students with experience of out of control flight. The aircraft was renowned for being easy to maintain with no work stands or ladders required for maintenance and large quick-open access doors for equipment, accessories and engines. However, while the airframe remained 'willing and able' to the end, the Buckeye's dated avionics and analogue cockpit have become obsolete and too far removed from the cockpit environment awaiting today's student pilots.

Richard Collens

VT-9 has now fully converted to the T-45C Goshawk for pilot training, but the Buckeye survives in the NFO and test pilot training roles.

shorter-range Su-24M strike aircraft, and work was also progressing to develop electronic warfare and reconnaissance variants. With a new cockpit layout, the Su-34 has received preliminary FRAF certification, and State Arms Programme-authorised procurement. An Su-34 would soon be received by the Lipetsk-based 4th TsBP i PLS, for further service evaluation and training.

NAPO also received an upgrade and life extension order for 12 Sukhoi Su-24 variable-geometry strike aircraft, some of which will be delivered as part of a contract with Algeria, and others to augment the active FRAF inventory.

In December, the C-in-C said that the FRAF would start to receive three Yakovlev Yak-130s, plus a few initial production MiG-AT advanced jet-trainers, by mid-2004. Having achieved FRAF service certification clearance, after over 80 evaluation flights, the MiG-AT would be operated "for experimental purposes", and offered to foreign customers, in conjunction with official French backing. Export potential for the MiG-AT to third-world countries is estimated by RSK MiG as 550-600 aircraft, worth $8-9 billion.

Mikhailov also announced production launch of the new Ilyushin Il-112 high-wing twin-Klimov TV7-117S turboprop-powered tactical transport, to begin replacing venerable RFAF Antonov An-24s and An-26s. Series production is also starting of heavy-lift Il-76MF transports at Russia's Voronezh plant, together with similar upgrades of current FRAF Il-76MDs.

New bombers

The Gorbunov Aviation and Production Association in Kazan, Tartarstan, has reportedly begun construction of three new Tu-160 strategic bombers for the Russian AF. The plant will also repair and upgrade the 15 operational Tu-160s that are still in service. The facility is also modifying a number of Tu-22MR reconnaissance aircraft.

South Africa

Arms package progress

A strengthened rand last year allowed the government to lower its overall cost estimate of South Africa's strategic arms package, which includes Gripen combat aircraft, Hawk jet trainers, submarines, corvettes and utility helicopters. The treasury's latest package cost assessment of R48.7 billion ($7.47 billion) compares with last year's estimate of about R53 billion ($8.14 billion), and the original R29.9 billion for the seven fiscal years from 2000-01, but package costs remain well over the original figure in current terms. This year, they represent about 40 percent of total defence spending, threatening allocations for operations and training. Nearly 30 percent more is earmarked for upgrading armoured personnel carriers, and the SAAF's 12 Denel AH-2 Rooivalk combat helicopters to Block 1E standards, plus acquiring new man-portable air-defence missiles (MANPADS).

South Korea

AEW&C RFPs issued

Three international military aircraft contractors were expected to respond to requests for proposals (RFPs) issued in March by the Korean Defence Ministry for four AEW&C aircraft, to meet the RoKAF's long-standing $2 billion E-X programme requirements. Boeing had previously announced its intention to offer its B737-700-based AEW&C system, equipped with Northrop Grumman's multi-role electronically-scanned array (MESA) dorsal planar radar, already ordered by Australia and Turkey. Israel Aircraft Industries is teaming with South Korean companies to integrate Elta's phased-array radar and Israeli communications systems in a twin-turbofan Gulfstream 550, to achieve the required RoK industrial offsets of 30 percent of overall programme costs. As the third potential E-X bidder, L-3 Communications has been considering an Airbus-based AEW&C solution. Selection is expected by the year-end, for deliveries from 2009.

RFPs were also expected earlier this year for the Korean multi-purpose helicopter (KMH) programme. Funding may be a problem for this project, involving planned purchases of 400-500 utility and attack helicopters, to replace current RoKAF Bell AH/UH-1s and McDonnell Douglas Helicopters MD-500s in 2010-2012.

Taiwan

Orion procurement approved

Long-discussed plans for Taiwanese acquisition of 12 upgraded ex-USN Lockheed P-3C Orion maritime patrol aircraft, eight US diesel-engined submarines and three Lockheed Martin/Raytheon Patriot Advanced Capability-3 (PAC-3) SAM/ABM defensive missile systems were finally approved in June. Earlier Taiwanese interest in 12 new P-3Cs, before production ended in the mid-1990s, waned with their reported $4 billion acquisition costs. With new advanced mission systems avionics, and structural upgrades, Taiwan's refurbished P-3C package is costed at $NT53 billion ($1.57 billion). Most of the work involved will be farmed out by Lockheed Martin to local AIDC and airline support companies.

United Arab Emirates

Cougars for UAE

The United Arab Emirates has placed an order for 25 AS 532 helicopters with Eurocopter. The first example will be delivered in 2005 and the final aircraft will follow in 2007. The UAE already operates the earlier AS 332 Super Puma and SA 330 Puma.

United States

MPA aircraft ordered

Lockheed Martin recently signed a $87.4 million contact with EADS CASA, on behalf of the US Coast Guard, that covers the purchase of two CN-235 MRSMPA (Medium Range Surveillance Maritime Patrol Aircraft). Based on the CN 235-300M airframe, the MRSMPA was selected to fulfil the medium-range portion of the Coast Guard's Integrated Deepwater Systems (IDS) programme. The contract, which could be worth up to $300 million, includes options for spares, integrated logistics support and six additional aircraft.

New Joint STARS aircraft

Northrop Grumman delivered the 16th E-8C Joint Surveillance Target Attack Radar System (JSTARS) aircraft to the USAF's 116th Air Control Wing (ACW) at Robins AFB, Georgia, recently. Although it is a component of the Georgia Air National Guard, the wing is composed of both active component and ANG personnel and is referred to as a 'blended wing'. The aircraft, which is known as P-16, is built to the current Block 20 configuration but features an upgraded satellite communication (SATCOM) capability, and is the first production JSTARS equipped with a new Global Air Traffic Management (GATM) capability.

Argentine 'specials'

***Right:** Operated by Brigada Aérea III at Reconquista, this Pucará retains the scheme it received in 1999 to celebrate 25 years of Pucará operations.*

***Below:** This FMA/Beech B-45 Mentor was given this scheme in 2002 to celebrate the 90th anniversary of the Escuela de Aviación Militar at Cordoba.*

***Below:** Another aircraft to retain its special scheme for several years is this MS.760 Paris, which was painted up in 1999 to celebrate 40 years of the type in FAA service.*

C-130J delivered to AETC

The first new C-130J for the USAF's Air Education & Training Command (AETC) was formally delivered to the 314th Airlift Wing at Little Rock AFB, Arkansas on 16 April 2003. The aircraft, which is assigned to the wing's 48th Airlift Squadron, actually arrived in Little Rock on 19 March 2004. Serial 02-0314 was the first of five aircraft that will be delivered during 2004. The stretched C-130J is the first to be assigned to an active USAF unit. The 48th AS, which will serve as a formal training unit for the C-130J, is scheduled to receive seven additional aircraft by December 2005. The wing will eventually receive 31 C-130Js.

Encore for the Marines

The US Army has placed a $30.9 million order with Cessna Aircraft for four UC-35D Citation Ultra aircraft on behalf of the US Navy. The aircraft will be operated by the USMC.

Global Hawk news

Northrop Grumman delivered the second production RQ-4A to the USAF on 1 March 2004. The delivery of air vehicle P-2 (serial 02-2009) marked the completion of the first low-rate initial production (LRIP) lot. The next lot includes four air vehicles, three integrated sensor suites, two electro-optical/infrared sensors, and one launch and recovery element (LRE) for the USAF. In addition, two air vehicles, two integrated sensor suites, one mission control element (MCE), and one LRE will be delivered to the US Navy.

The contractor has also been awarded a $50.7 million contract covering advance procurement associated with LRIP Lot 4 production. Lot 4 comprises four RQ-4B air vehicles with enhanced integrated sensor suites and clip-in sensor, a mission control element (MCE), a launch recovery element (LRE), support equipment and spares.

The US Navy UAVs will be stationed at NAS Patuxent River, Maryland, where the Global Hawk Maritime Demonstration Project will be conducted, beginning in spring 2005. Although based on the USAF configuration, the RQ-4s will be equipped with modified radar and electronic support measures tailored for maritime operations.

Naval Hawks ordered

Sikorsky Aircraft has received a $164.7 million modification to an existing contract covering the purchase of 13 MH-60S helicopters by the US Navy. The contractor has also received a $84.4 million modification to an existing contract that covers the purchase of four MH-60R low-rate initial production (LRIP) Lot 2 helicopters. Delivery is expected in March 2006.

Venezuela

MiG-29 interest reported

Government requirements were reportedly being finalised earlier this year, after negotiations with RSK MiG, for up to 40 upgraded MiG-29SMT multi-role fighters and 10 two-seat MiG-29UB combat-trainers. With new digital avionics, AAMs and precision-guided ASMs, these were costed at up to $2 billion. FAV C-in-C Gen R.A. Espin had made a formal request to the government for 50 MiG-29s, following late 2002 demonstrations in Venezuela of the prototype MiG-29M2 MRCA. RSK MiG was then quoting initial delivery dates within 18 months of contract signature. The MiG-29s would supplement Venezuela's 22 long-serving GD F-16A/Bs, for which the FAV has been seeking mission system upgrades.

Vietnam

More Russian equipment

Contract finalisation was recently confirmed of a Vietnamese People's army air force (KQNDV) follow-on $110 million order, through Russia's Rosoboronexport, for four KnAAPO-built Sukhoi Su-30MK two-seat air defence fighters, plus eight options. These will supplement five Su-27SKs and one two-seat Su-27UBKs received in 1994-95 from a $380 million contract, plus two Su-27SKs and four Su-27UBKs delivered from 13 January 1998, from a $120 million December 1996 order.

The Su-30MKs are similar to those being delivered to China, from Sukhoi's Komsomolsk-na-Amur factory, with digital Russian and western avionics, but lacking the canard foreplanes and thrust-vectoring of India's Su-30MKIs. Their delivery will increase the overall KQNDV complement to 24, to equip a complete fighter regiment. About 30 percent of the contract funding will come from trade offsets, with the rest in cash.

Zimbabwe

More Chinese fighters

Procurement was announced in mid-June by Defence Minister Mutsekwa of an arms package from China, which included 12 new combat aircraft costing $220 million, and 100 military vehicles, plus personnel training. Although unspecified, the AFZ's new fighters are considered likely to comprise current production variants of the Chengdu F/FT-7, to supplement a half-dozen surviving earlier versions from 10 delivered in 1986.

AIR ARM REVIEW

India

Indian armed forces expansion

Further increases from the 2003-04 Indian military budget total of Rs653 billion ($14.38 billion), some 16.6 percent more than the previous year's Rs560 billion ($12.33 billion) total, are implicit from IAF C-in-C ACM Krishnaswamy's forecast of expansion from 39 to 60 squadrons by 2014, although following LCA programme delays, the IAF currently faces reductions in its fighter fleet to 32 squadrons by 2006. Many of the IAF's 300 or so early MiG-21FL/Ms, 100 MiG-23BNs and 120 MiG–27Ms will then be retired, with additional replacements planned to include 126 Mirage 2000-5s costing some $7 billion, following current negotiations with Dassault. India is also considering nine 1997 Mirage 2000-5EDAs and three two-seat -5DDAs offered by Qatar.

HAL is well-advanced with its $626 million joint programme with ANPK MiG and Sokol to upgrade 125 IAF MiG-21bis to MiG-21-93 Bison standards, using multi-national digital avionics, and with upgrading many other Indian military aircraft. Following recent crashes of three MiG-21 Bisons, however, provisionally attributed to engine fuel control problems, and raising doubts concerning HAL's quality checks, the IAF C-in-C decreed that each new Bison squadron must spend six months at the MiG factory in Ojhar before service deployment. The Bison's new Phazotron Kopyo multi-mode radar also reportedly performed "below expectations".

EC 1/5 'Vendée', based at Orange-Caritat, celebrated 200,000 hours in the Mirage 2000 by painting this 2000C in fine style.

The Spanish AF's new helicopter team 'Aspa' performed in public for the first time at the Tablada Air Show, Seville, on 16 May 2004. The new team is formed by five Ala 78 Eurocopter EC 120 Colibris, all of them crewed by Armilla-based EdA helicopter school instructors. For the moment the helicopters maintain their standard training colours, although it is planned that they will receive an appropriate colour scheme in the near future.

Su-30 teething troubles

Initial deliveries of 18 IAPO Irkut-built Sukhoi Su-30K (MK-I) and 22 Su-30MKI (MK-III) two-seat multi-role fighters, the latter with canards, thrust-vectoring AL-31FP turbofans, new Russian, Indian, and western digital avionics, plus associated weapons, were recently completed to the IAF from a 1996 $1.8 billion contract. Media reports in December of IAF refusal to accept the last 12 Su-30MKIs when airlifted into their Lohegan base in Pune, because of earlier reliability problems with their Lyulka AL-31FP turbofans, were circulating in Delhi, where further payments to Russia were reportedly suspended.

Although the AL-31FP has a nominal life between (major) overhauls (TBO) of only 300 flying hours, the IAF has apparently had to change most of the Su-30MKI's engines well before this total. It seems that Russia's quoted engine TBO applies only to "normal flying conditions", which apparently exclude the Su-30MKI's extreme thrust-vectoring/FBW manoeuvres. The IAF then sought compromises with its Russian contractors, to cover further R&D costs, for the Su-30MKIs to achieve their full combat manoeuvrability potential. HAL recently received its first Su-30MKI kits from Irkut to assemble up to 10 by the year-end.

Su-30Ks were among the 25 IAF aircraft, which also included upgraded MiG-21 Bisons, -27s, and -29s, plus Mirage 2000s, involved in the first Indo/USAF joint exercise for over 40 years, at Gwalior air base in central India in February. Six PACAF Boeing F-15C/Ds, with a supporting Lockheed C-5, C-130 and a Boeing KC-135 tanker/transport, flew in from Elmendorf AFB in Alaska for Exercise Cope India '04, involving over 150 dissimilar and joint simulated air combat sorties.

Currently equipping only two IAF squadrons at Lohegan air base, IAF Su-30 deployment is also planned from Halwara, in Punjab province, to replace current MiG-23BN ground attack fighters, in India's strategic Western Air Command. Su-30MKIs will also operate from Bareilly air base, currently housing Mach 2.8 IAF MiG-25s, and Chetak helicopters, in Central Air Command.

AEW advances

IAF Su-30 potency will be further multiplied from 2003 deliveries from Uzbekistan of six Ilyushin Il-78MK three-point tankers, plus orders for an initial three Russian-supplied Il-76-based Beriev A-50EhI AEW&C aircraft. These will also be Uzbekistan-built, although fitted with 157-kN (35,300-lb) Aviadvigatel/Perm PS-90A-76 engines to Il-76MD-90 standards in Russia, and equipped with IAI Elta EL/2075 D-band phased-array radar with a 400-km (216-nm) operating range, in Israel. IAF CAS Air Chief Marshal Krishnaswamy said A-50 deliveries would start in 2006, with up to five received by 2008. IAI is receiving a $350 million advance payment, following January signature of the $1.1 billion A-50 contract.

In December, a senior Defence R&D Organisation (DRDO) official announced revival of an indigenous AEW&C aircraft programme, abandoned following the 1999 crash of an Avro/HAL 748 research aircraft. DRDO Centre for Airborne Systems Director K.U. Limaye said that a next-generation 200-km (108-nm) range fixed active phased-array radar, rather than the Avro 748's rotating assembly, would be installed on a smaller aircraft, such as the twin-turbofan EMBRAER EMB-145. Budget allocations of $154 million were made in December 2003 for four 14-seat IAF and one Border Security Force (BSF) EMB-145s, for government transport and surveillance patrols.

Iraq

Plans for new air force disclosed

Plans for re-establishing the Iraqi AF were announced by the Coalition Provisional Authority and the Ministry of Defence during a briefing held on 17 April 2004. Once it is activated the new Iraqi AF will be an integral part of coalition efforts and will work closely with ground, maritime and air units. Its roles will include policing 5633 km (3,500 miles) of Iraq's international borders and surveillance of its national assets. It will also be capable of rapidly deploying the Iraqi Army. The air force will initially be equipped with a pair of C-130Bs, provided by Jordan, by October 2004. The airlift fleet, which will be based at Baghdad, will grow to six aircraft by April 2005. A squadron of six UH-1H helicopters will initially be operational at Tadji AB by July 2004, however the number of helicopters will increase to 16 by April 2005. In addition, a squadron of light reconnaissance aircraft will become operational this summer at Basrah. It will initially be equipped with two SB7L-360 Seekers and tasked with infrastructure and border security duties. The squadron's fleet and its operating locations will eventually be expanded. Currently more than 100 members of Iraq's former air force are currently undergoing flight instruction with the Royal Jordanian AF in Amman. A Major General will command the new organisation.

United States

Texan II for Pax River

The first T-6A was delivered to the Naval Air Warfare Center at NAS Patuxent River during a formal acceptance ceremony on 11 March 2004. The Texan II joined Air Test and Evaluation Squadron VX-20's test fleet. VX-20 will be responsible for test programmes associated with both the US Navy and USAF T-6s. Initially delivered to the Navy in November 2002, BuNo. 165958 was the first production example built for the Navy. The service plans to purchase 328 aircraft to replace the T-34 in the primary training role and has already ordered 49 examples. Thirty Texan IIs are currently assigned to Training Air Wing 6 (TAW-6) at NAS Pensacola, Florida, where they support Naval Flight Officer (NFO) training.

Virginia guard could relocate

State officials have been briefed on a plan that could see the Virginia ANG's 192nd Fighter Wing/149th Fighter Squadron relocated to Langley AFB, where it would become an associate unit to Air Combat Command's 1st Fighter Wing. Under this proposal the 192nd, which has been stationed at Richmond International Airport in Sandston since 1947, would share the F/A-22A fighters at Langley.

Fighter Associate Programme

Air Combat Command and Air Force Reserve Command have jointly established the Fighter Associate Program (FAP), which will increase fighter pilot experience levels in active-duty flying squadrons. Under the programme a detachment of four experienced AFRC personnel, comprising a full-time detachment commander and three

A Northrop Grumman/Boeing E-8C J-STARS flies over the 128th Air Control Squadron ramp at Robins AFB, Georgia. The unit is assigned to the 116th ACW.

Recent C-130J deliveries to the USAF have included aircraft for the active-duty 314th AW's 48th Airlift Squadron, which acts as the OCU. This stretched C-130J is newly assigned to the 815th AS, part of Air Force Reserve Command's 403rd AW at Keesler AFB, Missouri.

traditional reservists are being assigned to specific active duty fighter squadrons (FS), where they will serve as instructor pilots. Additionally, two full- and four part-time maintainers, and one full-time administrator are being assigned to certain detachments. The FAP reservists are assigned to the 307th Fighter Squadron, which was activated at Langley AFB, Virginia, on 1 August 2003. The squadron, which reports to the AFRC's 10th Air Force, maintains administrative control of the AFRC personnel assigned to the FAP. Initially the 307th will establish four associate detachments that will fly and maintain F-15C, F-16C (Block 30, 40 and 50 series) and A-10A aircraft.

307th FS, Langley AFB, Virginia

Det. 1	Shaw AFB, South Carolina
Det. 2	Hill AFB, Utah
Det. 3	Nellis AFB, Nevada
Det. 4	Eglin AFB, Florida

Similar programmes already exist at Luke AFB, Arizona, where AFRC personnel assigned to the 301st Fighter Squadron (FS), support the 56th Fighter Wing (FW) and at Tyndall AFB, Florida, where instructors from the Florida ANG support the 325th FW. The 56th and 325th, respectively, serve as Air Education & Training Command's formal training units (FTU) for the F-16C/D and F-15C/D.

Under the active associate portion of the programme, ACC will embed three active-duty pilots, comprising one instructor pilot and two inexperienced pilots who have recently completed fighter upgrade training, within AFRC fighter squadrons. The active component detachments will be assigned to the following reserve units: 301st FW (NAS Fort Worth JRB, Texas), 419th FW (Hill AFB, Utah), 442nd FW (Whiteman AFB, Missouri), 482nd FW (Homestead JARB, Florida) and 926th FW (NAS New Orleans JRB, Louisiana).

FAP is an outgrowth of the earlier Fighter Reserve Associate Program (FRAP) and the Total Force Absorption Program (TFAP). Both were created in an attempt to provide new pilots with the 500 flight hours needed to become fully qualified and increase the number of qualified pilots assigned to the active component annually from 302 to 330-380. Established at Shaw AFB in 1997, FRAP offered 20th FW pilots, who were leaving active duty, the chance to join the reserves and continue flying with the wing. The TFAP sent inexperienced active-duty pilots, or those with limited experience, to AFRC and Air National Guard units. Although the latter programme, begun in 2000, reduced the burden placed on the active-duty squadrons, it still required that the pilots report to an active-duty unit quarterly for additional training. FAP combines the best aspects of both earlier programmes and will be fully implemented by mid-2004.

USAF may reactive Lancers

The commander of the USAF's Air Combat Command recently disclosed that the service is considering returning seven or eight of its retired B-1B bombers to service. Over the past two years the USAF retired 23 of the bombers, which were placed in storage at AMARC or put on display in museums. As a result of the aircraft's successes during Operation Iraqi Freedom, the US Congress allocated $18 million to return the aircraft to service, however the ACC commander indicated that the actual cost would be more than $3 billion. The reactivation of a small number of B-1Bs would provide ACC with a larger number of aircraft for training, test and an attrition reserve.

The S-3B Viking retirement process began on 16 April 2004 with the deactivation at NAS North Island of two Pacific Fleet squadrons: VS-29 'Dragonfires' and VS-38 'Red Griffins' (CAG-bird illustrated). Formal deactivation occurred on 30 April. In April the S-3s assigned to test units VX-1 and VX-20 were also retired. VS-21 is the next unit scheduled for disbandment, and under current plans the Viking will have left the fleet by January 2009, with VS-22 slated to be the last unit.

'Blended' Predator units

The USAF has announced plans to create integrated Predator teams comprising personnel from active-duty, ANG and AFRC units that will operate the MQ/RQ-1 Predator Unmanned Aerial Vehicles (UAV). Initially Air Combat Command will integrate assets from the California and Nevada ANG into the 11th and 15th Reconnaissance Squadrons (RS) at Indian Springs Air Force Auxiliary Field (AFAF), Nevada, which already have active duty and AFRC assets assigned. The moves will mark the first time that ANG personnel from different states have been assigned to the same units.

In related news Air Combat Command has terminated environmental studies that were examining Edwards AFB, California, and Holloman AFB, New Mexico, as possible homes for Predator UAVs. It also announced that the entire fleet will be based at Indian Spring AFAF. As a result of this decision, 443 additional personnel and 48 additional MQ-1 Predator A and MQ-9 Predator B air vehicles will be assigned to the facility. Approximately 40 MQ/RQ-1s are already assigned to the base. In addition to assigning 12 MQ-1s and four MQ-9s to the 17th RS, the decision clears the way for creating separate Force Development Evaluation (FDE) and Formal Training Units (FTU). The former organisation will be equipped with four MQ-1 and four MQ-9s, while the latter will have 12 of each variant assigned. The 11th and 15th RS currently carry out these tasks. The turboprop-powered MQ-9 is due to achieve initial operational capability (IOC) in 2005.

AETC update

The 12th Flying Training Wing's (FTW) 560th Flying Training Squadron (FTS) at Randolph AFB, Texas, took delivery of its first T-38C early in 2004. A crew from the AFRC's 420th Flight Test Flight (FTF) delivered serial 67-14921 to Randolph from Williams Gateway Airport, Mesa, Arizona, where it was modified to the latest configuration. The 560th, which trains instructor pilots, is expected to complete its conversion from the T-38A by August 2004. In related news, Boeing delivered the 200th T-38C to Randolph on 2 February 2004.

The 47th FTW at Laughlin AFB, Texas marked the midway point in its transition from the T-37B and T-6A trainers, when the 84th FTS conducted its final T-37B sortie on 7 January 2004. The flight also marked the completion of that squadron's transition to the Texan II. At that time the wing's mixed fleet included 25 T-37Bs and 25 T-6As, however the former's numbers continue to dwindle as the 85th FTS converts to the new aircraft. The 85th's first students began training with the T-6A in March 2004 and the last of the T-37Bs should be gone by the end of 2004.

The Texas ANG's 149th Fighter Wing, one of three Air National Guard F-16 formal training units (FTU), recently began providing instruction with the LANTIRN targeting pod as part of the F-16 basic course. Prior to the delivery of 10 targeting pods, the student were required to transfer to Luke AFB, Arizona, after completing the basic course.

Guard rescue wings reorganised

The California and New York ANG's 129th and 106th Rescue Wings (RQW) recently underwent a reorganisation that resulted in the activation of four new squadrons. Previously the wing's operations groups (OG) each controlled a single 'composite' rescue squadron (RQW), which had HC-130s, HH-60s and pararescue jumpers (PJ) assigned. As a result of the reorganisation, which parallels the structure of USAF and AFRC units, each of these assets is now assigned to individual squadrons. The two wings were reassigned from Air Combat Command to Air Force Special Operations Command on 1 October 2003.

106th RQW
Suffolk County ANGB/Gabreski Airport, Westhampton Beach, New York

106th OG	101st RQS	HH-60G
	102nd RQS	HC-130H/P
	103rd RQS	PJs

129th RQW Moffett Federal Airport, Sunnyvale, California

129th OG	129th RQS	HH-60G
	130th RQS	HC-130H/P
	131st RQS	PJs

More powerful Eagles

The Florida ANG's 125th Fighter Wing recently accepted its first F-15A fighter to be retrofitted with Pratt & Whitney F100-PW-220E engines. Initiated in 1997 the programme, which is commonly referred to as the -220 equivalent kit (220 E-kit) upgrade, modifies the fighter's F100-PW-100 engines to the more powerful configuration. Supplied by Pratt & Whitney and installed by the Oklahoma City Air Logistics Center (OC-ALC) at Tinker AFB, the 'E-kit' provides the USAF with -220 equivalent engines at less than half the cost of new ones. By 2006 each of the wing's 19 fighters will be updated.

New C-20A operators

The US Navy recently took delivery of a former USAF C-20A Gulfstream III. Serial 83-0500, which last served with the 76th Airlift Squadron at Ramstein AB, Germany, is assigned to the Executive Transport Detachment at NAS Sigonella, Italy, replacing a VP-3A Orion.

The US Army Corps of Engineers has taken delivery of serial 83-0502, which has been assigned the civil

End of the line for VX-9's Tomcats

Having conducted its final Test Detachment to NAS Key West in April and concluded all its remaining Test programmes, the long standing VX-9 'Vampires' F-14 Detachment at NAS Point Mugu held its formal disestablishment ceremony at the base on 22 June 2004.

VX-9 was formed in 1993 when VX-4 'Evaluators' and VX-5 'Vampires' were consolidated into a single operational test and evaluation squadron at NAS China Lake, with a permanent F-14 Detachment established at NAS Point Mugu, California. VX-9's wider mission has grown to include the operational testing (OT) of all attack, fighter and electronic warfare aircraft, weapons systems and equipment, and to develop tactical procedures for their employment in the US Navy and Marine Corps. Commanded by Captain Wade Tallman, VX-9 operates over 20 aircraft, including the F/A-18E/F Super Hornet, F/A-18A/B/C/D Hornet, EA-6B Prowler, AV-8B Harrier and AH-1 Cobra. Typically, aircrews are qualified in more than one of these aircraft types, increasing their versatility and providing broader based expertise to be applied to each test project.

VX-9's operational chain of command is through the Commander Operational Test and Evaluation Force (COTF) and administratively the squadron reports to the Commander, Naval Air Force, US Pacific Fleet. COTF receives direction on which projects to undertake directly from the Chief of Naval Operations (CNO). However, the Commander, Naval Air Forces Pacific provides the aircraft and parts support, while the majority of test funding is supplied by various commercial sponsors.

The F-14 Detachment at Point Mugu has operated as an autonomous unit for over 10 years, providing Operational Test support for the US Navy's Tomcat Fleet. In recent times the Tomcat operations at Point Mugu of VX-30 'Bloodhounds' and VX-9 have been pooled with VX-30 providing all the maintenance and line support. The two units aircrew have also 'cross operated' where operational requirements have required.

One of VX-9's three remaining F-14D aircraft at the time of the Point Mugu detachment's disestablishment was BuNo. 164599 '254', which had around 2,900 flight hours and was flown to AMARC on 23 June 2004.

Because of their role many of the VX-9 jets carried test instrumentation and they are incompatible with current fleet aircraft. Therefore, despite their relatively modest airframe flight hours few, if any, of the F-14s will continue flying and the majority are heading for retirement in the desert. Four of the unit's remaining F-14B/Ds have or will be retired to AMARC at Davis Monthan AFB, one F-14D has been reallocated to VF-101 at NAS Oceana, and F-14D BuNo. 164604 – the world famous 'Vandy 1' and the last F-14 ever built – is also headed for NAS Oceana but for static display.

Richard Collens

registration N65CE. It will be based at Omaha, Nebraska, replacing a Gulfstream I. The USAF originally took delivery of three C-20As, including serial 83-0501, which is currently assigned to the NASA Dryden Flight Research Center at Edwards AFB, California.

Reserve airlift news

The Tennessee Air National Guard's 164th AW has begun its transition from the C-141C to the C-5A and will retire its last Starlifter in May 2004. The unit is stationed at Memphis IAP. The Hawaii ANG's 204th Airlift Squadron, at Hickam AFB, is preparing for the delivery of its first C-17A in early 2005. The unit currently operates the C-130H in support of PACAF. Air Force Reserve Command's 934th Airlift Wing at Minneapolis-St. Paul IAP, Minnesota, recently transitioned from the C-130E to the C-130H. The move leaves the 913th AW at NAS JRB Willow Grove, Pennsylvania, as AFRC's last C-130E unit.

Edwards moves

The USAF reactivated the 445th Flight Test Squadron (FLTS) at Edwards AFB, California, on 11 March 2004. The squadron, which had previously been deactivated on 30 November 2001, operates the T-38 and F-15 aircraft assigned to the 412th Test Wing.

Navy changes

The Navy retired the first of two HH-1N helicopters that provided search and rescue (SAR) coverage for NAS Brunswick, Maine. The second aircraft will follow in July. In addition to SAR duties the aircraft supported the Navy's cold weather survival school.

VAQ-128 'Fighting Phoenix' held a deactivation ceremony at NAS Whidbey Island, Washington, on 7 May 2004. The squadron, which operated the EA-6B as one of four joint services expeditionary electronic warfare squadrons, will be formally deactivated on 30 September 2004. It recently returned from a deployment to Japan in support of Marine Air Group 12.

The 'Warhawks' of VFA-97 have finally begun transitioning from the F/A-18A to the F/A-18C. The squadron, which is based at NAS Lemoore, California, was the last active duty Navy squadron to operate the initial variant of the Hornet.

Strike fighter squadrons VFA-22 and VFA-27 will be the next two squadrons to transition from the F/A-18C to the F/A-18E Super Hornet. Both units will convert during 2004 at NAS Lemoore. The Atlantic Fleet's Super Hornet fleet readiness squadron will receive its first Super Hornets in March 2005, when the first aircraft are assigned to VFA-106 at NAS Oceana, Virginia.

VFA-204 'River Rattlers' at NAS New Orleans JRB, Louisiana, have transitioned from the F/A-18A to the F/A-18A+ variant. VFA-203 'Blue Dolphins' at NAS Atlanta, Georgia, were deactivated on 30 June 2004 and the squadron's remaining F/A-18As were transferred to VFC-12 at NAS Oceana, Virginia. The latter squadron currently serves as an adversary unit but will activate a fleet response unit.

Two Boeing 737-200 series airliners that were used by Fleet Air Reconnaissance Squadron Seven (VQ-7) as flight trainers by E-6B Mercury crews at Tinker AFB, Oklahoma, were recently replaced by a single B737-6Z9. Unlike the earlier -200s, the new aircraft, which wears the civil registration N743NV (c/n 30137/526), is powered by a pair of CFM56-7 turbofan engines.

Army guard news

The Missouri ARNG recently broke ground for a new army aviation support facility at Waynesville Regional Airport in Fort Leonard Wood. The new facility will house 10 UH-60 helicopters operated by 1-114th Aviation. The Florida ARNG's B/1-111th Aviation was recently called to active duty. Although the unit is normally based at Cecil Field Airport in Jacksonville, its personnel relocated to Fort Carson, Colorado, before deploying to Afghanistan. The company's AH-64As were not deployed. The Louisiana ARNG's 812th Medical Company (Air Ambulance) is currently converting from the UH-1H/V to the UH-60L. The unit is located at Alexandria-Esler Regional Airport in Pineville.

As part of its ongoing modernisation efforts the Army National Guard has activated several new heavy helicopter units equipped with CH-47D Chinooks. The aircraft will allow the guard to better support its state and homeland security missions and to augment active component units. Each company/detachment is divided across state lines and will operate 5-7 aircraft. In addition to the newly activated units, several existing units were redesignated or realigned under other units.

D(-)/113th AVN (HH) Reno-Stead Airport, NV
Previously Det. 1 G/140th AVN
Det. 1 D/113th AVN (HH) Eastern Oregon Regional Airport, Pendleton, OR
Previously Det. 1 E/168th AVN
G(-)/137th AVN (HH) Akron-Canton Regional Airport, OH
Det. 1 G/137th AVN (HH) Rochester IAP, NY
Det. 1 E/168th AVN (HH) Buckley AFB, CO
Assigned to E(-)/168th AVN in Washington
Det. 1 G/185th AVN (HH) Selfridge ANGB, MI
H/189th AVN (HH) Helena Regional Airport, MT

In April 2004 the Swedish Air Force introduced two Saab Tp 100C (340B) aircraft (right) to replace Beech Tp 101s as wing support aircraft at F17 Ronneby and F21 Luleå. Luleå's King Air 200 is seen above just days before its retirement.

RAH-66 Comanche

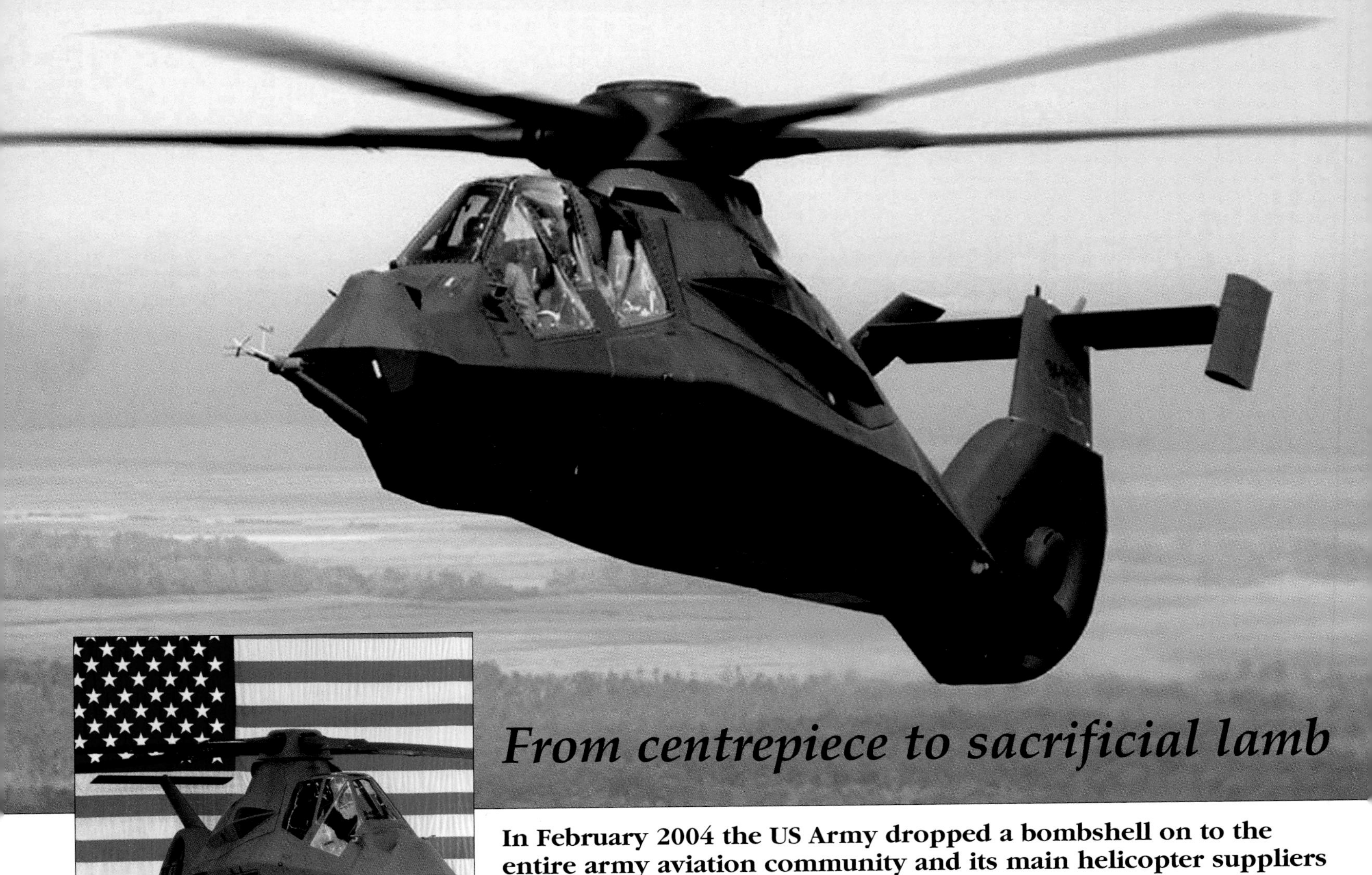

From centrepiece to sacrificial lamb

In February 2004 the US Army dropped a bombshell on to the entire army aviation community and its main helicopter suppliers – after already sinking $7 billion into the programme, the service was cancelling the RAH-66 stealthy attack helicopter. Here, the Comanche's development history and technological merits are reviewed, before the costliest and most ambitious helicopter programme ever launched recedes into distant memory.

In April 2000 the US Army unveiled its Aviation Force Modernization Plan (AFMP) and announced plans to retire its fleet of AH-1, UH-1 and OH-58 attack, utility and scout helicopters. The Boeing-Sikorsky RAH-66A Comanche served as the centrepiece of the AFMP, which would have resulted in a complete redesign of the army's aviation branch. Relying heavily on Comanche, the AFMP envisaged the purchase of 650 examples of the stealthy scout/attack helicopter. Less than four years later, the acting secretary of the Army and the chief of staff abruptly announced the cancellation of the Boeing-Sikorsky RAH-66A Comanche scout helicopter. This occurred on 23 February 2004 and coincided with the announcement of yet another reorganisation of the Army Aviation Branch. Just weeks before the announcement, the Army had released its fiscal year 2005 (FY05) budget request, which included $1.3 billion for research, development and acquisition associated with Comanche.

After its roll-out at Stratford on 25 May 1995 (above left), the prototype RAH-66 (94-0327) was displayed on the Pentagon helipad in June. It did not make a first flight until 4 January 1996, after having been trucked to West Palm Beach in Florida. Here it is seen on its maiden flight, with Sikorsky's Russ Stiles and Boeing's Rob Gradle at the controls. Because of a mechanical problem, it did not fly again until 24 August.

Despite the investment of $7 billion in Comanche since 1984, recent studies – which were not complete when the budget was developed – convinced officials that the new money could be spent more effectively on other projects. Although the Comanche project was moving forward it was both behind schedule and over budget. Through the life of the project the helicopter had been transformed from a simple, yet stealthy, scout to a near replacement for the more heavily armed AH-64D. Also, the programme's most recent restructuring had reduced production by nearly a half and pushed its in-service date back three years.

Development

The Boeing-Sikorsky RAH-66 Comanche was the first helicopter developed specifically for the role of armed reconnaissance; however, it was originally intended as a multi-platform replacement for the Army's AH-1 Cobra, UH-1 Iroquois and OH-58 Kiowa helicopters. When fielded, the Comanche would have been assigned to air cavalry troops and attack helicopter companies, and as the primary attack asset within cavalry squadrons, and light infantry and airborne divisions.

Originally developed to perform the light scout/attack role, the Comanche grew in sophistication until it emerged as a 'stealth Apache'. In its final incarnation the RAH-66 was intended to scout for the AH-64D within the operations of 'heavy' armoured and mechanised forces, or provide organic attack and reconnaissance for 'light' infantry formations such as airborne forces.

Development began in 1982 when the Army issued a request for Light Helicopter Experimental (LHX) design concepts. The service was then looking for two specific variants capable of carrying out light-utility and scout attack (SCAT) missions, and it expected to buy more than 4,500 examples. The systems employed by the two variants were expected to demonstrate 70 percent commonality. Despite their different missions, both variants were expected to weigh approximately 8,000 lb (3629 kg). Whereas the light-utility variant was to be designed to carry six troops or 2,000 lb (907 kg) of cargo, the SCAT model was intended to carry advanced sensors and weapons.

In June 1985 Boeing Helicopters and Sikorsky Aircraft joined forces to bid for the LHX contract, and by April 1986 Bell Helicopter and McDonnell Douglas Helicopter had similarly formed a team. After determining that a single platform could not fulfil both missions, the army issued a revised proposal for a single scout and attack aircraft in 1987. Accordingly, the number of helicopters required by updated request was reduced to 2,096. The service issued a formal request for proposals (RFP) in June 1988 and, by October, the two competing manufacturing teams, Boeing Helicopters/ Sikorsky Aircraft and Bell Helicopter/ McDonnell Douglas Helicopter, were awarded 23-month demonstration/validation (DEM/VAL) contracts valued at $158 million. The advanced designs proposed by both teams incorporated low-observable (stealth) characteristics.

Whereas the Bell/McDonnell Douglas team offered a platform equipped with a bearingless four-bladed main rotor and NOTAR (no tail rotor), the Boeing and Sikorsky team proposed a bearingless five-bladed main rotor and shrouded tail rotor that was known as a fantail. Both designs were to be powered by the LHTEC T800 engine that was the subject of a separate competition. Following a period of evaluation, the Boeing/Sikorsky team was declared the winner in April 1991 of the renamed 'LH' competition.

Comanche debut

The first prototype RAH-66 (serial 94-0327) was rolled out at the Sikorsky Aircraft facility in Stratford, Connecticut, on 25 May 1995 and its maiden flight took place on 4 January 1996 at the contractor's flight test facility in West Palm Beach, Florida. The aircraft, which was powered by two 1,432-shp (1068-kW) T800-LHT-800 turboshafts, was primarily dedicated to verifying the helicopter's structural design and flying qualities, and as development progressed numerous modifications were made to the airframe.

Among them were the installation of a new empennage, lowered exhaust doors, an alternate main rotor pylon, main rotor hub fairing and an aerodynamic representation of the mast-mounted radome for the fire control radar (FCR). The first flight with the new configuration took place on 18 December 2000. Installed in an attempt to resolve buffet problems, the reconfigurable empennage featured vertical and horizontal stabilisers, and vertical endplates were mounted on the horizontal structures.

Boeing and Sikorsky were partnered in what was known as the 'First Team' for the LHX competition . Between them, the two partners have become the primary Army helicopter suppliers of recent times, responsible for the AH-64, CH-47 and UH-60.

With gear down, the no. 1 Comanche leads no. 2 during a photo-shoot. The two flew together relatively briefly between no. 2's first flight in March 1999 and no. 1 being laid up for modifications in early 2000.

Power for the Comanche came from two of LHTEC's compact and powerful T800 free-turbine engines. In its initial T800-LHT-800 form, the engine developed 1,601 shp (1194 kW) at emergency rating, for a dry weight of just 310 lb (140.6 kg) and dimensions of 33.2 x 21.6 x 26 in (84.3 x 55 x 66 cm).

Eventually, more powerful T800-LHT-801 engines, rated at 1,563 shp (1166 kW) were also installed. The newer model provided nearly a 10 per cent increase in power over the earlier engines. The -801 also allowed the RAH-66 to meet the requirement to demonstrate a 500-ft (152-m) per minute vertical rate of climb at a temperature of 95ºF (35ºC) and an altitude of 4,000 ft (1219 m). It also provided the aircraft with a 165-kt (306-km/h) cruise speed, at a mission gross weight of more than 12,000 lb (5443 kg), in hot and high environments.

First flown on 1 June 2001, the engines were followed by new longer-span rotor blades that featured down-turned tips. Initially tested on 20 July 2001, the anhedral added to the rotor blades was intended to improve the aircraft's hover performance and reduce its infrared signature. Ultimately, all of these modifications would have been included in the production configuration. The prototype completed the last of 318 sorties on 30 January 2002, accumulating 387.1 flight hours over a 6.5-year period.

Above: An early milestone for the RAH-66 was achieved on 20 September 1996 when the Comanche was flown for the first time by an Army pilot (CWO John Armbrust). The RAH-66 prototype flew with its gear retracted for the first time on 25 February 1997.

Nicknamed 'the Duke' in honour of the late actor John Wayne, the second prototype (serial 95-0001), flew on 30 March 1999 at West Palm Beach. The aircraft initially served as a back-up for the No. 1 prototype and it accumulated 103.5 flight hours during 93 test flights before being removed from flight status in May 2001. Always intended as an avionics test bed, after being taken off flight status the 'Duke' underwent a modification programme that installed the T800-LHT-801 engines and the full Comanche mission equipment package (MEP) and software. The MEP was an integrated targeting, communications and navigation system. The Comanche returned to the flight test programme at the conclusion of integration testing, making its first flight in support of MEP development on 23 May 2002.

Design and features

The Comanche's design incorporated a number of stealth features, presenting a radar cross section (RCS) reported to be smaller than that of a Hellfire missile. Its stealth features included retractable landing gear, a stowable cannon and internally carried weapons. Constructed entirely using composite materials, the fuselage featured faceted sensor turrets, flat-plate canopies and flared fuselage sides. All were intended to deflect radar signals away from their source.

While the five-bladed composite main rotor and fantail reduced the aircraft's acoustic signature, its infrared signature was reduced by locating heat-producing engine components within the fuselage. Additionally, hot engine exhaust gases were mixed with ambient air and ducted out under the tail. As a result, the Comanche was said to radiate 75 percent less heat than military helicopters then in service.

Designed for ease of maintenance, its weapons bay doors, which doubled as maintenance work stands, comprised approximately 50 percent of the aircraft 'skin'. Most of the electronic equipment aboard the RAH-66 was installed in two primary avionics bays running along the sides of the fuselage and a third located in the nose. The triple-redundant fly-by-wire flight control system included sidestick controls for the pilot and co-pilot.

It was equipped with a built-in diagnostic system that made it more maintainable in the field and, rather than a traditional three-level maintenance system, Comanche's maintenance philosophy consisted of only two. Requiring only organisational (line- or field-level) and depot-level maintenance, Comanche had no requirement for traditional intermediate-level (shop-level) repairs. Keeping Comanche in service was expected to require only 2.6 maintenance man-hours per flight hour (MMH/FH) compared to 4.5 for the OH-58D and 8.6 for the AH-64D.

Turning the aircraft around at a forward army and refuelling point (FARP) was to have been accomplished by just two troops in 15 minutes. These features would have allowed the RAH-66 to be maintained by fewer personnel and with significantly less support equipment than those helicopters now in service. The aircraft was also designed for air transport

Right: The two Comanches are seen during flight trials, along with a Sikorsky S-76 which was used for test and support duties.

A Comanche demonstrates the retractable I-RAMS weapons doors, each of which could be fitted with three hardpoints. For heavy attack and long-range missions the RAH-66 could also mount non-retractable EFAMS stub wings with additional hardpoints.

The RAH-66's unusual shape was driven by the need for stealth. All major surfaces, including the kinked rear fuselage and canted fin, would reflect radar returns away from the illuminating radar. The rotor hub was shrouded by a radar-defeating fairing.

in a C-130, versus larger strategic transports, and could be made operational within 90 minutes of arrival.

Manufactured by the Light Helicopter Turbine Engine Company (LHTEC), a joint venture between Rolls-Royce (Allison) and Honeywell Aerospace (formerly Allied Signal/ (Garrett), a pair of T800-LHT-802 turboshaft engines would have powered the production RAH-66A. After competing for the contract against a team comprising Avco Lycoming and Pratt & Whitney, LHTEC was selected to develop the T800 in 1988. Besides the two T800s, each of which was rated at 1,680 shp (1253 kW), the Comanche was equipped with a small subsystem or secondary power unit (SPU) developed by Williams International. Designated the WTS124, the gas turbine engine provided electrical power and compressed air for engine-start and powered the no. 3 hydraulic system during normal operation of the aircraft. It also provided bleed air for the Comanche's environmental control system/ nuclear, biological and chemical protection (ECS/NBC) equipment.

Comanche cockpit

Identical front and rear crew stations each included an 8 x 6-in (203 x 152-mm) colour multifunction display that could show the digital map display, TAS imagery and system management data. Also installed were 7.8 x 5-in (198- x 127-mm) multi-purpose monochromatic touch-screen displays. Additional equipment included a wide-field-of-view, helmet-mounted display (HMD) system, and the avionics incorporated GPS/INS and defences. The latter included laser and radar warning receivers, and radar and infrared (IR) jamming equipment.

Compared with the systems installed in the AH-64D, the Comanche's advanced mission equipment package (MEP) provided a 40 per cent increase in stand-off range and an 80 per cent reduction in acquisition time. Equipment associated with the MEP included a passive electro-optical sensor system (EOSS) comprising two independent turreted sub-systems. Known as the electro-optical targeting and designation system (EOTADS), the target acquisition system (TAS) included an advanced targeting forward-looking infrared/low-light television (FLIR/LLTV) and a laser designator/ range-finder (LDRF). Both systems were equipped with three fields of view: 1.5º by 2º narrow (NFOV), 6º by 8º medium (MFOW) and 30º by 40º wide (WFOV).

Separately, a night-vision pilotage system (NVPS) included a second independent FLIR and an image intensifier television (NVPS/ I^2TV). The system provided a 40 per cent increase in range and 31 per cent increase in FOV over those in the AH-64D. The NVPS had a 30º by 52º FOV and could be slaved to either crewmember's HMD. The aircraft would also have been equipped with an updated version of the AH-64D's AN/APG-78 Longbow millimetre-wave fire-control radar (FCR). As on the AH-64D, the radar antenna would have been installed atop the rotor mast, albeit in a smaller, redesigned low-observable radome. Rather than equipping each individual RAH-66A, FCRs would have been installed on a total of 412 aircraft.

The Comanche's three-barrel General Dynamics/GIAT XM301 20-mm cannon was equipped with 500 rounds of ammunition stored in a rotary reel located below the pilot's cockpit. Firing at rates of 750 and 1,500 rounds per minute, the XM301 would have provided coverage from +15º to -45º in elevation and +/-120º in azimuth. To further reduce its infrared and radar footprint, the cannon could be rotated through 180º and stowed beneath the fuselage, facing aft at 2º elevation.

As well as the gun, the helicopter was also capable of carrying AGM-114 Hellfire air-to-surface-attack missiles, AIM-92 Stinger air-to-air missiles and 2.75-in (70-mm) rockets. The RAH-66A's stealthy configuration allowed a mix of Hellfires and Stingers, or rockets, to be carried internally on integrated-retractable aircraft munitions system (I-RAMS) panels. Each I-RAMS was equipped with three hardpoints capable of accommodating a single AGM-114, pairs of AIM-92s or rockets in four-tube composite pods. The I-RAMS panels were also designed to act as maintenance stands. The installation of an external, non-retractable enhanced fuel and armament management system (EFAMS) further increased the weapons load. The EFAMS, which could also be fitted with 230-US gal (871-litre) or 430-US gal (1628-litre) external fuel tanks, could be installed in just 15 minutes. With external fuel tanks installed, Comanche was capable of 'self-deploying' and flying a maximum distance of 1,260 nm (2333 km).

Operations

When deployed in its heavy configuration and assigned to armoured and mechanised infantry divisions, Comanches would primarily have been used as scouts for the larger AH-64D Apache. In light infantry and airborne divisions, however, the RAH-66A would have served both

RAH-66 no. 2 initially served as back-up to the no. 1 aircraft in aircraft trials, including Longbow radome trials (illustrated). From May 2001 it was grounded to have mission avionics fitted, flying again in May 2002.

RAH-66A specification

Maximum take-off weight: 13,000 lb (5896 kg)
Empty weight: 9,022 lb (4092 kg)

Overall length: 46 ft 10.25 in (14.27 m)
Fuselage length: 43 ft 3.75 in (13.2 m)
Main rotor diameter: 40 ft 0 in (12.20 m)
Fantail diameter: 4 ft 6 in (1.37 m)
Height: 11 ft 0.75 in (3.3 m); 14 ft 11 in (4.8 m) with FCR fitted

Powerplant: Two LHTEC T800-LHT-802 turboshaft engines, each rated at 1,680 shp (1253 kW) and driving a five-bladed main and eight-bladed shrouded fantail rotor
Maximum fuel capacity: 2,525 lb (1145 kg)

Maximum speed: 175 kt (324 km/h)
Service ceiling: 14,000 ft (4267 m)
Maximum range: 842 nm (1559 km) or 1260 nm (2333 km) with external fuel tanks
Mission radius: 150 nm (278 km) mission radius

Missions: reconnaissance, close air support, deep strike
Armament: XM301 20-mm cannon, AGM-114 anti-tank missiles, AIM-92 air-to-air missiles, 2.75-in (70-mm) rockets carried internally on I-RAMS and externally on EFAMS
Mission equipment: Night vision pilotage system (NVPS); electro-optical target acquisition and designation (EOTADS) system comprising FLIR, low light level television (LLLTV), laser designator/range-finder (LD/RF); and Fire Control Radar (FCR)

Left: An RAH-66 comes into land during field trials. The Comanche was designed for rapid refuelling and rearming.

Below: No. 2 Comanche (foreground) was employed to test the conical, stealthy radome for the Longbow FCR. No. 2 received the mission avionics in early 2002, returning to the air in May.

as a scout and an attack platform. The three primary missions earmarked for Comanche were armed reconnaissance, attack and deep strike, and the weapon configurations differed for each mission. Whereas the weapon load for the armed reconnaissance mission included four Hellfire and two Stinger missiles, carried on I-RAMS, for the attack mission EFAMS would be installed along with 10 additional Hellfires. The deep strike configuration would also make use of EFAMS but featured a pair of 230-US gal external fuel tanks rather than missiles, along with eight Stingers and just two Hellfires on the I-RAMS.

Reshaping the programme

By the time the second prototype made its initial flight in 1999, the Comanche programme had undergone another restructuring. The revised plans called for Boeing-Sikorsky to build 13 pre-production prototypes (PPPs), beginning in 2003. Included were five engineering & manufacturing development (EMD) aircraft and eight low-rate initial production (LRIP) aircraft. While the EMD examples would be used for development tests and participate in digital warfighting experiments, the eight so-called EMD 'production-representative' aircraft were destined to support initial operational test and evaluation (IOT&E), as well as user evaluations.

Before the Defense Acquisition Board (DAB) would allow the programme to enter the EMD phase, numerous criteria needed to be met. These included the completion of tests into ballistic vulnerability, radar cross-section (RCS), infrared signature and FLIR range, and a demonstration of the specified vertical rate of climb. Ultimately, the Boeing-Sikorsky team completed the last of the so-called Milestone II criteria, and on 4 April 2000 the DAB gave its approval for the Comanche to enter into EMD. Subsequently, the Army awarded the team a $3.1 billion contract on 1 June 2000. At that time the first RAH-66A built under the EMD programme was scheduled to fly in 2004. The purchase of the first lot of LRIP Comanches was then planned for FY05 and initial operational capability (IOC) would be achieved in 2006. Additionally full-scale production was planned for FY07.

In 2002 the programme was restructured yet again, with development being broken down into a series of blocks. The EMD programme was extended, adding 3,000 flight test hours, and an additional $3.4 billion was earmarked for development. In addition, the IOC date was pushed out to 2009. The extension of the EMD programme came at a cost, however, and procurement was reduced from 1,213 aircraft to

Sporting a dummy Longbow turret, the first RAH-66 prototype lifts off from West Palm Beach for its first flight with the redesigned empennage on 18 December 2000, after a 10-month lay-up. The flight lasted for 1.4 hours and tested the new tail feathers at speeds up to 165 kt (305 km/h) and bank angles up to 45°. The first prototype was retired from flight test duties on 30 January 2002.

Endplate fins were added to cure buffet problems. They were canted slightly to deflect radar energy upwards when the helicopter was being 'painted' from the side – from where most radar energy would come when the RAH-66 was operating at low level.

just 650. Although this change reduced the programme's cost from $39.3 billion to $26.9 billion, the unit cost rose from $29 million to $32.2 million. When the development costs were included as well, the unit cost rose to a staggering $60 million per aircraft.

Under this restructuring the service planned to buy 121 aircraft between FY04 and FY11 at a cost of $14.6 billion. Eventually, the initial Block I aircraft would be equipped with baseline systems including the FCR, which had originally been earmarked for Lot 6 production, and partial implementation of the tactics expert function (TEF). Among other things, the latter would support mission-planning, cockpit information management, and weapons selection and flight profile management. The initial aircraft would be fielded to cavalry and light infantry units as armed reconnaissance aircraft in FY09. Block II aircraft would be fielded around FY11 in the heavy attack configuration, and Block III would include the EFAMS, air-to-air missile, sensor fusion and full TEF implementation.

Always a true child of the Cold War, many of the Comanche's design requirements could no longer be justified as the Army attempted to restructure itself as a lighter, more mobile force. Although the RAH-66 would have easily fitted into this new force, the expense of developing its sophisticated systems was considered too expensive when weighed against their benefits, especially in the 21st Century battlespace.

In a Cold War scenario the RAH-66 would have been operating over battlefields bristling with sophisticated SAMs and radar-directed gun systems. In the real 21st Century world – in Iraq during 2003 – the US Army found itself mostly flying against small groups of irregular forces armed with shoulder-launched SAMs, rocket-propelled grenades (RPGs) and heavy-calibre machine-guns. As the service was to discover all too painfully, these low-tech systems could be no less deadly to its helicopters, and no amount of stealth or expensive electronics would make them any safer.

As the lessons of Iraq and other recent conflicts sunk in and the crunch time came, the Army was faced with a dilemma: its existing rotary-wing fleet was badly in need of updating, but it was equally clear that, if it proceeded with Comanche, it could not afford any of these modernisation programmes. Accordingly, the service took the difficult decision to cancel the RAH-66.

Reallocating the funds

$14.6 billion had already been earmarked for the production of 121 RAH-66s through 2011 and the Army will now reprogramme those funds to update its existing fleet and to replace aging helicopters operated by the Army National Guard and reserve. The service will also undertake yet another redesign of its aviation force structure, the seventh in 25 years. The funds diverted from the Comanche will be used to purchase 80 additional UH-60 Blackhawks, update the existing fleet of Blackhawks and field the Block III version of the AH-64D Apache Longbow. The service also plans to purchase nearly 800 new aircraft, including 303 light utility helicopters (LUH), 368 armed reconnaissance helicopters, 20 new CH-47 Chinooks and 25 fixed wing transports, as well as unmanned air vehicles (UAV). None of the three new aircraft types have yet been identified.

When the programme was terminated, only the two RAH-66 prototypes had been completed, although production was under way on the five EMD aircraft at Sikorsky facilities in Connecticut. The army is hoping to use several of the Comanche's systems in other platforms, allowing their development to continue and to recoup some of the money that has already been spent on them. These systems could find their way into the UH-60 and AH-64, especially the Block III Apache.

Tom Kaminski

What might have been – the futuristic Comanche was a technological step too far in today's austere budgetary environment. A fair proportion of the work done – principally that involving the avionics suite – will find application to other types, notably the Block III AH-64.

Jordan's special forces and upgrades

SOFEX 2004

The RJAF now has two EADS CASA C-295s in service. The first (352) was delivered in early 2003 and a second example (353) arrived in early 2004. A ceremony was held at SOFEX 2004 to formally mark the hand-over of these aircraft to No. 3 Sqn.

From 26 to 29 April Jordan hosted the latest in its biennial international special forces exhibitions, SOFEX 2004. His Majesty King Abdullah II is a former commander of Jordan's own army special operations unit and he has done much to see this element of the defence forces expand and improve its unconventional warfare capabilities. The army now boasts a well-trained, well-equipped and well-respected Special Forces Brigade, and a dedicated Special Operations Command. Jordan's SF troops have trained with other elite units around the world, including Britain's SAS. Much of the Jordanian tactics, techniques and procedures have evolved in co-operation with the SAS, and so the public demonstrations held at SOFEX – especially those of the counter-terrorist unit CTB-71 – provided a very rare glimpse into how such sensitive operations are conducted.

While much of Jordan's SF development has been focused on ground units, it has a growing air support element that was much in evidence during SOFEX 2004. The exhibition itself is held at Amman Marka Air Base (King Abdullah I) – a military facility with a civil terminal, right on the edge of the capital city's suburbs. One of the based units at Amman Marka is No. 14 Sqn of the Royal Jordanian Air Force (RJAF). This squadron is tasked with the special forces support mission and also provides troop-lift for Jordan's airmobile army units.

No. 14 Sqn has another important role, one that is highly classified and little-known. Hidden in the hangars at Marka are two Schweizer SA 2-37A high-altitude surveillance aircraft. Delivered sometime around 2000/2001, the SA 2-37As operate on sensitive border patrol missions around Jordan. So sensitive are these aircraft that they were flown out of Marka before SOFEX commenced, and secured at another location while their base was in view of visitors.

Until recently, No. 14 Sqn relied on its venerable UH-1Hs for troop transport and SF insertion. Beginning in mid-2003 the Hueys were joined by eight brand-new Eurocopter EC 635T1s, speedily acquired by Jordan from a cancelled Portuguese order. The RJAF intends to acquire a second batch of eight EC 635s to round out its fleet. It is also planned to equip these aircraft with FLIR systems, for full night/adverse weather capability.

Although the RJAF was the first military operator of the EC 635, the type is clearly well-established in service. The centrepiece of the Jordanian special forces demonstration at SOFEX 2004 was a full-scale hostage rescue conducted against a moving bus that had been 'hijacked' by terrorists. In a dramatic set-piece, the terrorists' vehicle was pursued and halted by ground units. With split-second timing, a team from Counter-Terrorist Battalion 71 (CTB-71) was dropped onto the roof of the bus from an EC 635, to blast its way in – using smoke and blank rounds to simulate live ammunition and stun grenades. No. 14 Sqn also participated in a simulated airbase capture using a group of four 'slick' UH-1Hs, escorted by a pair of AH-1F HueyCobras, to rapidly deploy troops that fought their way across the ground at Marka – all in front of the Royal review stand.

Left: The latest type to enter service with No. 14 Squadron is the Eurocopter EC 635T1. The first of these aircraft were delivered to Jordan in May 2003 and, during SOFEX, they were used to deliver the six-man assault teams of Counter-Terrorist Battalion 71 (CTB-71).

Jordan's industry

Elsewhere at SOFEX, Jordan's fledgling aerospace industry was on show. Jordan Aerospace Industries (JAI) and the King Abdullah Design and Development Bureau (KADDB) are working on a range of UAV systems. This includes the I-Wing mini-UAV – with a wingspan of just 1.25 m (4 ft 1 in) it can be carried (in pairs) by JAI's larger Trans Arrow UAV. The 50-km (31-mile) range piston-engined Trans Arrow can fly out to the target to release the covert I-Wings and then relays their sensor imagery back to the operator station.

Jordan is also developing several lightplanes with an eye on the military market. All are modest ultralight aircraft that have yet to enter series production. The most advanced programme is the Seabird Aviation Jordan

Jordan took delivery of 11 AS 332M-1 Super Pumas from 1986 and 10 remain in service with No. 7 Sqn today. Several of these helicopters are fitted with a winch for search-and-rescue work. The white panel on the cabin door allows for the quick application of Red Crescent markings when the aircraft is flying in a medevac role.

The Royal Flight element of the RJAF has taken delivery of two UH-60M Black Hawks that have not previously been reported. Both of these aircraft, dedicated transports for King Abdullah II, are fitted with weather radars and one has an undernose FLIR. These additional Black Hawks join three S-70A-11s already in service with the Royal Flight.

AH-1Fs led the special forces airfield assault demonstration, escorting troop-carrying UH-1Hs. Jordan has two squadrons of HueyCobras with about 20 aircraft in service.

Lockheed Martin and the RJAF are negotiating a novel upgrade package for the AH-1F fleet that would add a vastly improved weapons and sensor capability.

SB7L-360 Seeker. This is a development of an Australian-design that dates back to the mid-1980s. In co-operation with Seabird (Australia), JAI and KADDB are positioning the Seeker as a low-cost surveillance platform for border and EEZ monitoring. Powered by a Lycoming O-360-B2C 160-hp (119-kW) piston engine, the prototype Seeker was demonstrated at SOFEX with a FLIR Systems U7500 FLIR, although this had not then been fully cleared for export. A handful of Seekers have been ordered by government agencies in Jordan. In June news emerged that the Seeker had been ordered by Iraq's Coalition Provisional Authority to meet an urgent need for patrol and surveillance aircraft there. An initial buy of two FLIR-equipped Seekers will be followed by 14 more.

A second JAI project is the GulfBird Observer. Unveiled for the first time at SOFEX, it is based on the Ukrainian GulfBird X32-T ultralight. The Jordanian version is being marketed as an alternative to helicopters for border patrol and law enforcement tasks.

It emerged at SOFEX that the RJAF is discussing a major upgrade package for its Bell AH-1Fs with Lockheed Martin. This would add the weapons and systems that Lockheed Martin has developed for the US Marine Corps' AH-1Z SuperCobra – but would leave Jordan's AH-1Fs in their original single-engined configuration. Lockheed Martin is proposing to integrate the AAQ-30 Hawkeye Target Sight System and add an enhanced weapons capability that would include the laser-guided AGM-114 Hellfire. This would be the first time that Hellfires have been carried by a single-engined AH-1. The upgrade plan has been under negotiation for nearly two years and it is hoped that a final approval will be given early next year. One sticking point is that most of the RJAF's modernisation budget has been earmarked for its planned F-16 upgrade. It is expected that the final F-16 configuration will be decided soon, as the RJAF debates various options based on either the European MLU or US CCIP upgrades.

Robert Hewson

The Seabird Aviation Jordan SB7L-360 Seeker (below) made its public debut at SOFEX 2004. The Seeker is a militarised version of the original Australian-designed Seabird lightplane, optimised for surveillance and patrol missions with a new FLIR Systems U7500 fitted for trials (under the nose section). Note the LCD screen (to the right) for the observer's FLIR system (right)

Air Serv International is a US-based humanitarian aid agency that provides aviation support for several international organisations, including the UN. It operates a fleet of three Beech 1900Ds on daily services from Jordan to Iraq. Air Serv has been flying into Iraq since 1 May 2003, and was the first non-governmental air service to do so, following the end of major hostilities. It is the only civilian, non-commercial operator into Baghdad International Airport.

Throughout SOFEX 2004 Amman-Marka remained a fully operational airfield. This US civil-registered L-100-30 Hercules (N2189M) was one of two that undertook daily shuttle missions to destinations unknown – but probably in Iraq and Afghanistan, under US government contract. It is operated by Rapid Air Trans Inc. – a company with no profile but with an office registered in Maryland, close to Washington DC. It is most likely to be supporting US covert operations forces.

Upgrading the Mi-24/35 'Hind'

Still a feared weapon over the battlefield, Mil's 'Crocodile' serves in large numbers around the world. With no obvious replacement in sight, and with many of its operators facing severe budgetary restraints, the Mi-24/25/35 family is a prime candidate for modernisation. A number of companies are offering upgrade packages.

Above: A Ukrainian Mi-24V lets fly with rockets during a training flight. Combined with the nose gun and wing-mounted gun pods, rockets give the Mi-24 a powerful area attack capability.

Right: Rostvertol's Mi-35M demonstrator is currently the most sophisticated of the Russian upgrade proposals. It has a GOES-342 sensor turret, as also used in the Mi-24VM (VK-2) and PM (PK-2) upgrades for the VVS, and a GSh-23V cannon as well as airframe improvements.

The 21st century close air support/anti-armour and anti-insurgency focus of Russian army aviation, which since January 2003 has been under the control of the VVS (Voyenno-Vozdushniye Sily – Russian Air Forces), relies on the tried Mi-24V/P as its main workhorse. It is likely to soldier on with its principal operator for at least 15 more years – much longer than originally expected – due to the absence of credible successors.

It could be said that the Mi-24 is enjoying a prominent position simply because there is no real prospective, affordable replacement available in the required numbers, nor is there likely to be soon. The prospects of Mi-24 successors Ka-50/52 and Mi-28/28N entering large-scale production look rather bleak, because in 2000 the Russian Ministry of Defence redirected a significant proportion of the funds previously allocated for new hardware acquisitions to the launch of the Mi-8 and Mi-24 fleet-wide upgrade and life-cycle extension programmes. Apart from being considerably less costly, these upgrades were deemed much more urgent.

Today's 'Hind' still lacks any usable adverse weather and night operating capability and also, to some extent, weapons with useful stand-off range.

There is no credible and cost-effective alternative to an Mi-24 upgrade, and an array of options is now on offer in Russia and other parts of the world. Even the Central East European states which were granted NATO membership in 1999 now look committed to launching ambitious 'Hind' upgrade programmes; such investments are likely to turn the former Cold War combatant into a valuable NATO close air support/anti-armour (CAS/AA) asset. Non-European states such as India and Algeria have already elected to pump substantial finances into comprehensive operational capability enhancement programmes for their Mi-24 fleets, successfully carried out by non-Russian companies working as mission avionics/weapons system integrators.

Hybrid attack craft

Nicknamed 'Crocodile' in Russia and known as the 'Hind' in the West, the Mi-24 is considerably larger and heavier than its American cousin, the Bell AH-1 HueyCobra, as well as its projected successors the Mi-28 and Ka-50. It was originally designed in the late 1960s as an attack helicopter with a considerable assault transport capability, although the latter feature is very rarely used; moreover, the main cabin imposes a significant weight penalty and limits manoeuvrability and agility, especially in 'hot-and-high' operating conditions.

Unlike the Western concept of anti-armour helicopter deployment, which calls for predominantly ambush tactics, the 'Hind' has been used by the Russian/CIS air arms and client states around the world much more like a modern-day equivalent of the famous World War II Ilyushin Il-2 Shturmovik low-level attack aircraft. Ingressing at a slightly slower speed over the battlefield , Mi-24 pilots are taught to use the same tactics as their World War II predecessors: approaching the target area at high speed and treetop altitude, attacking from different directions and, if threat level permits, circling over targets for additional firing passes in two-, four-, six- or eight-ship formations – the so-called 'wheel of death' attack pattern.

There are up to 700 Mi-24V/Ps (and several dozen Mi-24R/K specialised NBC reconnaissance/artillery fire correction and battlefield reconnaissance derivatives) in the Russian Air Forces inventory today, although fewer than 200 or so are believed to be serviceable at any time. The older Mi-24D 'Hind-D' was reported to have been formally withdrawn from use in Russia in 2000. The Ukrainian and Belarussian air arms are other major Mi-24 operators among the CIS states, with inventories numbering around 200 and 75 'Hinds', respectively, although fewer than 60 and 35 are thought to have been in an airworthy condition in 2001/02.

Current airframe service life limits, authorised by Mil during the early/mid-1980s, call for 3,000 hours and/or 20 years, whichever is reached first. Time between overhauls (TBO) was initially set at seven to eight years, but later it was extended to 10 years, and in 2000 to 14 years for the Russian Air Forces fleet. TV3-117 engine TBO is set at 750 hours.

Most of the Mi-24 airframes in Russia, Ukraine and Belarus, as well as those with the air arms of Poland, Czech Republic, Slovakia, Hungary, Bulgaria and India, are well below the 2,000-hour mark. According to Mil Moscow Helicopter Plant (as the once-famous Mil Design Bureau is now officially known), if they are well maintained these airframes, all manufactured after 1980, could be good for 3,500 to 4,000 hours and 30 to 35 years of reliable service. Such significant activity would require extensive airframe structural wear points repair, rewiring and integrity/corrosion control efforts. By extrapolation, the life-extended 'Hinds' manufactured between 1980 and 1990 would be good – at least in theory – for use until 2010 to 2025, assuming that in the medium and long terms no troubles arise relating to unexpected airframe integrity, wiring and weapons/weapons control system reliability (due to ageing), and spare parts support.

Mi-24's apparent shortcomings

It is often claimed that the Mi-24's stressed-skin airframe structure, with extensive armour protection, can survive direct hits from small-calibre AAA projectiles. This is clearly an exaggeration: as real-world operations have shown, the Mi-24 (including powerplant, rotor system, transmission and systems) is vulnerable even to 12.7-mm (0.5-in) armour-piercing bullets, and some parts – especially the rear bottom fuselage and tailboom – to 7.62-mm (0.3-in) and 5.45-mm (0.2-in) bullets. Cockpit glazing is known to be a particularly weak point and is easily penetrated by small arms bullets. As well, the bladder fuel tanks, hydraulics and main gearbox oil system all demonstrated remarkably low resistance to combat damage in most local conflicts – especially those in Afghanistan (1979-1989) and Chechnya (1993-1995 and 1999-2000) – where Russian Army, Gendarmerie and Federal Border Service helicopters were involved in a great many combat and support operations.

Mi-24 fleets worldwide are known to suffer frequently from unreliable main rotor blades. Helicopters manufactured during the 1980s and early 1990s can be described as lacking any modern flight safety and combat survivability features such as further improved and refined armour protection, crash-resistant fuel systems, energy attenuating/armoured crew seats, self-defence aids and new digital avionics suites. Although the 'Hind' is, in general, considered to be an armoured platform, in today's typical threat environment its flight crews have to rely exclusively on good pre-mission planning and suitable tactics to keep out of harm's way.

It should be noted that, in complete contrast to the 1950s through 1970s, just 25 per cent of a modern attack helicopter's operational effectiveness comes from the airframe itself. Real battlefield effectiveness now comes from an integrated avionics suite for day/night operations, self-defence aids, and the man-machine interface.

Many flight safety problems during the most recent large-scale conflict for the Russian military, the 1999/2000 war in Chechnya, were caused by the Mi-24's poorly maintained and obsolete analog avionics suite. This was particularly true of the insecure communications gear (VHF/UHF and HF radios), radar altimeter and Doppler sensor. In addition, the unreliable main rotor blades essentially limited fleet availability. The basic Mi-24D/V/P is known to have only limited and somewhat untrustworthy autonomous navigation capability, something that is considered vital in any 21st Century operational environment. The original 'Hind' can use only Doppler navigation, which restricts a helicopter's usefulness in conditions of limited visibility, or when flying over featureless terrain or sea where map navigation could be difficult if not impossible.

As well, the Mi-24's vibration suppression is described as being particularly poor, with excessive pitch, roll and vertical vibrations generated by the far-from-perfect five-blade rotor system design at speeds below 80 km/h (43 kt) and above 300 km/h (162 kt), making weapons employment impossible when in the hover due to vibration-induced sighting problems. This particular inherent 'Hind' shortcoming does not allow Western-style tactics that rely on masking an attacker's presence behind available terrain and obstacles – so-called nap-of-earth (NoE) combat flying. To survive when operating *en masse* against targets with dense air defence cover, Mi-24 pilots are taught to use speed and surprise while approaching their targets at ultra-low altitude, thus limiting exposure to enemy air defences, then to execute a steep last-moment pop-up followed by a firing pass in a shallow dive.

The real-world experience gathered since 1979 clearly demonstrates that the Mi-24 often cannot absorb heavy punishment in combat and is particularly vulnerable to new-generation man-portable SAMs such as the Russian Igla and American FIM-92 Stinger. A usable stand-off engagement range was achieved only by introducing new sighting-observation systems, combined with the Mi-24V/P's existing 9M114 Shturm-V supersonic ATGM (anti-tank guided missile) or, better, with its considerably improved 9M120 Ataka derivative in the case of the Mi-35M. (The ATE-proposed 'Super Hind' Mk III upgrade would use the Kentron Ingwe,

'Hinds' have been exported to over 30 nations and many remain in service. New-build aircraft are available from Rostvertol, the Rostov-on-Don factory which is closely allied with the Mil design bureau. This is an export Mi-35 'Hind-E', carrying 9M114 Shturm-V missile tubes, B-8V20 rocket pods and UPK-23-250 cannon pods.

Mi-24s serve with 11 VVS regiments/air bases and two independent squadrons, plus two test units. Well over half of the 700-strong fleet is in storage. This Mi-24P is being followed by an Mi-8MTKO, a night-capable 'Hip' version with GOES-321 FLIR turret that was rapidly fielded for target-spotting operations in Chechnya.

The standard 'Hind' cockpit is a mass of dials and switches, with many more to the pilot's sides. These remain in most of the low-cost Mi-24 upgrades.

and the Israeli-proposed 'Hind' upgrade would use the Rafael Spike-ER.) The 1970s-vintage Shturm-V has a maximum range of 5 km (2.7 nm), but the new sighting-observation hardware to be integrated during upgrades would enable the crew to pick out and destroy targets from a greater distance, which would be safer in operations against the majority of the low-level battlefield air defences.

This is true for the Mi-24V and Mi-25, but the earlier Mi-24D 'Hind-D', still widely used in eastern Europe, and its export derivative the Mi-25, have been severely handicapped by the stringent employment limitations of their primary anti-tank weapon, the 9M17P Falanga-P (AT-2). This ATGM has a low speed, limited range (up to 4 km/2.2 nm) and relatively low probability of hit/kill; furthermore, stocks of the aged Falanga-P now have an

An armourer loads 12.7-mm (0.5-in) armament into the USPU-24 turret of an Mi-24. The standard gun of the Mi-24D and Mi-24V is the Yakoushev-Borzov YakB-12.7.

unacceptably low reliability due to their time-expired and old-element base electronics.

'Hind' in the Second Chechen War

During the outset of the second Chechen campaign (1999-2000), Russian army aviation had a fleet of 32 Mi-24V/Ps deployed in the region, assigned to three groups of the Russian Army advancing to occupy Chechnya. There were also a certain number of helicopters at the main base at Mozdok, for use as reinforcements on an as-needed basis. Their main use was within the tactical aviation teams, which comprised two to four Mi-24s and two or three Mi-8s, and were often tasked to support the advance of mechanised rifle regiments and to operate closely with ground or airborne forward air controllers (FAC). Another type of mission, without FAC assistance, was 'free hunting' or the search for targets of opportunity in certain designated areas (kill boxes) deep in rebel-held territory. In late 1999 and early 2000, free hunting accounted for some 30 per cent of all combat sorties flown by Russian Army Mi-24s.

As of late August 2000, when the main battles and anti-insurgency area clearing operations were over, it was reported that these Mi-24s had fired a total of 1,708 9M114 Shturm-V ATGMs (with only nine failures recorded), more than 85,000 S-8 80-mm (3.15-in) rockets and 89,850 rounds of 12.7-mm (0.5-in), 23-mm (0.9-in) and 30-mm (1.18-in) ammunition. During this time, 11 Mi-24s were lost and more than 40 suffered combat damage but were eventually returned to serviceable condition by the Russian Army's Mozdok-based field repair facility. The main causes of combat damage were 7.62-mm (0.3-in) and 5.45-mm (0.2-in) bullets fired from various Kalashnikov assault rifle and machine-gun derivatives, and on some occasions from 12.7-mm (0.5-in) and 14.5-mm (0.57-in) heavy machine-guns. 23-mm (0.9-in) ZU-23 truck-mounted twin-barrelled guns and ZSU-23-4 Shilka self-propelled systems with four-barrelled 23-mm (0.9-in) guns were also encountered occasionally, but no man-portable SAMs were launched against Mi-24s during this period. Nearly half of the write-offs during the combat operations in 1999 and 2000 were attributed to pilot error, mainly during final landing approach, and collisions with terrain resulting from inadequate pilot training for ultra-low-level operations, coupled with high levels of physical and mental stress.

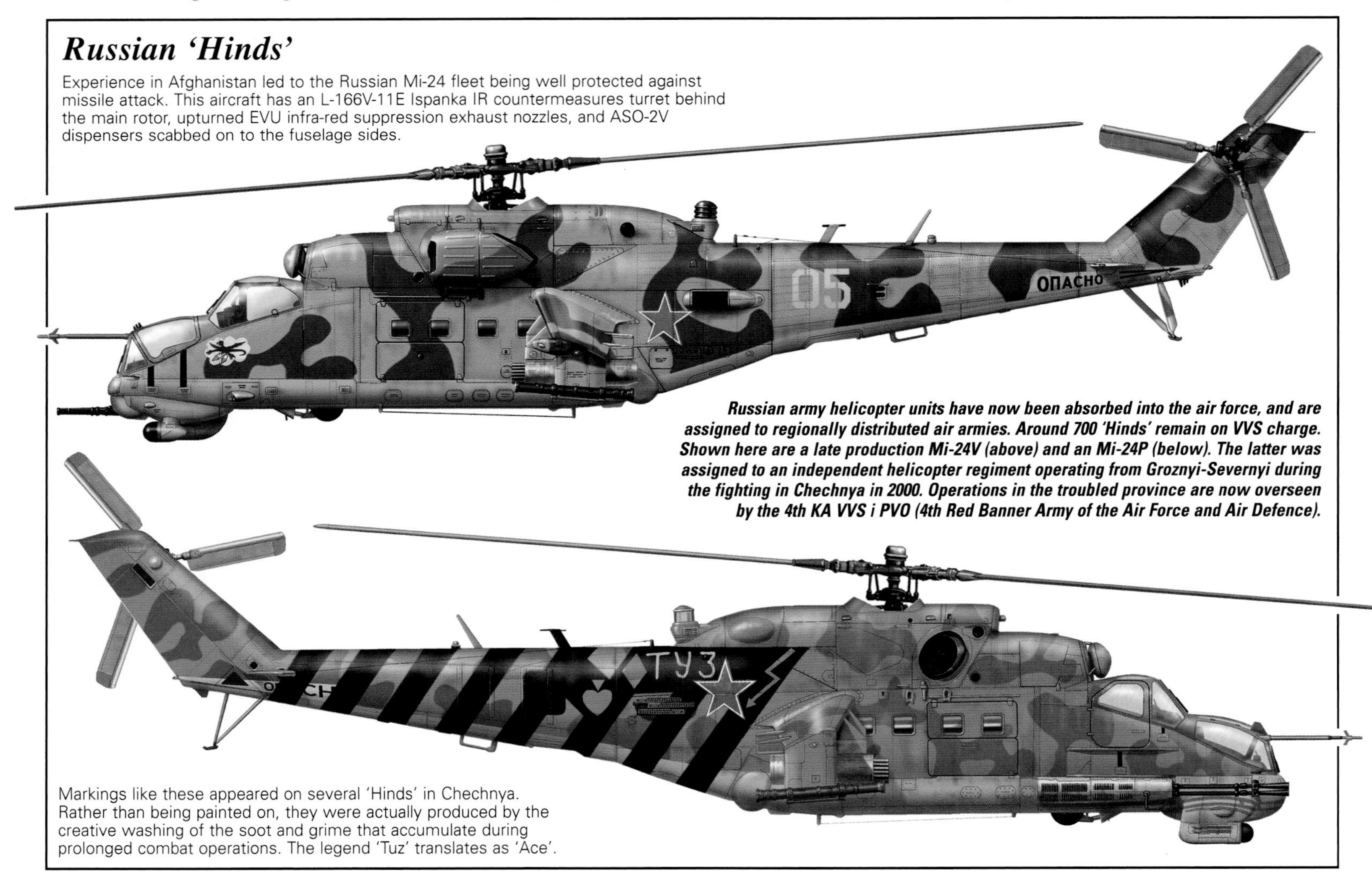

Russian 'Hinds'

Experience in Afghanistan led to the Russian Mi-24 fleet being well protected against missile attack. This aircraft has an L-166V-11E Ispanka IR countermeasures turret behind the main rotor, upturned EVU infra-red suppression exhaust nozzles, and ASO-2V dispensers scabbed on to the fuselage sides.

Russian army helicopter units have now been absorbed into the air force, and are assigned to regionally distributed air armies. Around 700 'Hinds' remain on VVS charge. Shown here are a late production Mi-24V (above) and an Mi-24P (below). The latter was assigned to an independent helicopter regiment operating from Groznyi-Severnyi during the fighting in Chechnya in 2000. Operations in the troubled province are now overseen by the 4th KA VVS i PVO (4th Red Banner Army of the Air Force and Air Defence).

Markings like these appeared on several 'Hinds' in Chechnya. Rather than being painted on, they were actually produced by the creative washing of the soot and grime that accumulate during prolonged combat operations. The legend 'Tuz' translates as 'Ace'.

The final production version of the Mi-24 for the Russian forces was the Mi-24VP, with GSh-23L twin-barrelled cannon in the nose. Only 25 were built, this example being assigned to the 'Berkuty' (eagles) display team from the Torzhok-based 344th TsBPiPLS AA.

Upgrade blocks

The block approach offered for the upgrade of Russian and export Mi-24 fleets has been designed by Mil and Rostvertol (the primary manufacturing organisation for the 'Hind') and encompasses a package of five 'building blocks' or modules that were promoted publicly for the first time in early 1999. All the airframe life extension and operational capability enhancement efforts offered by Mil Moscow Helicopter Plant for domestic and export customers are spread over Blocks 1 through 5. These can be carried out independently or together, depending on a customer's operational requirements and allocated budget.

Block 1 includes a comprehensive service life extension programme (SLEP) based on thorough inspection of the condition of each individual airframe. Main problems to be encountered include corrosion in certain places due to a low standard of maintenance and untidy storage, as well as fatigue cracking due to the very nature of inflight loading of the Mi-24's semi-monocoque fuselage structure.

Block 2 covers rotor system replacement, including the Mi-28's main rotor hub and composite blades as well as an X-shaped low-noise tail rotor. There is an option in this block to introduce the uprated Klimov TV3-117VMA (VK-2500) turboshaft, rated at 1790 kW (2,400 shp) in take-off, 2014 kW (2,700 shp) in emergency (OEI mode) and 1641 kW (2,200 shp) in maximum continuous mode; this would give additional performance improvements in 'hot-and-high' operating conditions.

Block 3 encompasses a host of airframe refinements such as shortened stub-wings with new DB-3UV pylons, and non-retractable landing gear. The latter is motivated by safe crash-landing requirements during low-level operations. Actual combat experience demonstrated that in an emergency Mi-24 pilots often do not have enough time to implement the undercarriage extension that is required for a safe landing. As a result, they often prefer to fly and fight at ultra-low level with the landing gear down, but such a configuration imposes severe speed restrictions. The fixed undercarriage proposed in Block 3 causes only an 11-km/h (6-kt) reduction in maximum level speed. The total weight reduction from all airframe refinements offered by Mil is said to be some 600 kg (1,320 lb), which, together with Block 2 enhancements, increases service ceiling by 300 m (985 ft) to 3100 m (10,170 ft) in standard conditions, and rate of climb to 12.4 m/s (40 ft/s).

Block 4 incorporates a host of weapons suite enhancements, the main one being the replacement of the 9M17P Falanga-P and 9M114 Shturm-V semi-automatic command line-of-sight ATGMs with the latter's much improved 9M120 Ataka-V (Attack) derivative, which was purposely developed for the Mi-28 in the 1980s. The supersonic 9M120 missile has a maximum range of 5.8 km (3.13 nm) and is equipped with tandem warheads having an armour-penetration capability of up to 850 mm (33 in), compared to figures of 4 km (2.16 nm) and 500 mm (19.7 in) for its predecessor. Up to eight ATGMs can be carried on an APU-8/4-U launcher, attached to the new DB3-UV weapons pylons which are equipped with built-in hoists. The upgraded Mi-24 can carry 575-litre (126-Imp gal) underwing fuel tanks borrowed from the Mi-28.

For self-defence, anti-UAV and anti-helicopter operations, the 9M39 Igla-V (SA-18 'Grouse') air-to-air missile is offered in twin launcher packs. It has a range of up to 5.8 km (3.2 nm) and combines a very sensitive cooled IR seeker with a lethal warhead. An Ataka-V ATGM derivative, the 9M220O, equipped with a proximity fuse and blast-fragmentation warhead, is also offered in the air-to-air role, for use against helicopters and UAVs.

It is possible to replace the Mi-24V's YakB-12.7 four-barrelled machine-gun with a GSh-23L twin-barrelled cannon mounted in the NPPU-23 turret. Mil promotional materials indicate that the GSh-23's 23-mm (0.9-in) projectiles are 1.4 to 1.6 times more effective than 12.7-mm ammunition against individual targets. During area suppression fire, the turreted GSh-23 gun can saturate an area 2-2.5 times larger than the Mi-24V's YakB-12.7 can cover. The new weapons control system has a laser rangefinder, and a BVK-24 weapons control digital computer greatly enhances the precision of unguided weapons; the upgraded helicopter retains the original VG-17 reflector gunsight, fed by the new digital ballistic computer. The upgraded 'Hind' can use the large S-13 122-mm (4.8-in) rockets, carried in five-round pods.

Block 5, the final and most expensive set of improvements offered by Mil, introduces day/night and adverse weather operating capability through NVG-compatible cockpit illumination, an advanced display system, precise navigation system (digital map is optional) and gyro-stabilised optronics system, integrated into a digital weapons control system.

The first 'Hind' prototype, featuring only the new rotor system, shortened stub-wings and fixed undercarriage, was flown officially for the first time on 4 March 1999. According to Mil, by 2001 a trio of upgraded Mi-24s was involved in the flight test programme. The first was tasked with testing the airframe and rotor system improvements; the second was used to test the uprated VK-2500 turboshafts; and the third machine, designated Mi-24VK-1, was involved in testing the new day/night observation-sighting hardware.

Mi-24PN – the first night proposal

The Mi-24PN is the most widely-publicised 'Hind' upgrade for the Russian Air Forces. A variant of the Mi-24P 'Hind-F', it is the first production version to have night ATGM capability while retaining the 30-mm (1.18-in) GSh-30-2 gun. It was in tests with Mil and Rostvertol between 2000 and 2003. Designated Mi-24PN (N for *nochnoy*, night capable), this simplified/low-cost upgrade features NVG-compatible cockpits and a BREO-24 avionics suite integrating one colour LCD screen in each

Under the designation Mi-35D it was proposed to equip the 'Hind' with the Vikhr tube-launched missile used by the Ka-50 'Hokum'. This impression shows the Mi-35D2 version, based on the Mi-24P but with the Ka-50's Shkval-V targeting system mounted in the nose. It has six Vikhr tubes under each wing and two Shturm-V tubes on the wingtips.

TMM 1410 display for FLIR imagery and navigation information, plus a VH 100 head-up display (HUD) and SMD LCD screen in both cockpits. However, no customers were found

Special Operations Unit, subordinated to the State Security authority, was reported to have acquired a pair of late production Mi-24Vs from Russian sources. For most of these new

of newly-built airframes fitted with scavenged components and systems from old Mi-24s, as this proved to be a cheaper option for most African customers.

On 28 January 2004 the first five Mi-24PNs (this is the

When Mil displayed this night-capable Mi-24 upgrade mock-up at Zhukovskiy in 1999 it was described as an 'Mi-350' (obozreniye = view). It was fitted with a GOES-321 turret as used by the Mi-24VK-1.

During the preceding decade, the newly exported 'Hinds' saw much use in many local conflicts, peacekeeping efforts and anti-insurgency operations around the globe. Most conflicts were essentially low-technology clashes (such as those in Sri Lanka and Sierra Leone) in which the Mi-24s' targets often lacked any sophisticated air defence, and thus the 'Hind' demonstrated fairly good results in numerous force projection missions. Weapons of choice were the inexpensive 57-mm (2.24-in) and 80-mm (3.15-in) rockets, 23-mm (0.9-in)/ 30-mm (1.18-in) gun pods/guns, and nose-mounted 12.7-mm (0.5-in) Gatling machine-gun, with little or no need for ATGM employment. Night operations were the exception rather than the rule during these conflicts.

Undoubtedly, the main driving force behind the renewed interest in the well-tried armoured gunship and the steady demand for the 'Hind' in Africa is the type's satisfactory warfighting capability in a Third World operational environment, coupled with a low price. The remarkably low price tag for newly-built or refurbished 'Hinds' from Rostvertol is possible due to the extremely low fixed manufacturing costs in Russia which result from a combination of very cheap labour, an old and inexpensive manufacturing technology base, ready availability of scavenged components (main gearbox, avionics and accessories) from retired Russian Air Forces or foreign Mi-24s, and an enormous production run in the preceding two decades with its impressive attendant economics of scale. Newly-built Mi-35Ps are reported to have been offered for $US4-5 million each, while 10- to 15-year-old second-hand examples fresh from major refurbishment are going for around $US1.5 million a copy, and in some cases even less. Seven Mi-35s, newly-built in 2002 and 2003 by Rostvertol for Czech Republic and delivered in lieu of part of Russia's trade debt, apparently had a total price of $US30 million, or $US4.3 million each. In September 2004, a pair of newly-built Mi-35Ps was delivered to Indonesian army aviation and follow-on orders are now expected to equip a full attack helicopter squadron.

Belarus and Ukraine are other major sources for cheap, second-hand Mi-24s – both refurbished and non-refurbished – that have sold well to Third World customers and some European states under UN embargo. These two CIS nations are thought to have exported more than 80 Mi-24s. Ukraine alone is believed to have sold in excess of 60 'Hinds': up to 20 Mi-24V/Ds to Croatia in the mid-1990s, 12 to Macedonia in 2001 (10 Mi-24Vs and two Mi-24Ks), and 14 to Algeria in 1999. Kyrgyzstan is another 'Hind' exporter and reportedly sold as many as 15 Mi-24s to India in 1995, most if not all of which were later refurbished by Rostvertol.

Israeli upgrade proposals

The wake-up call to Mi-24 export operators (as many as 600 Mi-24/25/35 export types are deemed suitable for various scales of upgrade) came from Israel Aircraft Industries. Its Mission 24 Mi-24 upgrade proposal was selected by the Indian Air Force in an order comprising 25 upgrade kits under a $US20 million contract signed in 1998. It proved to be a good advertisement with which to convince potential customers that a cost-effective alternative upgrade proposal exists, and was viewed as highly competitive to those offered by Russian and French companies. With an affordable and rapid integration of observation/sighting, navigation, self-defence and self-protection equipment proven in real-world operational conditions, the helicopter could perform well in the demanding CAS/AA role in a 21st century battlefield, including at night and in adverse weather.

IAI's Tamam electro-optical division was quick to offer an affordable upgrade package for the Mi-24 that required reduced development time and risk. The Indian Air Force contract covered prototype manufacture and testing in Israel, with production conversion to be undertaken at the customer's facilities. No airframe, flight control system, autopilot, powerplant, transmission or rotor system changes have been made by Tamam due to the

Mi-24PM/PK-2

When the 'full-spec' Mi-24M systems upgrade with GOES-342 turret is applied to the Mi-24P, the result is known as the Mi-24PM (air force) or Mi-24PK-2 (design bureau). The VVS favours upgrading Mi-24Ps as they are for the most part younger than the Mi-24Vs which remain in service.

Above: An important option for the Mi-35M is the replacement of the YakB-12.7 machine-gun by the much harder hitting GSh-23V water-cooled cannon.

Mil/Rostvertol Mi-35M/Mi-24VK-2

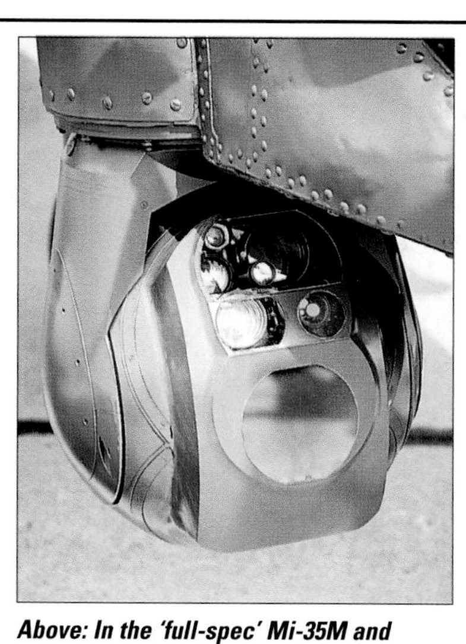

Above: In the 'full-spec' Mi-35M and Mi-24VK-2/PK-2 upgrades the sensor turret is the GOES-342.

Both front and rear cockpits have IV-86-2 multi-function displays – two in the rear (above) and one in the front (below). The Mi-35M export model is offered with HOCAS controls.

Right: The avionics system of the Mi-35M and Mi-24VK-2/PK-2 is designed to support the employment of the 9M120 Ataka-V anti-tank missile, of which eight can be carried under each shortened stub wing. Another important weapon is the 9M39 Igla-V short-range air-to-air missile, two of which are seen here in their launching tubes under the inboard pylon. Note also the upturned heat-suppressing exhaust box.

complexity and costs of such rework and any subsequent required qualifications.

The Mission 24 upgrade package as sold to the Indian Air Force is built around a Mil-Std 1553B digital databus. The heart of the upgrade is a single mission computer developed with IAI MLM; it is a derivative of the model used in the US Air Force T-38 upgrade programme, in which IAI is the principal subcontractor. Mission 24 utilises Tamam's proven helicopter multi-mission optronic stabilised payload (HMOSP), which weighs around 30 kg (66 lb). It is a turret ball-mounted derivative of the combat-proven IAI Tamam night targeting system installed in the US Marine Corps' AH-1W Super Cobra and Israeli DF/AF Cobra attack helicopters. It provides day/night observation and targeting through TV and FLIR sensors with variable FoV (between 2.4º and 29.2º on the FLIR). The HMOSP can incorporate two types of FLIR: a scanning array of 4 x 480 cadmium-mercury-telluride detectors operating in the low-wavelength band, and a 320 x 240-element indium-antimony focal-plane array functioning at middle wavelengths. Monochrome or colour CCD TV cameras are included, together with a laser rangefinder, designator and pointer, plus a built-in auto-tracking unit that uses centroid and edge-tracking techniques.

The cockpits are NVG-compatible, and both crew members have the option to use IAI's advanced NVG set with built-in monocular display on which all necessary navigation and targeting information can be presented. Both cockpits feature a single multi-function display (MFD) for TV, FLIR and targeting information, in addition to a keyboard and display unit for navigation and communication control. The CPG has control grips borrowed from the AH-1W, with all necessary sensor/weapons control switches and knobs; such devices, when combined with the MFD, can dramatically change work practices and reduce workload.

The HMOSP has been integrated with the Shturm-V ATGM SALOC guidance system through an IR goniometer and interface unit designed by IAI Tamam engineers, which has fully replaced the original old and bulky Raduga-F sighting/ATGM optical tracking system. The unit weighs more than 200 kg (440 lb). The Raduga-F's role for targeting and tracking both the target and missile was taken over by the HMOSP, with guidance commands being produced and transmitted to the missile through the existing equipment. Russian sources indicated in 2002 that Tamam experienced some guidance problems during Shturm-V test firings using the HMOSP system in place of Raduga-F. In order to solve the sensitive missile guidance problems, IAI contacted the Russian design authority for the Shturm, the KBP company of Tula, which provided important technical assistance. The contact was arranged through the Russian arms export agency Rosvoorouzhenie (the predecessor of Rosoboronexport, the only current Russian arms export agency). KBP expertise is believed to have been instrumental in solving the software/hardware, guidance and control problems associated with the HMOSP/Shturm-V assimilation on India's upgraded Mi-24s.

There are other types of ATGMs on offer for the Mission 24 system, such as Rafael's Spike-ER which has a maximum range of 7 km (3.8 nm) and employs 'fire-and-update' fibre-optic guidance. However, the Indian Air Force, which has ample stocks of Shturm-V missiles, preferred to limit the guided weapons integration work on its upgraded helicopters to the relatively cheap and well-proven Russian-made supersonic ATGM. The Shturm-V is still considered to be highly effective against older generation main battle tanks, such as the Chinese-made Type 59 and Type 69 – the chief potential target for Indian 'Hinds' in a future war with Pakistan. Integration of the indigenous Nag ATGM is known to have been earmarked for IAF Mission 24 machines at a later stage.

Navigation improvements introduced by IAI include a GPS receiver integrated into the existing DISS-15D Doppler sensor, and a three-dimensional digital map display. Both the HMOSP and YakB-12.7 gun are slaved to the pilot's line of sight through the use of a helmet-mounted sensor; the machine-gun can also be slaved to the HMOSP. A self-defence capability has been provided by IMI chaff/flare dispenser units and Elta radar/laser/missile warning systems. The total weight of these new systems is about 50 kg (111 lb).

It was reported that the production phase of the Indian Air Force Mi-35 upgrade was successfully running in 2001 or 2002. During the contract implementation phase, the IAF was tight-lipped about the upgrade details, and IAI has also been reluctant to disclose any details about the launch customer for its Mission 24

package. The upgraded helicopters, in overall light grey camouflage, were displayed publicly for the first time during the Aero India 2003 air show held in February 2003 in Bangalore.

Elbit Systems of Haifa, Israel, has also proposed a number of 'Hind' upgrades. The company has comprehensive expertise and experience in the upgrade business in general, and has one of the most aggressive marketing approaches in today's rather crowded defence upgrades arena. As identified by its engineering development team, which possesses enormous experience in battlefield helicopter upgrades, the main trends in any future Mi-24 upgrades are very short time to market, low risk, maximum integration of systems, and maximum utilisation of off-the-shelf hardware.

Elbit's proferred upgrade emphasises pilot situational awareness improvement, survivability enhancement, reduced aircrew workload with maximum 'head-out' operation, and maximum HOCAS controls employment. Such concepts were introduced and proved during the late 1990s on the Romanian Air Force Puma SOCAT upgrade, optimised for operations in a medium-to-high threat environment. Elbit's basic modular avionics package for a rotary-wing platform provides a day/night and adverse weather capability for the Mi-24 and enables its use in roles such as close air support, armed reconnaissance, air defence, anti-armour, escort and Combat SAR, in addition to the more conventional ones of assault transport, medevac and cargo lifting. It is centred around dual Mil-Std 1553B databuses and could include Elbit's own MIDASH (Modular Integrated Display And Sight Helmet) advanced helmet-mounted display and targeting system, highly capable TopLight or El-Op's COMPASS optronics systems. The latter weighs 45 kg (99 lb) and incorporates a gyro-stabilised, steerable turret with colour CCD TV camera, a third-generation FLIR operating in the low-wavelength band, and laser designator/rangefinder. Other components include fully NVG-compatible cockpit, hybrid GPS/INS navigation system, Western-standard navaids (VOR, ILS and DME), highly efficient self-protection suite, new radios, digital mapping module, and weapons interfaces. The MIDASH variant of Elbit's helmet-mounted sight and display family was designed especially for helicopter pilots

'212' is one of two Mi-24Ks serving with the Macedonian air force's 201 POHE (badge right). The reconnaissance equipment has been removed and they are now used as gunships. A modern reconnaissance capability may be restored as part of a wider plan to upgrade the Macedonian fleet, for which Elbit is seen as the most likely integrator.

and has a standard helmet body to which image intensifiers are attached to give a 50º x 40º FoV, providing flight and weapon-aiming information, including see-through binocular night imagery.

Elbit Systems adopted a unique approach to introducing night ATGM capability to the Mi-24, choosing to upgrade the Raduga-F sighting-missile guidance system rather than apply an all-new guidance system, thereby saving both development time and money. Shturm-V ATGM capability at night became possible due to the embedding of a miniaturised and relatively cheap CCD camera into the Raduga-F sighting system, and the upgraded system was dubbed 'Raduga by Night'. The CCD camera and a laser rangefinder are integrated with the upgraded helicopter's mission computer, enabling the processed and enhanced IR image of the target (from the Raduga-F) to be displayed on the CPG's display in the centre instrument panel. Raduga by Night's range is enough to provide a clear image of a tank-sized target from more than 11 km (6 nm).

Like its rival IAI, Elbit offers a number of Israeli and Western weapons systems for the Mi-24, including the Rafael Spike-ER ATGM and 70-mm (2.75-in) rocket pods.

An unconfirmed news release in January 2004 indicated that Elbit Systems has concluded (or is near concluding) a contract for upgrading some if not all of the Sri Lanka Air Force's Mi-24Ps, four of which were delivered newly-built in 1998, followed by three more (second-hand) examples in 2000. As well, it was reported that a contract for upgrading two Macedonian Air Force Mi-24Vs with high attack capability was near completion in January-February 2004.

South African 'Super Hinds'

Another successful upgrade, in both commercial and engineering terms, is that offered by the South African company Advanced Technologies and Engineering (ATE). Its upgrade packages, 'Super Hind' Mk II and Mk III, were publicly revealed during the Africa Aerospace and Defence 2002 exhibition at Waterkloof in September 2002, where ATE's Mk III demonstrator ZS-BOI was on display.

The basic avionics/weapons modernisation is being offered by ATE in two different versions built around common avionics and weapons systems but featuring different levels of equipment sophistication to suit customers' operational requirements and, of course, budgets. ATE developed its Mi-24 upgrade package on the basis of the valuable experience and expertise gained during the Denel Rooivalk combat support helicopter development programme, for which the company was selected as avionics and system integrator.

ATE studied 'Hind' upgrade possibilities in 1996 and, understandably, it came to the same conclusions as the Russian and Israeli studies of the mid-/late 1990s. The chief possible developments outlined were night operation capability introduction, weapons accuracy and total firepower improvement, navigation and self-defence capability enhancements, and improvement to the logistics support and fleet management, which under the old Soviet/Russian service concepts were quite primitive. Like their counterparts at IAI Tamam, ATE engineers took the pragmatic decision to retain the Mi-24's basic analog avionics suite and to interface new digital mission avionics. However, if a customer wishes, a fully-digital system is an option.

The first Mi-24V to be used as a testbed was acquired at a bargain price from Ukraine in 1998 and received the civil registration ZU-BOI. An export order for 40 aircraft to be upgraded to the Mk III standard was signed in 1999 with

IAI Tamam Mission 24

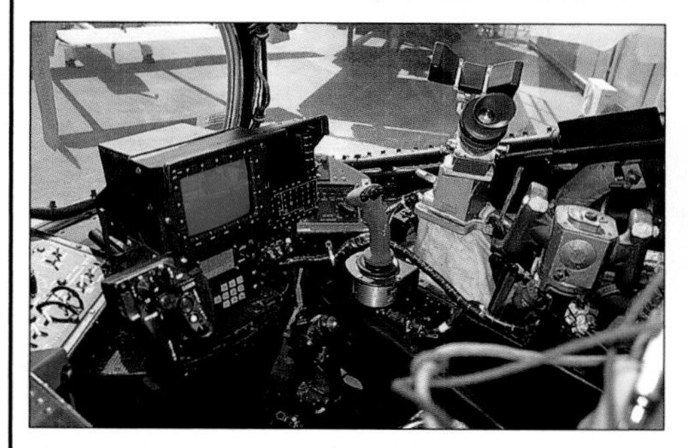

The front (pilot-operator's) cockpit has been extensively updated with AH-1W-style hand controls and a new screen display. Note also the monocular sight and control column.

The rear (pilot's) cockpit has a new MFD for targeting and tactical displays. The cockpit is NVG-compatible, and is equipped to support helmet-mounted sights.

Above: In the Mission 24 upgrade for the Indian Air Force (which supplied this airframe for the prototype conversion), the YakB-12.7 gun is retained, but it can be slaved to either the HMOSP or helmet-mounted sights.

Left: At the heart of the Mission 24's sensor suite is the HMOSP multi-sensor turret, fixed above the existing Raduga-F installation.

Right: Rafael's Spike-ER missile (centre pylon) is an option for Mission 24, alongside the existing Shturm-V (wingtip tubes).

Algeria and the first upgraded helicopters were delivered in 2000. A new order – for a smaller number of helicopters to be upgraded to the Mk III standard with additional airframe and equipment improvements, for an undisclosed customer – was announced by ATE in late 2002; it is believed that the customer for this derivative is once again Algeria, which intends to re-upgrade some of its earlier machines overhauled by ATE.

The ATE-developed 'Super Hind' core package is on offer in two main versions. The first, Mk II, aims to enhance the capability of the Mi-35P gun-armed helicopters (a certain number of which, new or refurbished, were sold to several African nations in the late 1990s and early 2000s); it retains the original Russian weapons. It could be a suitable and cost-effective Mi-35P upgrade solution for customers who have ample stocks of Shturm-V missiles and 30-mm (1.18-in) rounds. Mk II introduces only an observation/targeting and precise navigation package.

The second upgrade version, Mk III (as selected by the Algerian Air Force for 40-plus Mi-24D/Vs), features an entirely new weapons control system, replacement of the 12.7-mm (0.5-in) four-barrelled Gatling machine-gun by a 20-mm (0.787-in) gun, and replacement of the Falanga-P/Shturm-V ATGMs by the Kentron ZT35 Ingwe laser beam-riding missile. There is also an option for an agile Mk III derivative, dubbed 'Agile Hind', which would have much improved agility and manoeuvrability due to an extensive airframe/rotor system re-work.

The core system all three ATE upgrades is based on that developed by ATE for the Rooivalk, and comprises an ATE-produced mission computer that interfaces with various onboard systems and is responsible for all navigation/weapon delivery data processing; NVG-compatible cockpit and formation lights, as well as steerable infra-red landing light; and Kentron Cumulus Argos 550 gyro-stabilised multi-sensor system with TV-camera, FLIR, auto-tracker, and laser rangefinder/target designator. The nose-mounted Argos 550 system can rotate through 220º in the horizontal plane and +20º/-90º in elevation. Other elements incorporated into the core Mk II/Mk III systems include a head-up display; new NVG-compatible navigation displays in both cockpits, capable of providing position data as a Doppler fix, GPS fix or (more reliable and precise) a combined Doppler/GPS fix; new self-defence system with programmable Vinten chaff/flare dispenser units and control panel, installed in the pilot's cockpit; and a digital/optical rotor balancing and tracking facility to reduce vibration during low- and high-speed flight, thus improving comfort and weapons delivery precision as well as extending airframe/mission equipment life.

Archer sight and Armscor cannon

The 'Super Hind' Mk III package combines the core system with additional high-tech features. One is the Archer R2 helmet-mounted sight developed by Kentron, which is offered for both crew members and is useful for steering the IST Dynamics 20-mm (0.787-in) gun turret and cueing the Argus 550 multi-sensor system. Another is a high-rate-of-slew turret with the Armscor F2 20-mm gun (a licensed Giat Industries F2 model) with 840 rounds, housed in 'cheek' fairings protruding on both sides of the fuselage. The gun can be steered through 110º left/right and –50º/+15º in elevation, at a rate of 100º per second, and can be aimed through either the R550 system, Archer R2 HMS or the pilot's advanced HUD. The Kentron F2 20-mm gun has an impressive muzzle velocity of 1050 m/s (3,445 ft/s), a rate of fire of 750 rpm and an effective range of 3 km (1.6 nm). ATE claims that the combination of the Archer R2 HMS, robust turret and high-muzzle-velocity F2 gun can be deadly effective against air targets, making the integration of air-to-air missiles unnecessary.

One of the most important elements in the 'Super Hind' Mk III's operational capability enhancement package is the new Ingwe ATGM; a warload of eight missiles can be carried on two four-round launchers on the modified wingtips. This 127-mm (5-in) ATGM weighs 64 lb (29 kg), can be launched from 0.25 to 5 km (0.13 to 2.7 nm), and has an armour penetrating capability of up to 1000 mm (3.28 ft). Its laser beam-riding guidance method is highly resistant to countermeasures. In addition, the helicopter can carry a wide range of Russian- and South African-made rocket pods and free-fall bombs (with delayed fuses for a precise low-level delivery profile) – two or four weighing up to 500 kg (1,102 lb) each. The Mk III's enhanced package can also include a further improved and refined man-machine interface, digital HUD, new-generation Doppler sensor and, for NATO member states, a new IFF system. Also on the upgrade list are additional mission avionics components and systems, a digital autopilot/auto stabilisation system, and an advanced integrated communications suite.

The 'Agile Hind' is perhaps the most sophisticated Mi-24 upgrade package offered by ATE. It is aimed at producing a combat support helicopter with superior agility, manoeuvrability and power margins. Serge Vidal, the marketing manager of ATE's Helicopter Division, maintains that such a rotary-wing attack machine can easily be employed in accordance with

India displayed one of its IAI-upgraded Mission 24 'Hinds' at the Aero India show in 2003. The Indian Air Force has large numbers of Shturm-V missiles and these are retained as the primary armament for the modernised aircraft.

sophisticated Western attack helicopter tactics. Vidal points out that there is more to NATO compatibility than just the onboard IFF and radios. Interoperability, he says, can be truly achieved only if the helicopter can comply with NATO tactics. For the Mi-24, this means enough manoeuvrability to achieve nap-of-earth flight – and this is what ATE's follow-on upgrade is targeted at providing.

In order to achieve better agility, a comprehensive weight reduction programme was deemed necessary, so ATE stripped out an impressive 2000 kg (4,409 lb) of airframe empty weight. Company sources maintain that this can be achieved through replacing the old electrical, hydraulic and fuel systems with lighter and more reliable modern ones. As well, the existing armour plating can be replaced by lighter solutions that use composite and ceramic plating and by armoured crew seating. Additional weight reduction could be achieved through fixing the undercarriage in the down position, enabling the removal of surplus hydraulic systems and mechanical components.

ATE announced plans for new composite rotor blades (which in early 2004 were still in testing), new much more efficient and lightweight dust/sand engine intake filters, and IR exhaust suppressors. To prove the 'Agile Hind' airframe/rotor system improvements, in 1999 ATE acquired its second Mi-24V example from Ukraine for conversion as a testbed, which received the civil registration ZU-GAL.

ATE has designed a Western-style logistics support system, which focuses on TBO extension and transition to an 'on-condition' maintenance concept, new service documentation, spares, field services and expertise, as well as computer-aided training and maintenance.

SAGEM's upgrade proposal

Another Western company with a successful Mi-24 avionics upgrade is France's SAGEM, a major avionics and system manufacturer and integrator. The core system consists of an integrated modern suite of modular avionics including accurate and autonomous navigation, passive high-tech sensors and sights, 'glass' cockpit with map display and terrain avoidance system, advanced mission computers and night vision equipment.

At the 2001 Paris air show, the company displayed an Uzbekistan Air Force Mi-24P featuring a variety of digital systems borrowed from the Eurocopter Tiger and NH90 new-generation combat and support helicopters. Each cockpit is fitted with a 15 x 15-cm (6 x 6-in) colour display and has NVG-compatible lightning for use with SAGEM's CN2H night-vision goggles. The highly precise Sigma 95L navigation system feeds inertial and GPS sensor inputs to the Mercator mapping module. A turret beneath the nose houses the modern Strix system, which was developed for the French Army Tiger HAP and has second-generation FLIR, CCD TV and laser rangefinder-designator. A proposed ground segment includes the MARS (Mission Analysis and Restitution System).

It was widely reported that in 2001 SAGEM won a contract to upgrade 12 Uzbekistan Air Force Mi-24Ps, in co-operation with the local aircraft manufacturer TAPOiCh of Tashkent. To develop its Russian-made helicopter upgrade activity further, in June 2003 SAGEM teamed with Mil and Rosoboronexport to offer an upgrade package for the Mi-24; it is being advertised as fully interoperable with NATO army aviation assets, and primary potential customers are the states of the Visegrad Four group (Czech Republic, Hungary, Poland and Slovakia) and Bulgaria.

New players in the upgrade field

In 2001, BAE Systems and Eurocopter announced their intentions to become two more significant players in the already crowded Mi-24 upgrade market. They both offer NATO compatibility and weapons/sighting/observation system packages for the East European states' Mi-24 fleets.

Since late 2001/early 2002, BAE Systems has put much effort into promoting the battlefield mission capability for the Mi-24 upgrade and NATO interoperability programmes in Central East European countries. Its Mi-24 upgrade promotional material claims that investing in capability means investing in the avionics systems – enhanced sensors, increased avionics system incorporation, and integration of the platform and mission avionics. BAE Systems proposed its upgrade for the first time in September 2001, and a company press release stated that the company was going to offer only proven off-the-shelf equipment. Its modular upgrade package is based around open architecture avionics using a Mil-Std 1553B databus which would provide room for further growth,

Having been hawked round the air shows and used for ground tests/demonstrations for some years, Rostvertol's Mi-35M demonstrator 'Yellow 77' eventually flew in late 2003. Hopes for upgrade orders from traditional Russian 'client' states and former Soviet republics are high.

ATE 'Super Hind' Mk III

Key features of the ATE Super Hind Mk III are the Armscor F2 20-mm cannon and Argos 550 multi-sensor turret in the nose (right), and the Kentron ZT35 Ingwe laser-guided missile (below), of which eight can be carried. The 'cheek' fairings house the ammunition for the hard-hitting cannon. Algeria was the first, and to date only, customer for th ATE upgrade, although the company is active in promoting its modification packages to other 'Hind' operators.

i.e., new weapons integration. The proposed avionics package is designed for full ICAO/NATO interoperability and compliance with NATO STANAG 4555, and can be supplied on a 'pick and mix' basis.

One option is the advanced Striker helmet-mounted display, already selected for the USMC AH-1Z upgrade. It comprises a light-weight flying helmet fitted with a pair of detachable image-intensified CCD cameras that may be removed for daylight operation. The cameras weigh 260 g (9 oz) each, have a 40º FoV and project onto the helmet visor a very high-resolution raster binocular image that is electronically combined with targeting and navigation symbology.

Another element in the BAE Systems' proposed upgrade package is the Titan 385 or Titan 485 multi-sensor turret, incorporating cutting-edge technology developed from more than 30 years' experience in combat operations worldwide. The Titan 385 combines a large-format General III gyro-stabilised FLIR (operating in the mid- and low-wavelength bands), CDD camera for day operations and laser rangefinder, installed to port in the nose. The mission suite also includes hybrid GPS/INS navigation system, NATO/ICAO-compatible navigation aids (VOR/ILS/MB/TACAN), advanced IFF, weather radar, state-of-the art jammers, chaff/flare dispensers and radar/laser/missile approach warning receivers, which would considerably improve the Mi-24's survivability in a medium-to-high threat environment. Particularly for battlefield survivability, BAE System proposes its HIDAS – a fully-integrated Defensive Aids System which it says is a modern attack helicopter's essential (albeit very expensive) piece of equipment, already proven on the British Army WAH-64D Apache.

Several types of ATGM are proposed, the chief ones being the AGM-114 Hellfire and Spike-ER/LR. As well, BAE Systems suggested replacement of the Mi-24D/V's 12.7-mm (0.5-in) four-barrelled machine-gun with an Oto Melara 20-mm (0.787-in) TM-197 three-barrelled gun.

In 2001, BAE Systems bid with the WZL-1 maintenance facility of Poland for a Polish Army Mi-24 upgrade, but with no result. In mid-2003, BAE Systems was ranked among the prime bidders in the ongoing competition to upgrade 100-plus helicopters for the air arms of the Visegrad Four group nations. In order to demonstrate to potential customers its capability to upgrade the Mi-24, BAE Systems has acquired a former Russian Air Forces example from a private owner in the UK, which will be used as a company demonstrator vehicle. In mid-2003 the company started work on a prototype which is slated to be ready for the 2004 Farnborough air show. BAE Systems plans to invest some $US5 million in the prototype upgrade, which company sources believe will remove as much as 90 per cent of technology risk. Talks are known to have been carried out with MVZ Mil in mid-2003 regarding co-operation in the Mi-24 upgrade but, by February 2004, no definitive agreement had been signed.

East Europe's upgrade ambitions

In early 2002, the East European states of the so-called Visegrad Four (V4) group – Czech Republic, Hungary, Poland and Slovakia – decided to adopt a comprehensive joint upgrade and life extension of their 'Hind' fleets. On 30 May 2002, the defence ministers of the four countries signed a memorandum of understanding for the joint 'Hind' upgrade efforts, covering up to 106 Mi-24s (40 for Poland, 24 for Czech Republic, 28-32 for Hungary and 10-12 for Slovakia). Their original service life would

SAGEM's Uzbek upgrade

A 'prototype' conversion of SAGEM's Mi-24P upgrade for the Uzbekistan air force was displayed at Paris in June 2001. The French company has based much of the modification on sensors and equipment developed for the Eurocopter Tiger and NH90, including the undernose Nadia sensor ball which is mounted centrally under the chin (right). SAGEM is now officially linked with Mil and Rostvertol in offering NATO-compatible Mi-24 upgrades. This move overcomes one of the main stumbling blocks faced by other potential upgrade contractors, namely, withdrawal of support by the design authority.

BAE Systems upgrade

The 'Hind' upgrade package offered by BAE Systems focuses on NATO compatibility and integrated defences, in addition to improved navigation and combat effectiveness. This demonstrator (above) was displayed with a Titan multi-sensor turret ball, while a front cockpit model (right) shows the new screen display and hand controller proposed for the type. Weapons can include the Israeli Spike and the AGM-114 Hellfire. Aimed primarily at eastern European nations, the BAE upgrade was selected by Poland in February 2004 to proceed to prototype stage.

expire around 2006 and the helicopters were originally to be updated for another decade's service.

However, this rather ambitious programme, expected to exceed $US500 million, has proceeded very slowly. A technical agreement for the joint upgrade was signed in February 2003. This agreement authorised Poland to organise a tender for potential system integrators, which would have to co-operate with the WZL-1 maintenance facility of Lodz, Poland, to develop the prototype of the upgraded Mi-24. Companies which had expressed their willingness to participate in the tender in 2002 included BAE Systems, Eurocopter, Elbit Systems, IAI Tamam division, SAGEM and ATE. In the event of contract signature by the end of 2003, the prototype was to be test-flown in late 2004.

In early 2003, the four nations disagreed on the scale of the proposed upgrades in each country – and therefore on the eventual level of commonality, schedule of the project, and how to move from prototype to production scale.

Then, in June 2003 Poland decided to drop its requirement for upgrading 40 helicopters and opted for the NATO minimum of 16 units, abandoning plans for 24 Mi-24Ds. New plans foresee upgrading as many as 13 Mi-24Ws (as the Mi-24V is known in Polish Army service), which are slated to receive NATO-interoperable avionics and will be used as all-weather attack helicopters. Three more will be upgraded to serve as CSAR platforms. The upgraded Mi-24s are intended to remain in service until 2018. The Slovak Air Force reduced to 10 the number of its Mi-24s to be upgraded, and Czech Republic and Hungary dropped theirs to 12-18.

A major hindrance for the Mi-24 upgrade programme is reaching a working agreement with Russia over the politically sensitive service life extension and spare parts delivery issues. According to unofficial sources from the Polish MoD, in April 2003 Russia set conditions that were unacceptable to the V4 group. Mil and Rostvertol aspired to carry out the majority of the upgrade work themselves, which was impossible because the work would involve sensitive NATO communications, targeting and identification technologies to which the Russian engineers could not have access.

Furthermore, at the 2003 Le Bourget air show, Rostvertol and Mil officials declared that they will not support any Mi-24 upgrade programmes led by Western, South African or Israeli companies, and will ban the supply of spare parts to operators of such helicopters.

A meeting of the V4 Ministers of Defence on 27 June 2003 eventually freed each member to sign a separate agreement with the Russian Federation and Rostvertol for fleet service support and for technical support during the development and testing of the Mi-24 prototype. Anticipating serious programme delays and even cancellation of the Mi-24 upgrade project, in early 2003 the WZL-1 facility and the Polish Air Force Technology Institute initiated joint development work on a Combat SAR variant for Polish army aviation, which has a requirement for three or four examples to be ready for service in 2004. This activity is independent of the main Mi-24 upgrade activity. The Mi-24 CSAR programmes include the installation of an enhanced self-defence suite, FLIR, SAR-optimised direction-finder equipment, and other mission avionics. WZL-1 was awarded a $US1.25 million contract for prototype work. In December 2003 it was announced that WZL-1 was awarded the long-expected contract to upgrade two Mi-24Vs to the CSAR variant, complying with the requirements of NATO Standardisation Agreement G 4555 dealing with minimum equipment packages for battlefield helicopters for attack operations in a low-threat environment. The upgraded helicopters, known as Mi-24CSAR, are to be taken on strength by the CSAR Group based at Pruszcz Gdanski which is expected to be formed in late 2004.

In late 2003 it was clear that the V4 group's joint upgrade had collapsed, although officially it still existed. The Czech Air Force is thought to be interested in some small upgrades to bring its older Mi-24Vs to the same standard as the Mi-35s delivered as payment of Russian debts in 2003. Slovakia, however, is still considering ways to co-operate with Poland. Poland itself issued a contract to BAE Systems in February 2004 covering the production of two prototype upgraded aircraft.

Above: In 2003 the Czech air force received seven new 'Hinds' in the form of Mi-24V/35s delivered from Russia as part of a debt repayment deal. The Czechs are looking to upgrade around 12-18 'Hinds' to NATO-compatible standards, reduced from the 24 specified during the initial Visegrad Four discussions.

Left: Poland's 'Hinds' are being repainted in this all-over green scheme. Poland was the first to break from the V4 group by placing a contract for the prototype upgrade of two aircraft with BAE Systems. It is also modifying three aircraft for the CSAR mission.

There were some reports in 2002 that the Visegrad Four group attempted to invite Croatia and Ukraine to join the Mi-24 upgrade undertaking, thus sharing the benefits of the multinational approach and further reducing development expenses. However, V4 group expansion failed to materialise. The Croatian Air Force, whose Mi-24s were reported in early 2002 to have been grounded for three years, eventually elected to go ahead with only structural refurbishment and a minor avionics upgrade. Seven Croat Mi-24Vs were slated for the overhaul, which it was thought would most likely be carried out by the WZL-1 facility in Poland. However, in July 2003 Croatian Air Force Commander-in-Chief Brigadier General Victor Koprivnjak declared that the upgrade contract would go to Russian or Ukrainian companies, as only they had the necessary expertise and could offer an affordable price.

Bulgarian upgrade moves

In 2002/2003 the Bulgarian MoD prepared for an Mi-24D/V avionics upgrade and service life extension so that NATO-interoperable Bulgarian Air Force 'Hinds' could serve until about 2015. As many as 18 Mi-24s (12 late D variants and six V variants) were included in an upgrade that was estimated to cost between $US65 and $US70 million. BAE Systems, Eurocopter, IAI Tamam division, Elbit Systems (teamed with Lockheed Martin-Owego), ATE, Aviation Services (an Mi-24FM team from the Czech Republic with Flight Visions and Marconi), Eurocopter, SAGEM and Rosoboronexport (teamed with Mil and Rostvertol) all expressed interest in participating as avionics integrators and/or main contractors, but the tender procedure was delayed due to funding difficulties and organisational problems.

In early December 2003, the Bulgarian Minister of Defence was authorised to select a strategic partner to run this important project, valued in total between $US150 and $US200 million, without the need to organise an open tender. The leading contender appeared to be BAE Systems, which in 2002 and 2003 actively sought local partners for a large-scale offset programme. However, the Bulgarian Air Force raised strong objections to the BAE Systems' offer, said to be a 'paper upgrade' with no prototype so far being demonstrated, and the Ministry of Defence eventually decided to go to the tender procedure, which was scheduled for announcement in early March 2004.

ATE has also been very active in Bulgaria, the company having formed a temporary (one-year) consortium with local maintenance company TEREM to undertake private-initiative-funded Mi-17 and Mi-24 upgrades. These will be used as company demonstrators for potential customers – chief among them, of course, being Bulgaria. The Mi-24 prototype, a phased-out Bulgarian Air Force Mi-24D serialled 123, is expected to be ready for ground and flight testing in May 2005.

SAGEM, in its turn, is trying to penetrate the Bulgarian market along with Mil Moscow Helicopter Plant and Rosoboronexport. Mil is known to have presented an independent bid to Bulgaria based exclusively around new-generation Russian-made avionics, night-vision equipment and weapons. However, sources say the bid has been deemed unacceptable.

CIS upgrade market

Ukraine is believed to be currently considering a bid by IAI Tamam to upgrade 50 to 100 of its 200-plus Mi-24s. This country, the second-largest in the CIS, has already established a good working relationship with Israel in the military aviation area. The main local contributor to a Mil upgrade would undoubtedly be the Aviacon helicopter maintenance facility, based at Konotop. It is the largest of its kind in the CIS and is widely known among Third World Mi-24 operators as a source of readily available, cheap, refurbished Mi-24s and spare parts drawn from Ukrainian military inventory. There are also indications that SAGEM has been very active in Ukraine, promoting its avionics upgrade solutions. Budget constraints, however, could delay any upgrade process.

Bulgaria's delayed 'Hind' upgrade programme has attracted attention from BAE Systems, ATE and from the SAGEM/Mil/Rostvertol partnership. A clear and urgent requirement exists for NATO-compatible aircraft, but funds for such programmes are difficult to find.

Belarus is another CIS state operating a sizeable 'Hind' fleet. It is expected to remain loyal to Russia, preferring the Mil/Rostvertol-proposed Mi-24VK-1/VK-2/PK-2 packages for 20 to 30 of its 70 or so Mi-24s. The primary facility for production upgrades and airframe refurbishment in such a case will be the Baranovichi-based 557 ARZ Company.

The same option – embracing Mil/Rostvertol or SAGEM Mi-24 upgrade proposals – may be taken up by other CIS states such as Turkmenistan, Kyrgyzstan and Kazakhstan. Other important Mi-24 upgrade markets which can be described as lying firmly within the Mil/Rostvertol/Rosoboronexport sphere of influence are those in Libya, Syria, Sudan and North Korea.

Alexander Mladenov

Ukraine maintains a sizeable 'Hind' fleet and is seeking to upgrade a number of its aircraft. In the light of ongoing joint efforts with Israel to upgrade L-39 trainers, the IAI Tamam Mission 24 'Hind' modernisation is the most likely candidate.

Italian air force training is built around three types: the Aermacchi SF-260 (left) for elementary training, Aermacchi MB-339 (main photo) for basic, advanced and weapons training, and the Nardi-Hughes NH 500E (below) for rotary-wing instruction. The latest MB-339CD has many operational systems, such as RWR, and has a 'glass' cockpit to provide an excellent lead-in to Italy's modern fighters such as the Lockheed Martin F-16 and Eurofighter Typhoon.

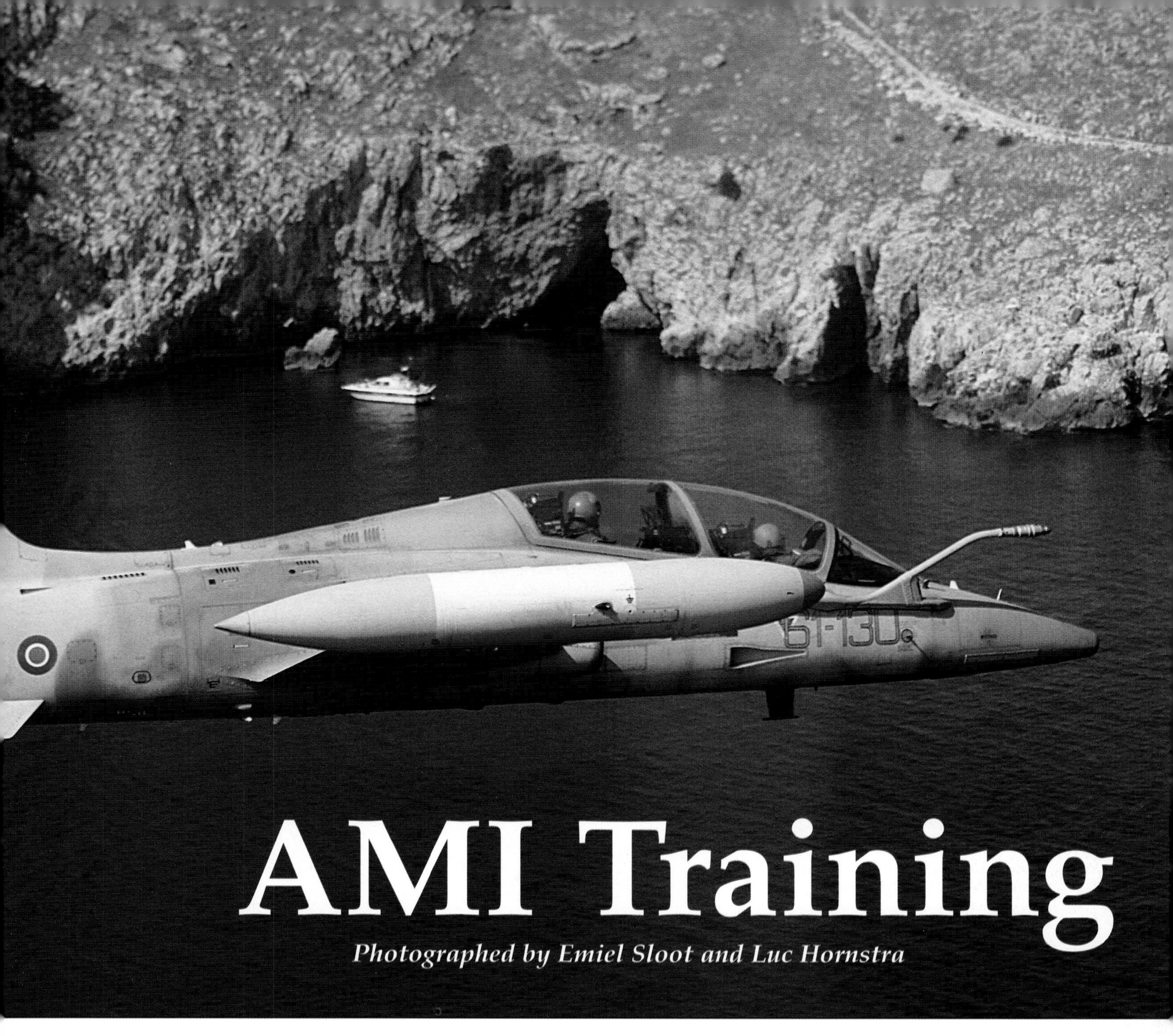

AMI Training

Photographed by Emiel Sloot and Luc Hornstra

Headquartered at Roma-Guidonia, the Comando Generale delle Scuole (General Training Command) of the Aeronautica Militare is responsible for pilot training for the Italian Air Force, from *ab initio* until the trainee is ready for operational type conversion. The CGS controls the three different units involved in this programme: 61°, 70° and 72° Stormi.

61° Stormo, Lecce-Galatina

Graduates from 70° Stormo at Latina will pass on to Lecce. Here, the 61° Stormo is responsible for basic and advanced jet training, while the unit also has a limited tactical task. 61° Stormo consists of three flying units: 213° Gruppo Scuola di Volo Basico Avanzato su Aviogetti (213th basic and advanced jet flying training squadron), 212° Gruppo Pre-operativo (212th pre-operational squadron) and 214° Gruppo Istruzione Professionale (214th flight instructor squadron). All squadrons operate the Aermacchi MB-339.

Trainee pilots coming from Latina enter 213° Gruppo, equipped with the MB-339A, for their first jet experience. The whole jet training course takes 167 flying hours in 152 sorties, and is divided into three phases. The first is mainly concerned with general aircraft handling, ending with a few solo flights, the first one generally being after 11 sorties. The second phase focuses on instrument training, with the trainee in the backseat covered under a hood. This phase is supported by two simulators for IFR and emergency procedures training. The last part of the course is the advanced phase, which includes formation and low-level flight, as well as the first tactical manoeuvres in both air-to-air and air-to-ground regimes. With the 213° Gruppo syllabus successfully completed, the graduate is then posted to an operational conversion unit, depending on the needs of the air force at the time. Generally, the best graduates are destined for jets, to preserve their skills. Three classes of around 10 pilots each start their training with 213° Gruppo annually.

If the trainee pilot is to transfer to one of the fighter OCUs, he or she will first join 214° Gruppo for a pre-operational course, also known as the weapons school. The course involves 40 sorties in the squadron's MB-339CDs. For training purposes, the aircraft can be armed with a warload of up to 2000 kg (4,409 lb) under six hardpoints, consisting of gun pods (either 12.7-mm or 30-mm), rockets (70-mm/2.75-in), and Mk 76 or Mk 106 practice bombs. The latter are dropped at a range north of Lecce, while for strafing and rocket-firing the squadron flies to Decimomannu in Sardinia.

In 2003 214° Gruppo painted up this MB-339A as part of the 100th anniversary of powered flight celebrations.

214° Gruppo is responsible for training new instructor pilots for Lecce, operating both the MB-339A and CD. All new instructors volunteer for this task. Pilots with previous operational experience in fighter units are preferred, however. The instructors of 212° and 213° Gruppi also have a limited tactical role. The new MB-339CD can be armed with two AIM-9L Sidewinders and is suitable as an interceptor against slow-moving targets, a role that grew in importance in the wake of the events of 11 September 2001. CAP flights were carried out during the EU summit in Rome during October 2003. In times of war, these two squadrons could be added to NATO in this role, as well as in the CAS and CSAR support missions.

Four different versions of the MB-339 are operated by 61° Stormo. Of 106 MB-339As delivered to the air force since 1979, some 50 are still based at Lecce. Others are in use with the 'Frecce Tricolori' or the various base flights around Italy. Presently, most remaining airframes are undergoing a mid-life update at the Venegono plant. All flight controls are revised where necessary, while GPS, an ELT and electro-luminescent strips for night formation flying are fitted. By the end of 2004, the last MLU will be redelivered. In addition to the A-model, the air force bought 15 MB-339CDs with deliveries starting in 1997. The MB-339CD features a mission computer, three MFDs, HUD, GPS, RWR and ECM. Furthermore, a refuelling probe is added for training with Boeing 707T/T or Tornados equipped with a buddy-buddy refuelling pod. Twelve MB-339CDs are in use with 61° Stormo. From 2002, some 15 second-series MB-339CDs are being delivered to Lecce. These aircraft have a moving map display and an embedded threat simulation. Virtual air and ground threats can be programmed in along the flightpath, to which the trainee pilot has to respond.

Above are the badges of the 61° Stormo and its three constituent Gruppi. Aircraft wear the wing badge on the fin, inherited from the SVBIA (Scuola Volo Basico Iniziale Aviogetti) which was formed at Lecce in 1962 to fly the Macchi MB.326.

Above: Two MB-339CDs practise the slow-moving intercept mission. The lowered gear is an internationally recognised signal for the 'interceptee' to follow the aircraft and land.

The rounded nose identifies the MB-339A advanced trainer. This aircraft has undergone the mid-life update, as evidenced by the formation lighting strips, ELT antenna on the fin and GPS receiver aft of the cockpit.

Below: Wearing markings of the AMI's RSV test centre, this aircraft is the first of the second series (2° lotto) MB-339CDs. Among the improvements is the addition of a moving map.

70° Stormo, Latina

Screening and elementary flying training is done at Latina air base. Here, the 207° Gruppo Scuola di Volo Basico (207th basic flying training squadron), as part of 70° Stormo, operates a number of Dayglo-sprayed SIAI-Marchetti SF-260AM piston-engined trainers. In total, 46 of these aircraft have been delivered from 1976 in three different batches, although all are identical. Some 40 are still in use for the syllabus at Latina, during which mainly VFR training including aerobatics is taught, beside some basic IFR skills as well. From late 2003, gradual replacement by the SF-260E has started, of which 30 are on order. This version has improved flying characteristics (wider operational envelope) and a new canopy design.

The first screening takes some 15 sorties. If successful, the new pilot continues training for four years at the academy, while at the end of each year, 15 to 20 training sorties will be completed. Others, who only sign a temporary contract (for 10 years), undergo the whole syllabus of 60 flying hours in one instalment.

Instructors from 207° Gruppo maintain their operational proficiency using one of the MB-339A aircraft of the 674° Squadriglia Collegamenti, the base flight at Latina.

Stelio Frati's classic SF-260 trainer design first flew in July 1964 but, despite being evaluated in the mid-1960s, it was some time before the AMI adopted the type to replace ageing Piaggio P.148 trainers. The first SF-260AMs were delivered to 207° Gruppo in 1976, initially for pilot screening duties but soon adding the basic training task. The SF-260 is a highly regarded trainer with excellent aerobatic handling and the instrumentation necessary to allow initial IFR instruction. Pilots who complete the 70° Stormo course usually move on to the MB-339 at Lecce, but some train in the US under the ENJPPT scheme at Sheppard AFB, Texas.

72º Stormo, Frosinone

Some of the graduates from Lecce will learn to fly helicopters. At Frosinone, the 208° Gruppo Scuola di Volo Elicotteri (helicopter flying training squadron, part of the 72° Stormo) is equipped with 48 Breda-Nardi NH-500E helicopters. All air force, army, navy, police, Carabinieri, forestry service and treasury guard trainee pilots come to 208° Gruppo for rotary-wing training. In 2002, five pilots from Albania joined 208° Gruppo for training. The squadron has also been involved in international operations. A few helicopters were detached to KFOR in Kosovo as part of Operation Joint Guardian, while two NH-500Es sprayed white (right) were detached to UNMEE in Eritrea, for patrolling the border with Ethiopia. Armed with rockets and door-mounted machine-guns, the squadron's helicopters even have a limited CSAR capability. Air force pilots who successfully complete their training will pass on to 85° Gruppo at Pratica di Mare, for conversion to the HH-3F or AB 212. Two float-equipped NH-500Es (below) allow water landings to be practised.

Panavia Tornado

Rebirth of the MRCA

Following its first flight 30 years ago in 1974, the first Panavia Tornado Interdictor Strike Variants (IDS) entered service from 1980 with the partner nations that made the concept become a reality. Since that time the Tornado has evolved to fulfil many roles in an ever-changing world. This review examines the Tornado's in-service record since the first Gulf War of 1991 and how it has adapted over the last decade through a number of significant upgrade programmes.

Seen in its natural element, a Tornado GR.Mk 4A of No. II(AC) Squadron, RAF, transits over the English countryside. The Marham-based unit is one of two operational squadrons to operate the RAF's reconnaissance version of the Tornado, the other being co-located No. 13 Squadron. Following the end of the Cold War all three European Tornado nations have reduced their Tornado IDS fleets, yet, through a series of mid-life updates, the aircraft has added a host of new capabilities, ensuring it remains at the forefront of NATO's ground-attack and reconnaissance assets for another two decades.

The Aeronautica Militare Italiana (AMI) maintains a fleet of some 85 Tornado IDSs/ECRs. The 36° Stormo is the sole operational unit to operate both the IDS and ADV variants from its base at Gioia del Colle. The crew of this IDS from 156° Gruppo, 36° Stormo, prepares for a maritime attack training sortie in early 2004 – the only unit tasked in this role. As the Italian mid-life upgrade (MLU) programme gathers pace in 2006 the unit will cycle its aircraft through modification, with all operational aircraft achieving IT-Full MLU standard by 2010.

The Panavia Tornado project was initiated in 1968, in the middle of the Cold War, as a collaborative project for the UK, Germany and Italy. The Tornado multi-role combat aircraft, known as the Interdictor Strike (IDS) variant, was designed to penetrate highly-defended Warsaw Pact airspace at low-level under all weather conditions for attacks on high-value targets.

The advanced and integrated avionics system of the Tornado, using a digital main computer and advanced navigational systems, provides the accuracy needed for a single-pass weapon delivery. In adverse weather conditions this excellent navigational precision is also the basis for target identification using the Ground Mapping Radar (GMR) and for low-level penetration using the Terrain Following Radar (TFR).

The Tornado IDS (GR.Mk 1 in RAF service) made its combat debut in Operation Desert Storm in 1991, marking a definitive phase of the Tornado IDS's career.

Alongside the IDS, the UK's development of the Tornado Air Defence Variant (ADV) saw the existing airframe modified to take on this new role. Both types have seen busy and varied service careers with the air forces that fly them.

RAF Tornado at war

One of the undoubted success stories of 'Gulf War I' was the operational deployment and combat debut of the RAF Tornado GR.Mk 1 force under Operation Granby. The deployed aircraft began the war at low-level, flying high-risk missions against Iraqi AF airfields in this highly-demanding regime. By the end of the conflict, working in tandem with RAF Buccaneers, the Tornados had switched to medium-level operations using Laser-Guided Bombs (LGBs), marking a total change in operating disciplines. Wing Commander Mark Roberts, Officer Commanding No. 12(B) Squadron in 2003, was a junior pilot on No. 16 Squadron operating from Tabuk during Granby. He commented; "In Gulf War I we arrived expecting to do the whole thing at low level. When we then went to medium-level it was a completely new skill for the Tornado".

Following Granby, the successes of the deployed squadrons was tempered by a swathe of cutbacks in the operational force. Even as the Tornados headed into combat over Iraq, plans were already in place to disband the RAF Germany (RAFG) Tornado Wing at RAF Laarbruch. Nos XV, 16 and 20 Squadrons had been disbanded by May 1992, and the four remaining RAFG

Although the Eurofighter Typhoon represents the next generation of combat aircraft, and will possess a significant air-to-ground capability in Tranche 2 form, it will not replace, but operate alongside upgraded Tornado IDSs in UK/German/Italian inventories. Here the first single-seat Eurofighter development aircraft (DA1) conducts air-to-air refuelling trials with a Tornado from the Wehrtechnische Dienstelle für Luftfahrtzeuge 61 (WTD 61) test and evaluation unit, fitted with a centreline-mounted Sargent Fletcher 28-300 buddy-refuelling system.

Tornado squadrons at RAF Brüggen (Nos IX(B), 14, 17 and 31 Squadrons) became part of No. 1 Group, headquartered at RAF High Wycombe.

Operation Desert Storm and a changing operational environment dictated that the IDS had to evolve to meet worldwide crises reaction scenarios. New requirements called for an aircraft that could meet offensive counter air requirements at low and medium altitudes. High survivability against opponents operating with a highly integrated air defence system, with a possible mixture of western and eastern equipment, an ability to adhere to strict Rules of Engagement (ROE) and the capability to strike with precision accuracy was required.

A Cold War requirement for enhanced capabilities for the RAF's GR.Mk 1 fleet had been issued under Staff Requirement (Air) 417. This was designed to allow the Tornado to make a more covert penetration of enemy airspace, with greater use of new stand-off weapons. In addition, the RAF sought to extend the operational career of the Tornado GR.Mk 1 towards 2020 in order for it to complement the Eurofighter Typhoon. The Mid-Life Upgrade (MLU) began to take shape under the Tornado GR.Mk 4 programme.

The complete 'wants list' for the GR.Mk 4 was ambitious, and went as far as to suggest a need for a fuselage stretch, all new stealthy intakes, a composite taileron, Eurojet EJ200 engines, a stealthy gold canopy coating, and a new covert terrain referenced/following navigation system known as TERPROM (TERrain PROfiling and Matching). All of these were subsequently dropped due to budgetary restrictions, as were plans for 26 new-build examples. As the succession of post-Cold War defence cuts bit harder it became clear that the Tornado IDS fleet would lose out on any radical new upgrades. As decision-making delays threatened the entire programme it finally became clear that the MLU would comprise of a series of refinements and additions for 142 existing airframes, born more out of necessity in order to keep the RAF Tornado GR.Mk 1 in line with its counterparts.

Ongoing operations

The 1990s saw the RAF heavily committed to UN operations over the former Yugoslavia as well as covering ongoing peacekeeping duties over Iraq over the southern 'No-fly' zone. The rotational deployments to Al Jaber and Ali al Salem in Kuwait under Operations Jural and (from 1998) Bolton, respectively, by all operational GR.Mk 1 squadrons saw the aircraft regularly employing precision-guided weapons as Iraq flouted various UN Resolutions.

The US-led Operation Desert Fox in December 1998 came about as a result of Saddam Hussein's continued obstruction of UN Weapons Inspectors. This marked the most substantial air campaign against Iraq since Desert Storm, with (appropriately due to its Fox's head unit badge) No. 12(B) Squadron crews from the Operation Bolton detachment at Ali al Salem attacking a number of targets in southern Iraq with Paveway II/III laser-guided bombs (LGBs). Following Desert Fox, the RAF Tornado GR.Mk 1 force remained in situ at Ali al Salem and continued to patrol the southern 'No-fly' zone. However, with the GR.Mk 4 programme consuming an ever-increasing number of GR.Mk 1s, the two Gulf-region detachments started to cause problems. The GR.Mk 4 was insufficiently

The Tornado remains one of the world's most capable low-level interdiction aircraft, offering good-range, a large weapons load and excellent penetration speed. Since the closure of the Tri-Tornado Training Establishment in 1998, an expanded No. 15(R) Squadron is now responsible for all RAF Tornado training, as demonstrated by this GR.Mk 1 in 2001 (above). In the reconnaissance role Nos II(AC) and XIII Squadrons have now equipped with the GR.Mk 4A, an example from the latter seen in Low Flying Area 13 in 2003 (below).

Having participated in Southern Watch operations since the end of the first Gulf War, RAF Tornado GR.Mk 1s were called into offensive action again in December 1998 during Operation Desert Fox. Unlike during the earlier conflict, the Tornados were able to self-designate using the GEC-Marconi TIALD Srs 400 laser designation pod. Manned by personnel from No. 12 Squadron, a total of 12 Tornados, taken from different units, was dispatched to Ali al Salem AB, Kuwait, conducting strike missions against airfields, Iraqi Republican Guard barracks and suspected WMD sites. Here a GR.Mk 1A, carrying No. II(AC) Squadron markings, departs Ali al Salem carrying two 2,000-lb (907-kg) GBU-24 Paveway III laser-guided bombs – the first time RAF Tornados had used the weapon in combat. In total 28 sorties were flown with 48 Paveway IIs and four Paveway IIIs having being dropped.

prepared to be deployed on combat operations, and the RAF needed to reduce its strike force of Tornados in the region. The answer came in the shape of transferring the Jural deployment to Al Kharj, Saudi Arabia, and equipping it with Tornado F.Mk 3s. The two deployments to the south of Iraq eventually became known as 'Resinate South'. These continued to function until the start of Operation Telic in March 2003.

The RAF's Tornado GR.Mk 1 force had not been included in peacekeeping operations over the Former Yugoslavia, however, the GR.Mk 1 force was called upon for NATO's Operation Allied Force in 1999 over Kosovo. Aircraft from the Brüggen Wing flew missions from its home base before a detachment from Nos IX(B) and 31 Squadrons was sent to Solenzara, Corsica, to fly missions up to 31 May. In total, RAF Tornado GR.Mk 1s flew 160 sorties in support of Operation Allied Force.

British upgrade – the GR.Mk 4

One of the main driving factors behind the GR.Mk 4 upgrade was to correct the 'fleets within fleets' problem. The end of Allied Force saw the pendulum swing towards procurement of the GR.Mk 4. It was widely recognised that the RAF's Tornado fleet had become vastly disjointed in terms of capabilities, as modifications and adjustments were made to aircraft as operational requirements dictated. Consequently, one of the major aims of GR.Mk 4 was to standardise the force, principally allowing greater flexibility when organising logistical support and allocating operational taskings. The GR.Mk 4 was to be compatible with the entire inventory of equipment and weapons already available in the Tornado community, as well as new weaponry that was planned.

With BAE Systems (British Aerospace) as the prime contractor, programme SR(A)417 (announced in mid-1994) was to cover 142 RAF Tornado GR.Mk 1/1As to be rotated through a Return To Works (RTW) programme at BAE Systems Warton. The upgrade was designed to be implemented in several stages (or Packages) with the main hardware of the GR.Mk 4 upgrade to be followed by incremental refinements. A pair of Tornado GR.Mk 1s on the BAE Systems test fleet began trials for the proposed upgrades.

The first aircraft converted to full Tornado GR.Mk 4 standard for the RAF made its maiden flight from Warton on 4 April 1997. The first two Tornado GR.Mk 4s were handed over to the RAF on 31 October 1997 and an extensive trials period was followed by initial deliveries to front-line squadrons beginning in May 1998, as the programme got into full swing.

Advanced capabilities

The upgrade package for the GR.Mk 4/4A saw a couple of external modifications being made. A new GEC 10-10 Thermal Imaging Common Module System (TICMS) Forward-Looking Infra-Red (FLIR) sensor was incorporated in a large fairing adjacent to the existing Laser Ranger and Marked Target Seeker (LRMTS) under the nose. The new FLIR required the port Mauser cannon to be removed on the GR.Mk 4 (the GR.Mk 4A followed the GR.Mk 1A in having no cannons). The original IDS

Operation Allied Force

Launched in March 1999 in response to accusations of Serbian 'ethnic cleansing' in Kosovo, Operation Allied Force involved a huge array of NATO air assets, with Tornados from the RAF, the AMI and the Luftwaffe making a valuable contribution. RAF Tornado GR.Mk 1/1A operations were entrusted to RAF Brüggen-based Nos IX(B) and 31 Squadrons. Flying seven-hour missions from their home base, involving three air-to-air refuellings, the Tornado strike packages, usually consisting of six aircraft, employed Paveway II and Paveway III LGBs against high-value targets in Serbia. In late May 12 of the aircraft were dispatched to Solenzara, Corsica, to increase the sortie tempo.

The Luftwaffe had deployed to a combat area for the first time in the mid-1990s during Operation Deny Flight, operating both IDS and ECR variants from Piacenza, Italy, as Einsatzgeschwader 1. For Allied Force the Luftwaffe returned to Piacenza with eight Tornado ECRs from JBG 32 and six Tornado IDSs from recce-roled AKG 51, again operating as Einsatzgeschwader 1. The Tornado ECRs were heavily involved in the destruction of Serbian radar sites using HARM missiles. Pre- and post-strike reconnaissance was conducted by the AKG 51 aircraft carrying MBB/Aeritalia centreline reconnaissance pods.

As well as providing numerous bases for NATO aircraft, Italy committed significant air assets to the operation. Co-located with the German Tornados of AKG 51, Tornado ECRs from 155° Gruppo, 50° Stormo conducted SEAD missions carrying HARMs. Ghedi-based IDS Tornados, alongside 156° Gruppo examples, attacked barracks, armour and high-value targets using LGBs, and 36° Stormo conducted offensive CAPs using eight Tornado F.Mk 3s.

Above: RAF groundcrew prepare a 1,000-lb (454-kg) CPU-123 Paveway II LGB for loading on a No. 617 Squadron Tornado GR.Mk 1, 'lent' to the Brüggen Wing for Allied Force. A total of 11 strike packages was dispatched during the first 25 days of the operation. Re-location to Corsica allowed the RAF Tornados to increase their sortie rate.

Left: AMI Tornado's flew attack, recce, SEAD and combat air patrol missions during Allied Force. This 6° Stormo IDS was one of six examples that operated from Ghedi AB, equipped with the Thomson-CSF Combined Laser Designator Pod (CLDP). The CLDP was used to designate targets for GBU-16 LGBs, as the mission marks on this aircraft demonstrates.

procurement included 24 purpose-built Tornado GR.Mk 1As, which featured the internal Tornado Infra-Red Reconnaissance System (TIRRS). Naturally the reconnaissance bias of this variant was retained in the upgrade, thus becoming the GR.Mk 4A.

Advances in night capabilities had reached the RAF Tornado GR.Mk 1 force by the late 1980s, with Night Vision Goggles (NVGs) in limited use on selected aircraft. The GR.Mk 4's avionics package included full Night Vision Goggle (NVG) compatibility with equipment from Oxley Avionics. Revised cockpit lighting helped permit unrestricted NVG use by both pilot and navigator in all GR.Mk 4s and, complimented by the FLIR, gave the Tornado a radical enhancement in all-round night capability. In the past aircrews had to invent ingenious methods to enable the use of new equipment, most notably the NVGs. These usually involved utilising large amounts of sticky tape to mask out prominent light sources in the cockpit to enable effective NVG use. The improved NVG integration included auto separation from the helmet in the event of ejection, and also meant that the crew could test and fit the 'gogs' as and when required during flight, dramatically reducing the fatigue on long missions caused by the weight of the goggles. A new Ferranti wide-angle holographic Head Up Display (HUD) incorporated into the package allowed the FLIR imagery to be overlaid onto the pilot's forward field of view. Similarly this 'picture' could be presented on a new Smiths Industries Multi Function-Head Down Display (HDD), and on the left TV display in the navigator's position. The FLIR allowed the crew to pick out any relevant or useful 'hot-spots', which could be displayed as 'thermal cues' in order of perceived importance. These could be either a wingman or thermal emissions from a camouflaged target, vastly improving aircrew awareness. The fixed infra-red picture, coupled with the all-round views of the NVGs, permitted the pilot to fly manually in the dark. This reduced the need for the GR.Mk 4 crew to draw on active emitting resources such as the Terrain Following Radar.

In addition to FLIR imagery, the pilot's Multi-Function Display (MFD) can present a new moving digital map. This Digital Map Generator (DMG) replaced the old film-based moving map in a circular display, and allows the pilot to see his intended routing superimposed. The map can also be annotated to highlight specific points of interest *en route*, such as a SAM site avoidance area – the chosen safe radius of which is decided by the mission planner with an alarm prompt set to warn of any infringement.

Fitted with the standard BOZ chaff/flare pod beneath the starboard wing and a TIALD designating pod beneath the fuselage, a Tornado from No. 12 Squadron taxies past a hardened aircraft shelter at Ali al Salem, Kuwait, damaged during the first Gulf War. During Operation Desert Fox the Tornados worked in pairs with one TIALD pod and three or four Paveways between the aircraft.

Above: The RAF maintained constant Tornado detachments for Southern Watch duties from the end of the 1991 Gulf War until Operation Telic. Tornados have been based in Saudi Arabia and Kuwait (Operation Jural) and Turkey (Operation Warden) and, as more aircraft progressed through the RAF's mid-life upgrade programme, GR.Mk 1s were replaced by GR.Mk 4s. Here a No. 14 Squadron GR.Mk 4, adorned with mission markings depicting its role in attacking Iraqi targets with Paveway LGBs, undergoes modifications in the UK before another deployment for participation in Iraqi Freedom.

Right: A Paveway-armed No. II(AC) Squadron Tornado GR.Mk 1A awaits its tasking while on Southern Watch duties in the late 1990s. The addition of TIALD allowed the Tornado to attack targets from medium level, vastly reducing the risk from shoulder-launched missiles and anti-aircraft fire.

Prime contractor BAE Systems (then British Aerospace) used three of its own GR.Mk 1s as trials aircraft for the numerous proposed upgrades that would result in the GR.Mk 4. The upgrade had been initiated in 1989 with an investment and development contract, allocated to Panavia, as the Cold War was still raging. The wide-ranging upgrade was initially planned to include 'stealth' intakes, a fuselage stretch, new engines and a host of new avionics, systems and weapons. The collapse of the Soviet Union brought about drastic defence cuts, which reduced the scope of the upgrade and cut the procurement from 26 new-build and 165 conversions to a total of 80 conversions plus 62 options only. Nevertheless, BAE flew the first GR.Mk 4 in May 1993 and in mid-1994 was awarded the full contract worth £640 million. Here two of BAE's three development GR.Mk 4s formate for the camera at the height of the trials programme. After three years of testing the first RAF Tornado entered the upgrade programme at Warton in April 1996.

Flight planning was also revised to utilise a new solid state data 'brick', which replaced the old cassette tape that downloaded to the navigation computer via the cockpit voice recorder. The new system fed the sortie data more reliably and quickly into the GR.Mk 4's system via the Computer Symbol Generator. Navigation itself was enhanced with the incorporation of an Integrated Global Positioning System (GPS). This allowed the navigator to 'fix' the system to fewer than 10 metres in accuracy. In order for brisker performance, it was further decided to incorporate Turbo Union RB.199 Mk 103s to replace the original Mk 101 engines. The wet film that the GR.Mk 1 relied upon to record HUD, radar and left-hand navigator's TV display data was also replaced. A new Video Recording System (VRS) is employed in the GR.Mk 4 to record the HUD, radar, FLIR, audio transmissions and both navigator TV displays. This allows the navigator to leave the cockpit with the tape ready to utilise during debrief.

No. XIII Squadron at RAF Marham received initial deliveries of GR.Mk 4As from Warton in mid-1998 alongside No. IX(B) Squadron at RAF Brüggen, which received the first delivered GR.Mk 4 on 11 May of that year. After initial acceptance, senior aircrews at the squadrons embarked on a period of NVG and FLIR work-up. The first five pairs of crews from No. XIII Squadron were declared combat ready on the new night equipment during Tactical Leadership Training 99/1, held at RAF Leuchars in early April 1999.

Cuts in the fleet

As the GR.Mk 4/4A firmly established itself as the strike attack platform for the future, defence cuts saw the end of RAF Germany and the closure of RAF Brüggen. In 1999, No. 17 Squadron disbanded and Nos IX(B) and 31 Squadrons returned to the UK and took up residence at RAF Marham to join the reconnaissance specialists of Nos II(AC) and XIII Squadrons. Elsewhere, No. 14 Squadron

RAF Tornados were rotated through the Return To Works (RTW) line at BAE System's Warton facility. It took a total of seven years for the entire fleet to undergo conversion to GR.Mk 4 standard, with the final delivery to the RAF in June 2003. The programme consumed in excess of 1.6 million man hours. Rumours have persisted that BAE Systems has offered a GR.Mk 4-style upgrade programme to the Royal Saudi Air Force, but no order has yet been forthcoming.

moved to RAF Lossiemouth to join Nos 12(B) and 617 Squadrons as well as the National Tornado Operational Conversion Unit (NTOCU) No. XV(R) Squadron. The latter unit marked the combination of the RAF element of the Tri-national Tornado Training Establishment (TTTE) from RAF Cottesmore, that disbanded in 1999, and No XV(R) Squadron Tornado Weapons Conversion Unit (TWCU).

To the Four

The GR.Mk 4 modification facility at Warton accommodated 20 aircraft, with initial turnaround time for each aircraft put at 12 months, but later reduced to 7.5 months by aircraft 17. At its peak, throughput rate for the GR.Mk 4 upgrade was one aircraft every eight working days.

The implementation packages saw the baseline GR.Mk 4/4A receive incremental upgrades, added mainly at squadron level. Package 0 covered some minor software glitch fixes and was quickly followed by Package 1 (Part 1), which was again mainly software-orientated for the integration of Paveway III laser-guided bombs. Package 1 (Part 2) was initiated in April 2001 and involved extensive trials work in order to address a number of software-related issues with targeting and avionics equipment that were plaguing the programme. Successful completion of this phase permitted a full Military Aircraft Release (MAR) for the new software.

Most significant was full Thermal Imaging and Airborne Laser Designator (TIALD) and Paveway III integration. This paved the way for the first operational deployment for the GR.Mk 4, which took place in June 2001 when the first examples deployed to the Gulf region to take part in Operation Resinate South, patrolling the southern 'No-fly' zone over Iraq.

All GR.Mk 4/4As from number 61 in the programme onwards emerged from BAE Systems with a new pilot control column stick-top to allow HOTAS control of some functions and a Laser Inertial Navigation System/ GPS/GPWS (Ground Proximity Warning System). All pre-61 aircraft were subsequently retrofitted with the new stick-top at the earliest opportunity. In early 2002, around

As the GR.Mk 4 programme entered full-swing in the late 1990s, a new swathe of defence cuts saw a reduction and realignment of the Tornado fleet. In addition, some units worked-up on the GR.Mk 4 with other squadrons continued to operate the GR.Mk 1. Here a pair of No. 14 Squadron GR.Mk 1s conduct a sortie from their new base at RAF Lossiemouth in early 2000 following the withdrawal of the Tornado from Germany and the closure of RAF Brüggen.

No. IX(B) Squadron, based at RAF Brüggen, was given the honour of introducing the GR.Mk 4 into operational RAF service in mid-1998 alongside No. XIII Squadron, which received the first of the GR.Mk 4A reconnaissance variant. Initial work-up involved senior aircrew undergoing intensive training on the NVG and FLIR systems. Shortly after converting to the GR.Mk 4 No. IX Squadron returned to the UK and since that time has been based at RAF Marham. here a No. IX GR.Mk 4 conducts a low-level sortie in the Welsh mountains.

The GR.Mk 4 programme was devised to include incremental improvements as the technology became available. The latest stage is named Package 2 (Part 2), and was ongoing in spring 2004. This included the full integration of the MBDA Storm Shadow long-range cruise missile, giving the Tornado fleet a much-needed stand-off capability. Used for pre-planned attacks against heavily defended, hardened and high-value targets, Storm Shadow's GPS-guidance allows fire-and-forget delivery, and a high-resolution IIR seeker and Automatic Target Recognition (ATR) ensures terminal accuracy. No. 617 Sqn (above) debuted the weapon in combat under an Urgent Operational Requirement (UOR) for Operation Telic in 2003. In early 2004 BAE Systems was testing with a maximum load of four missiles (top).

75 per cent of the GR.Mk 4 fleet had received the HOTAS controls, with all completed by mid-2003.

As the RAF shifted towards a predominantly GR.Mk 4/4A fleet, the package upgrades continued. Package 2 (Part 1) was again software-orientated and also included installing a new Upgraded Main Computer (UMC). This Commercial Off The Shelf (COTS) Litef Power PC-based system replaced the archaic original and swamped computer in order to more efficiently drive the weapons bus.

Following lessons learned in Operation Allied Force, the RAF sought an all-weather precision guided bomb (PGB) capability. The requirement was announced in July 2000, with Raytheon selected to perform the upgrade to the weapons in November that year. The Raytheon Enhanced Paveway II and III GPS/laser-guided bombs (known as E-Paveway – EPW) in RAF service, fulfilled this Interim Precision-Guided Bomb (IPGB) requirement. The Strike Attack Operational Evaluation Unit (SAOEU) led trials with a Tornado GR.Mk 4 in September 2001 and cleared the weapon for service on 27 October 2001. The first GPS-guided bombs in the RAF's inventory, the 1,000-lb (454-kg) Enhanced Paveway II and 2,000-lb (907-kg) Enhanced Paveway III were fitted with external interface connectors and conduits, and enhanced guidance units comprising GPS/INS, an inertial measuring unit and two GPS antennas. Laser-designation can be used in conjunction with GPS when weather permits.

Introducing RAPTOR

The Package 2 (Part 1) element of the MLU also contained some functionality for the new Goodrich RAPTOR (Reconnaissance Airborne Pod for TORnado) system to make it a recognised store for the GR.Mk 4. Project RAPTOR came about under Staff Requirement (Air) 1368, with a requirement for target recognition at 40 nm (72 km) visually and 20 nm (36 km) with infra-red with a minimum slant range of 20,000 ft (6069 m). RAPTOR features an Electro-Optical (EO) and Infra-Red (IR) stand-off system that allows the Tornado to gather imagery of pre-planned or 'opportunity' targets, which can then be datalinked back to a ground station. RAPTOR is able to cover as many as 200 separate 'targets' in a single mission.

The RAF currently has eight RAPTOR pods and the system was brought into service by No. II(AC) Squadron at RAF Marham. Under trial 'Flote', the unit performed

The ALARM anti-radiation missile remains an important part of the Tornado's armoury and the weapon was used extensively by the GR.Mk 4 force committed to Operation Telic. Here a No. IX Squadron example carries a typical load-out of two ALARMs on the fuselage pylons.

No. IX Sqn has been at the vanguard of introducing the the GR.Mk 4 into service. Here one of the squadron's aircraft formates with a No. 101 Sqn VC.10 K.Mk 2 after topping up its tanks during a training exercise. The incremental addition of capabilities, especially the integration of the Storm Shadow and Brimstone weapons, has exhausted the capacity of the aircraft's main computer. The old Litef F computer is to be replaced by a hybrid Upgraded Main Computer (UMC) under a conversion contract awarded to EADS, BAE Systems and Litef. The UMC runs simultaneously on the old assembler and new Ada software. Object-orientated programming with Ada adds greater capacity and allows easy integration of future weapons and systems. In June 2003 the first GR.Mk 4 flew from Warton with the new computer. Deliveries to the RAF are due to commence in 2005.

service trials of RAPTOR, with the squadron working with pod manufacturer Goodrich to complete RAF acceptance trials. At that time, Officer Commanding No. II(AC) Squadron Wg Cdr 'Moose' Poole explained; "We have allocated four crews from the squadron for the trials. RAPTOR is a medium-level, stand-off, Electro-Optical/Infra-Red recce pod for the RAF Tornado. It's been proved that we can't rely on satellites and we need good tactical recce, and we know that all service branches will be queuing up to use our imagery. We will be able to perform both pre- and post-strike reconnaissance, with real-time datalinks to an RAF RIC (Reconnaissance Interpretation Centre). We will also be able to perform 'swing-role' whereby the pod is carried on a shoulder pylon; with weapons on other pylons".

The last of the 142 Tornado GR.Mk 4s was redelivered to the RAF on 10 June 2003, when the last example (ZA492) was handed over in a ceremony at Warton. On this date, BAE Systems announced it had been awarded the GR.Mk 4 Combined Maintenance Upgrade (CMU) contract. This will allow future refinements to be incorporated during major servicing to maximise fleet availability. Package 2 (Part 2) is the latest stage in the story as regards GR.Mk 4/4A mission capability and is ongoing in early 2004. This includes full functionality for the MBDA Storm Shadow stand-off weapon integration, despite the weapon having been temporarily released to service under an Urgent Operational Requirement (UOR) for Operation Telic in 2003. The package also includes full integration of the MBDA Brimstone anti-armour weapon,

Initial handling and clearance trials for the MBDA Brimstone anti-armour weapon commenced in December 1998. BAE Systems' Tornado GR.Mk 1 ZA328 (below) was used to conduct firing trials in early 2001, the missiles configured in three groups of four beneath the fuselage. The weapon system will be integrated into GR.Mk 4s under Package 2 (Part 2).

Right: The RAF's original procurement included 24 Tornado GR.Mk 1As featuring the Tornado Infra-Red Reconnaissance System (TIRRS). The reconnaissance element of these aircraft was retained with the GR.Mk 4A, which, like its predecessor, has no cannon armament, but retains full attack capabilities. No. II(AC) Squadron is one of two RAF squadrons equipped with the variant and operates from RAF Marham. The integration of the NVG cockpit has allowed the GR.Mk 4A to operate more effectively at night, especially in conjunction with the new RAPTOR system.

Above and below: No. II(AC) Sqn has been chosen to operate the new Raytheon RAPTOR recce pod. Eight pods have been delivered along with two ground stations. RAPTOR has a DB-110 electro-optical/infra-red dual-band sensor to offer real time imagery via datalinking. Although usually carried on the centreline pylon, it will also be cleared for carriage on the aircraft's shoulder pylons.

with trials of these two systems having been in progress since 2002. The RAF Air Warfare Centre's Strike Attack Operational Evaluation Unit (SAOEU) is understood to have conducted service test launches of Storm Shadow in late 2003 on the US Navy test ranges at China Lake, California. In February 2004, BAE Systems was testing handling characteristics with up to four missiles loaded. The GR.Mk 4 is also receiving Successor Identification Friend or Foe (SIFF) under Package 2 (Part 2).

Future GR.Mk 4

The RAF's Tornado GR.Mk 4/4A force is expected to remain operational until about 2018 when it will be replaced by the as yet unspecified Future Offensive Aircraft Capability (FOAC). As well as undertaking studies to extend the service life by up to 8,000 hours to meet these dates, other upgrades are planned.

Package 3 of GR.Mk 4 is not fully specified, however plans include a Tactical Information Exchange datalink capability in the form of Joint Tactical Information Distribution System (JTIDS) or Improved Data Modem (IDM), a collision warning system to conform to meet full CAA requirements, new Modular Defensive Aids Sub-system (MoDAS – see below) and main computer upgrades. Other options being evaluated include a new Missile Approach Warning System, new targeting pod (Litening reportedly seen as a preferred option) and reliability/structural upgrades also being evaluated.

BAE Systems was awarded an £82 million contract by the UK Defence Logistics Organisation (DLO) for a new cockpit Pilot MultiFunction Display (PMFD) to be integrated with a new radar and navigator's map processor known as Tornado Advanced Radar Display and Information System (TARDIS). Testing was scheduled to begin in July 2004 to meet an In-Service Date (ISD) of August 2006. The RAF also received bids from BAE Systems, Elta and Thales for a new Electronic Warfare Self-Protection System (EWSP) to replace the Sky Shadow ECM pod. The new Modular Defensive Aids Sub-system (MoDAS) will provide electronic, infra-red and radio countermeasures for the GR.Mk 4. BAE Systems, EADS and Elta were eliminated from the contest in September 2003, with remaining bids from ITT, Northrop Grumman and Saab in

the running for the new system to replace Sky Shadow from 2006.

At the Paris Air Show in June 2003, Raytheon was awarded a contract for a new 500-lb (227-kg) Precision Guided Bomb (PGB) to enter service from 2006, with the Raytheon Paveway IV the preferred option. In June 2004 Raytheon delivered three Paveway IV 'shapes' to the UK for initial dropping trials

Operation 'Telic'

At the start of 2003, it was clear that Iraq's continued breach of UN Resolutions was wearing thin in the eyes of US President George W. Bush and UK Prime Minister Tony Blair. Forces were massed in countries surrounding Iraq, for an impending invasion. The RAF deployed a force of approximately 32 Tornado GR.Mk 4/4As to Ali al Salem and Al Udeid in Qatar. This bolstered the existing Resinate South detachment of around eight aircraft. The Ali al Salem team was comprised of aircraft and crews from Nos II(AC), IX(B) and 31 Squadrons from RAF Marham and No. 617 Squadron from RAF Lossiemouth. The No. II(AC) Squadron crews (supplemented by crews from No. XIII Squadron), led by Squadron Leader Ian Gale, formed a time-sensitive targeting (TST) flight that became known as the 'Truffle Snufflers'. This involved flying low-level 'Scud' hunting missions in the western Iraqi desert. The No. 617 Squadron 'Dambusters' element at Ali al Salem was tasked with the combat debut of the MBDA Storm Shadow conventionally-armed stand-off cruise missile.

Prior to deploying, the GR.Mk 4/4As received a number of minor modifications under Urgent Operational Requirements (UORs). As well as the interim MBDA Storm Shadow integration, Defensive Aids System (DAS) upgrades included new hand-controllers in the rear cockpit and new BOL-IR Sidewinder launch rails, as well as repainted light grey nosecones to reduce visual signature at medium level. The Tornados were flying missions from the outset and the GPS-guided Enhanced Paveway III was also used in combat for the first time.

No. 617 Squadron debuted the Storm Shadow on 2 March 2003, following an urgent clearance process conducted by the squadron. The involvement of No. 617 Squadron with this new missile was particularly significant as it had been exactly 60 years since the squadron had been formed to undertake the famous attacks on the Ruhr Dams under Operation Chastise in World War II. "It was an historic mission for us", said Wg Cdr Dave Robertson, OC No. 617 Squadron, as he spoke to reporters in Kuwait after the dramatic first mission. Having led the first Storm Shadow mission over Iraq, Wg Cdr Robertson described how the sortie took the 'Dambusters' crews through heavy anti-aircraft fire. It transpired that one of his wingmen was forced to jettison fuel tanks in order to increase manoeuvrability as he was engaged by an Iraqi Surface to Air Missile (SAM).

As the war turned to a peacekeeping effort, the RAF maintained a valuable presence in the region, with the Tornado GR.Mk 4/4A deployment at Ali al Salem flying reconnaissance missions utilising the RAPTOR pod. In 2004, the detachment moved to Al Udeid, Qatar, and was again operated on a two-month rotational basis by each squadron.

Back in the UK, the GR.Mk 4/4A fleet is kept busy with a hectic training schedule and training deployments around the world under programmes such as Torpedo Focus – the now annual live bombing detachment to Davis Monthan AFB, Arizona. The RAF is studying the GR.Mk 4 replacement platform under the Future Offensive Aircraft Capability (FOAC), which is as yet undefined but could take on a number of guises from UAV to stand-off cruise missiles from around 2018.

RAF air defence

The evolution of the Tornado Air Defence Variant (ADV) came about as a result of a Cold War RAF requirement for a long-range interceptor, to replace the Lightning and then F-4 Phantom. The Tornado ADV was designed to work in tandem with an Airborne Early Warning (AEW)

The second of the RAF's GR.Mk 4A units is No. XIII Sqn. The main visible external difference between the GR.Mk 1/1A and the GR.Mk 4/4A (clearly seen on the aircraft at top) is the addition of the Thermal Imaging Common Module System (TICMS) Forward-Looking Infra-Red (FLIR) sensor housed beneath the aircraft's nose, adjacent to the Laser Ranger and Marked Target Seeker (LRMTS). This new system necessitated the removal of the port cannon from the GR.Mk 4. The standard Sky Shadow ECM pod, as seen on this No. XIII Sqn GR.Mk 4A (below), is due to be replaced by a new Modular Defensive Aids Sub-system (MoDAS) from 2006.

Above: RAF Tornado F.Mk 3s were heavily involved in Southern Watch operations from 1999, squadrons rotating to Al Kharj AB in Saudi Arabia. Squadrons deployed for nine-week tours, although this was negotiated between the squadrons, as was the order in which squadrons deployed. Here a No. 11 Squadron F.Mk 3 is seen transitting over Kuwait during Operation Jural.

Below: Tornado F.Mk 3s were used to defend high-value assets (HVAs) in the Goalkeeper role during Operation Jural, which ensured the Tornado's low operating altitude of around 20,000 ft (6096 m) did not conflict with strike and recce assets. The fear was that Iraq might launch a MiG-25 suicide mission to shoot down an HVA.

Below right: A No. 111 Sqn F.Mk 3 undergoes an engine change at Ali al Salem, Kuwait, in February 2001. The base was used as a diversion airfield, this example having shut-down an engine over Iraq. The high temperatures ensured afterburner had to be employed to maintain altitude on one engine.

platform, today being the E-3 Sentry. The ADV was never designed as a dogfighter, or to enter a close battle and punch above its weight. It was designed to loiter on CAP (Combat Air Patrol) and engage targets at long range.

Following the early problems of the original Tornado F.Mk 2 that entered service with the RAF in 1985, the subsequent Tornado F.Mk 3 was much improved and was a solid baseline on which to build through incremental upgrades, which have seen the aircraft develop into a highly capable and versatile platform.

Following the well-documented Stage One HOTAS-orientated and Stage One Plus Gulf War improvements to the baseline F.Mk 3, a series of minor 'tweaks' have been made to the AI.24 Foxhunter radar to make it a formidable primary sensor. As with most types, a number of minor modifications are incorporated on the fleet as required.

F.Mk 3 in combat

RAF Tornado F.Mk 3s are no stranger to combat operations. The RAF maintained an F.Mk 3 detachment at Gioia del Colle from April 1993, with squadrons deploying on a rotational basis to support UN Peacekeeping operations following the civil war in Yugoslavia. This became Operation Deny Flight – maintaining the ceasefire between the new states of Croatia and Serbia as well as to support the humanitarian aid effort into Bosnia. As the ground war intensified in late 1994, Serb ground forces regularly 'painted' NATO aircraft with SAMs, and F.Mk 3s were fired upon. This led to a rapid programme to upgrade the type's defensive aids systems (DAS), principally with the integration of the aerial Towed Radar Decoy (known inevitably as 'TURD'). With NATO action under Operation Deliberate Force drawing down, the RAF F.Mk 3 deployment to Gioia was wound up in February 1996.

The F.Mk 3 was not called upon for Allied Force in 1999, however, the peacekeeping operations over southern Iraq saw the F.Mk 3s employed on Operation Jural from February 1999 at Al Kharj (Prince Sultan AB). The deployment at Al Kharj continued through to the 2003 Operation Telic. Deployed F.Mk 3s, crewed by the Leuchars Wing, participated in the air war flying strike escort missions. However, the presence at Al Kharj ceased when the detachment returned to the UK at the end of hostilities. The crews had come under fire several times and are thought to have taken part in around 1,000 hours of flying during the combat phase of Operation Telic. On 11 April, the first Tornado F.Mk 3s, crewed by No. 111 Squadron, arrived back at RAF Leuchars.

Incremental upgrades

RAF Tornado F.Mk 3's have received a number of incremental upgrades following the first Gulf War package of enhancements. The Stage 2G radar upgrade was introduced from 1995 to give automatic target recognition and tracking, recognition of head-on targets through analysis of compressor discs, and introduction of Link-16 Joint Tactical Information System (JTIDS). Celsius Tech BOL (now Saab) integral chaff/flare dispenser and missile

Left: Operation Jural sorties were usually flown with four SkyFlash missiles beneath the fuselage and two AIM-9L Sidewinder (later ASRAAM) missiles on the glove pylons. A No. 11 Squadron F.Mk 3 is seen here in December 2002, when the squadron was heavily involved in the task of preparing the battlefield for the looming war. On a number of occasions the F.Mk 3s were locked by Iraqi radar systems and shot at with AAA. Additionally, an Iraqi MiG-23 entered the 'No-fly' zone in an attempt to shoot down a Predator UAV, but was chased off by the F.Mk 3s.

Left: No. 25 Squadron F.Mk 3 ZE808/XV is seen minus underwing fuel tanks and aerial Towed Radar Decoy (TRD) after defending itself against an Iraqi surface-to-air missile launch. The F.Mk 3s often came under threat from Iraqi ground-based units, particularly when conducting operations close to Baghdad, and were highly valued by the US commanders as they brought JTIDS to the theatre at a time when USAF F-15s were not equipped with the system.

launch pylons were also introduced during this period but not as part of Stage 2G. Stage 2G STAR and Stage 2H followed to improve radar processing as well as fine tuning of the primary sensor.

Sustaining the F.Mk 3

Similar to the GR.Mk 4 standardisation upgrade, a more concerted effort was launched in order to bring the F.Mk 3 up to a common specification as well as to introduce new technology. The Capability Sustainment Programme (CSP) recognised the fact that the 114-strong RAF F.Mk 3 fleet had vastly differing modification standards. The lack of a single baseline modification state was exaggerated with the phased introduction of different stages of the Foxhunter radar upgrades and limited integration of other more significant modifications, such as the TRD. Delays to the In-Service Date (ISD) of the Eurofighter Typhoon, the F.Mk 3's direct replacement, resulted in the need to implement a core standard. As a result, CSP was defined and chosen to be implemented on 100 examples.

With BAE Systems acting as prime contractor, it designed and tested the modifications before upgrading the first 24 aircraft at RAF St Athan. Following this, DARA (Defence Aircraft Repair Agency) teams set about modifying the remaining 76 examples using BAE-supplied kits, while 14 aircraft with No. 56(R) Squadron, the F3 Operational Conversion Unit, were not selected to be upgraded.

The CSP upgrade includes a MIL-STD-1553B digital databus link between the radar data processor and the fuselage weapons stations, a new main computer processor, a new Ada-based missile management system and improved cockpit displays. The new main computer processor facilitated improved cockpit display symbology and presentation but, more significantly, enhanced the system to permit full functionality during simultaneous use of radar and the Joint Tactical Information Distribution System (JTIDS). The radar upgrades coupled with the new systems in CSP led to crews boasting a 'second to none ability to track while scan'.

With the new weapons management system, the type was made compatible with the Raytheon AIM-120

Above: No. 5 Squadron F.Mk 3 ZE042 rests under a shelter at Al Kharj following the first operational MBDA ASRAAM sortie (seen on the glove pylon) on 14 November 2002. Initial aircrew experiences with the missile have reported much improved performance over the AIM-9 Sidewinder.

Left: A No. 5 Squadron F.Mk 3 rests on the tarmac at Al Kharj in 2002 during the unit's last operational deployment before the squadron was disbanded in 2003. The UK government's July 2004 announcement that a further F.Mk 3 squadron (No. 11 in October 2005) will be disbanded, will leave three operational squadrons and No. 56(R) Squadron F3 OCU.

***Right:* The RAF's F3 Operational Evaluation Unit (now part of the Fast Jet OEU) is tasked with developing the tactics for the ADV force as well as conducting trials to evaluate new equipment and systems. One of the unit's four aircraft is seen here during evaluation of the RAF's new Rangeless Airborne Instrumentation and De-briefing System (RAIDS) pod in 2000. This system, developed by MBDA in conjunction with IAI(MLM), and now in service with the RAF, allows air forces to train in any airspace, uses improved weapon simulations and provides real-time kill assessment and collision avoidance.**

***Right:* The AIM-120 AMRAAM medium-range air-to-air missile was selected to replace the venerable SkyFlash as the Tornado F.Mk 3's principle beyond visual range (BVR) weapon. Like the SkyFlash, the AMRAAM is loaded either as one or two pairs (as seen here) recessed into the lower fuselage. The missile was first fired from an F.Mk 3 in December 1999, but initial results were disappointing and a programme was initiated to introduce mid-course guidance. In early 2004, however, the SkyFlash remained the preferred weapon option.**

Advanced Medium-Range Air-to-Air Missile (AMRAAM), aimed at replacing the SkyFlash. The first AMRAAM launch from an F.Mk 3 took place on 16 December 1999 as part of firing and clearance trials. The Ministry of Defence revealed in mid-2002 that Tornado F.Mk 3s were also to undergo the AMRAAM Optimisation Programme (AOP), a £28 million project to introduce Mid-Course Guidance Capability (MCGC) to improve AIM-120 AMRAAM targeting and dramatically enhance its Beyond Visual Range (BVR) capabilities. Service release was expected during 2003, however this would appear to have slipped as the F.Mk 3 community continued to favour carrying the SkyFlash in 2003. This can be attributed to the fact that AMRAAM without the full MCGC capability is arguably less effective than the popular SkyFlash.

In addition to CSP, a further upgrade package was defined to improve equipment cooling for hot-and-high operations and an upgraded cockpit configuration to

Tornado F.Mk 3 turns Weasel

The Royal Air Force reportedly had an Urgent Operational Requirement (UOR) in 2000 for a limited number of RAF Tornado F.Mk 3s to be modified with an Emitter Locator System (ELS) for use as unarmed Suppression of Enemy Air Defence (SEAD) platforms. It had long been understood that the F.Mk 3 would make a highly suitable SEAD platform due to its tail and wing-glove mounted Radar Homing And Warning System (RHAWS) antennas – giving it an extremely effective Emitter Locator System (ELS). Sources suggest that the aircraft is able to pinpoint enemy radars with greater precision than dedicated SEAD aircraft like the Tornado ECR and F-16CJ. Furthermore, the Tornado F.Mk 3's Joint Tactical Information Distribution System (JTIDS) datalink provides key links to other strike assets, as well as to decision-makers.

Several SEAD demonstration sorties were known to have been flown over the Spadeadam electronic warfare range by 2002, with further reports that the F3 Operational Evaluation Unit (OEU) received one BAE Systems Thermal Imaging And Laser Designation (TIALD) pod, possibly for additional trials to gain a precision-guided munition designation role. In early 2003, it emerged that around 12 RAF Tornado F.Mk 3s had indeed been modified with the ELS under a hastily introduced new Urgent Operational Requirement (UOR) as part of the build up to Operation 'Telic'. These aircraft, unofficially referred to as Tornado EF.Mk 3Bs, had been further equipped to carry the MBDA Air-Launched Anti-Radar Missile (ALARM) on two underfuselage pylons, giving the aircraft a lethal SEAD/DEAD (Destruction of Enemy Air Defences) capability. The cockpit of the modified aircraft was also fitted with a new control panel for the SEAD mission, although details remain vague.

Aircraft assigned to No. 11 Squadron at RAF Leeming were initially given the new SEAD capability – giving the unit a dual-SEAD and Air Defence role. Despite being ready to deploy for Operation Telic, the Tornado 'EF.Mk 3B' played no part in the operation. However, the programme underlined the ability for the RAF to react quickly to urgent requirements, to use in-house resources and work in conjunction with those of QinetiQ and industry, to implement them effectively.

Although the SEAD UOR only lasted for one year, the RAF initially looked at retaining this capability for the type. Further reports suggested that the F.Mk 3 could receive expanded capabilities in the form of targeting pods for precision strike, or even the capability to carry a reconnaissance pod as a useful force multiplier. Indeed, during 2003 unconfirmed reports suggested that the ALARM fit was being extended to allow the aircraft to carry additional missiles on the inner wing stub pylons. However, by early 2004, it appeared that momentum in the SEAD/multi-role programme had been lost, with talk of yet further cuts on the table for the F.Mk 3 force.

Tornado 'EF.Mk 3' ZE763 from No. 11 Squadron carries two ALARM missiles in March 2003. The squadron began working with the missile in January of that year and was prepared to deploy for Operation Telic, before the withdrawal of Turkish bases left the SEAD F.Mk 3s with nowhere to go.

address ergonomics flaws (including some HOTAS-related refinements) and the provision of NVG-compatible lighting to common standard. This upgrade was known as Common Operational Value (COV) and was quickly combined with CSP under ADV2000 to streamline the two upgrade programmes. The first aircraft to undergo the combined procedure was the 41st CSP aircraft, which was completed in early 2000.

Under ADV2000, the F.Mk 3 is also receiving new Active-Matrix Liquid-Crystal Displays (AMLCDs) as well as an improved radio fit (which includes a double Have Quick secure module).

The baseline ADV2000 upgrade has also received additional enhancements since conception. Modification to the navigation system was completed as a priority in mid-2002. The new system includes a Honeywell 764GT LINS (Laser Inertial Navigation System) with embedded GPS fitted. A replacement for the present main computer is also desired with one or two Radstone Power PCs the suggested Commercial-Off-The-Shelf (COTS) alternative. These could be added to the Raytheon IFF 4810 Successor Identification Friend-or-Foe (SIFF) upgrade, which was also completed in mid-2002.

As well as the systems upgrades, the structural integrity of the airframe is also of major significance. At RAF St Athan, DARA not only runs the ADV2000 upgrade and routine deep maintenance, but is also involved in the Mid-Life Fatigue Programme (MLFP) to give complete centre fuselage overhaul and replacement. Fuselage sections are removed at St Athan and shipped to EADS at Manching, Germany, to be re-worked before re-introduction to the fleet.

New strengths

In September 2002 the RAF declared the Advanced Short-Range Air-to-Air Missile (ASRAAM) ready for operational use on the F.Mk 3 as a replacement for the AIM-9L Sidewinder. The MoD initially refused to accept ASRAAM from manufacturer MBDA in 2001 due to reported failures for the weapon to meet less than half of its key user requirements due to technical reasons. This included the ASRAAM's all-aspect acquisition and track, probability of kill, countermeasures resistance and off-boresight acquisition and launch characteristics. The full operational missile software was cleared in 2003 following high off-boresight launch test firings at Eglin AFB, Florida.

ASRAAM can be deployed in different operating modes, 'Lock before Launch' for normal front hemisphere engagements or 'Lock after Launch' for engagements with targets beyond the seeker head acquisition range with target data provided by the aircraft's sensors or a third party. Alternatively, for close-in combat, the pilot can fire the

A 'Treble-One' Squadron F.Mk 3, sporting an 85th anniversary special colour scheme, is seen carrying four MBDA ASRAAM missiles on the underwing pylons. Technical problems with the missile were ironed out during 2002 and the missile has now fully replaced the AIM-9 Sidewinder with all four operational squadrons.

Below: The RAF used Exercise Saif Sareea II, held in Oman in the autumn of 2001, to test its new equipment and tactics in a simulated combat environment. Nos 11 and 25 squadrons were deployed to the region and one of the former's aircraft is seen here releasing flares from beneath the rear fuselage during an air-to-air combat training sortie.

Right: As part of the UK's commitment to Operation Iraqi Freedom the RAF deployed a force of some 32 Tornado GR.Mk 4/4As to Ali Al Salem, Kuwait, and Al Udeid in Qatar. Here a quartet of GR.Mk 4s, devoid of all squadron markings apart from individual aircraft codes on the tail fin, transits over the Mediterranean at the start of the deployment.

Above: The MBDA Storm Shadow stand-off cruise missile was rushed into operational service in time for Operation Telic. The squadron chosen to conduct the first sorties was No. 617 Squadron, which had previously been involved in trials, using dummy Storm Shadows. Here, Wg Cdr Dave Robertson and Sqn Ldr Andy Myers taxi from their protective shelter at Ali al Salem during the first ever GR.Mk 4 Storm Shadow sorties on 21 March 2003.

missile 'off-boresight', i.e. over the shoulder, using the 'Lock after Launch' mode employing a Helmet Mounted Cueing System. However, the F.Mk 3 fleet is thought unlikely to receive helmet-mounted cueing systems, and will likely fire the weapon only in 'Lock before Launch' mode to be certain of accurate target acquisition. Full, digital, ASRAAM integration and the provision of the BAE Systems Striker Helmet-Mounted Sighting System (HMSS) would have given the F.Mk 3 a real off-boresight missile aiming capability. The full capabilities of the ASRAAM system will, however, be available for the Eurofighter Typhoon and F-35 Joint Strike Fighter.

The RAF is very pleased with the new missile. Commanding Officer of the F3 OEU, Wg Cdr Robin Birtwistle stated in 2003; "ASRAAM has an acquisition and track performance twice that of the AIM-9 Sidewinder and has a robust counter-countermeasures capability. The weapon will provide a first shot in most scenarios, and as standard equipment with the F3 is an enhancement to the aircraft's capabilities. It will give our crews an edge in obtaining air superiority".

Wg Cdr Birtwistle and the F3 OEU team embarked on a major programme of missile development and integration that culminated with successful trials in May 2002 at Eglin and at the US Navy facility at NAWS China Lake, California. Tornado F.Mk 3 squadrons have now completed training on the new missile and the two operational F.Mk 3 wings at RAF Leeming and RAF Leuchars are fully utilising ASRAAM in place of the AIM-9 Sidewinder.

JTIDS

Of all the upgrades to the RAF F.Mk 3, none have created as much enthusiasm as the Link-16 Joint Tactical Information Distribution System (JTIDS). The F.Mk 3 was always designed with a tactical information system in mind, in order for the type to work in close harmony with the RAF's E-3D Sentries.

JTIDS is well-documented as a high-capacity tactical information distribution system providing integrated voice and digital data communication, navigation and identification capabilities. It is designed to provide secure and jam-resistant real-time information transfer to enable all participants 'on the network' to be interconnected simultaneously. By direct interface of air defence surveillance platforms such as the E-3 Sentry, the co-ordination of resources can be accomplished more effectively, allowing the F.Mk 3 crew to have advantageous situational awareness of the battlespace, maximising effectiveness.

However, JTIDS is by no means new to the F.Mk 3. Flight testing of the system on the type began in 1986, however service trials with the F3 OEU continued into the early 1990s, with the MoD having awarded GEC-Marconi a US$70 million contract for the initial production terminals for the F.Mk 3 in 1991.

Following the disbandment of No. 29 Squadron at Coningsby in 1998, No. V(AC) Squadron followed suit in 2003, leaving No. 56(R) Squadron F3 OCU and the F3 OEU as the only resident Tornado units at the Lincolnshire base. Talk of F.Mk 3s being forward-deployed to bases in southern England in 2002 to act as a Quick Reaction Alert (QRA) force for the southern UK have failed to materialise. When the OCU moved out to

Below: RAF Tornados regularly relied on USAF tankers for mission support during Iraqi Freedom. Here, a GR.Mk 4A, armed with four RBL-755 cluster bombs, takes on fuel from a KC-10.

Above: Following the non-deployment of the SEAD-rolled F.Mk 3s, the responsibility for conducting RAF operations against Iraqi radar sites fell to the GR.Mk 4/4A force. The improved Air Launched Anti-Radiation Missile Mk 2 (ALARM Mk 2) was used in combat for the first time and a pair is seen here being loaded onto the under fuselage stations of an Ali al Salem-based aircraft. No. IX Squadron flew the first RAF offensive sorties of the war, carrying ALARMs.

Left: Armed with ASRAAM missiles and a GEC-Marconi TRD mounted in a modified BOZ pod on the port outer wing pylon, a Tornado F.Mk 3 takes on fuel from a USAF 117th Air Refuelling Wing KC-10 tanker during an Operation Telic sortie. RAF ground crew elected to name a number of the F.Mk 3s in honour of famous RAF/RCAF/RAAF aces. Names such as McCudden, Bishop, Collishaw, Bader, Stanford-Tuck, Caldwell and Deere and their number of 'kills' appeared on starboard nosewheel doors.

Below: Tornado F.Mk 3s flew escort sorties for a wide range of air assets during Iraqi Freedom. Here a pair of No. 111 Sqn aircraft keeps a watchful eye on a B-1B taking on fuel en-route to Baghdad.

RAF Leuchars in 2003 to join Nos 43(F) and 111(F) Squadrons, the RAF was able to rationalise the F.Mk 3 to just two operational stations – the other being RAF Leeming with Nos XI(F) and 25 Squadrons.

Under Case White at BAE Systems Warton, the RAF's Typhoon Operational Evaluation Unit, No 17(R) Squadron, started pilot training in late 2003. The Operational Conversion Unit, No. 29(R) Squadron, is due to stand-up initially at Warton, but will move to Coningsby in mid-2005. The first two operational Typhoon squadrons (Nos 3 and 11 Squadrons) should then stand up at Coningsby, followed by two units at Leeming and two more at Leuchars in 2009 – signalling the end for the RAF Tornado F.Mk 3. However, uncertainty over Typhoon procurement figures continues, and F.Mk 3 Out Of Service Dates (OSD) are still not set in stone.

German Tornados

In August 1995, as peacekeeping operations in the Balkans turned increasingly sour, the Luftwaffe deployed eight Tornado Electronic Combat Reconnaissance variants (ECRs) from JBG 32 and six reconnaissance-configured IDSs from AKG 51 to Piacenza, Italy, as NATO's Operation Deliberate Force came into effect. The deployment under Einsatzgeschwader 1 was significant in that it was the first operational deployment for the Luftwaffe Tornados. The deployment was reduced to a mainly reconnaissance-orientated capability from late 1996 and, in contrast to the RAF's GR.Mk 1 and the Italian IDS communities, the Luftwaffe did mot get its first taste of actual combat until Allied Force in 1999. This conflict over Kosovo saw the Luftwaffe again deploy JBG 32 ECRs and AKG 51's recce-configured IDSs. The ensuing conflict saw the ECR earning respect as a formidable SEAD platform, launching an astonishing 236 AGM-88 HARMs against Serb SAM sites.

Despite having to make way for Eurofighter Typhoon to a certain extent, the Luftwaffe Tornado IDS/ECR force is planned to remain operational until around 2020. Luftwaffe Tornado training transferred to Holloman AFB, New Mexico, when the TTTE disbanded at RAF Cottesmore in March 1999. The wings JBG 31 at Norvenich and JBG 33 at Büchel are planned to retire the Tornado IDS in favour of Eurofighter in around 2010 and 2015, respectively. The remaining Tornados (82 operational) are being upgraded under an incremental programme by EADS Military to keep these remaining aircraft operational after Eurofighter Typhoon deliveries are completed. Under current German defence planning (Luftwaffe Struktur 5), two wings will continue to operate the Tornado until around 2020-2025.

Below: Despite the availability of the AMRAAM, the RAF elected to use the SkyFlash missile combined with the ASRAAM for escort operations during Iraqi Freedom. Here a pair of Tornado F.Mk 3s patrols the vast desert of central southern Iraq.

Operation Telic nose art

Tornado GR.Mk 4/4A

Of the approximately 32 Tornado GR.Mk 4/4As deployed by the RAF for active duty during Operation Iraqi Freedom, the majority received decorative nose art.

ZD715 *Alarm Maiden*
ZA607 *Delightful Debs*
ZA554 *Born Fighter*
ZA542 *Danger Mouse*
ZA547 *Star Turn*
ZA589 *Deadly Nightshade*
ZG714 *Truffle Snufflers*
ZA606 *Big Deal*
ZD720 *Talisker*
ZD850 *Op Telic 2003*
ZA553 *Dishy Intel*
ZG775 *The Macallan*
ZA600 *Hot stuff*
ZA614 *It's Show Time*
ZD714 *Johnny Walker*
ZG792 *Glen Moray*
ZG727 *Look'n For Twouble*
ZD740 *Desert Raven*
ZG777 *Craigellachie*
ZD793 *Tamdhu*
ZG707 *BABS*
ZG794 *Glenfarclas*
ZG726 *Kylie*
ZG729 *Mean One/Grinch*
ZG711 *Oh Nell*
ZA400 *Scud Hunters*
ZA560 *Brave Coq/Benromach*
ZA559 *Aberlour*
ZA449 *Strathisla*

***Top:* GR.Mk 4 ZG777/TC** Craigellachie *from No. 15(R) Squadron is seen here with seven Paveway mission marks.*

***Above left:* GR.Mk 4 ZA542/DM** Danger Mouse *from No. 31 Squadron.*

***Above:* GR.Mk 4A ZG727/L** Look'n for Twouble *from No. 13 Squadron.*

***Above right:* GR.Mk 4 ZA560/BC** Brave Coq/Benromach *from No. 14 Squadron.*

Combat Air Wing badge on ZA400/T Scud Hunters.

GR.Mk 4 ZD714/AM Alarm Maiden *from No. 31 Squadron.*

GR.Mk 4 ZA614/AJ-J It's Show Time *with Storm Shadow from No. 617 Squadron.*

GR.Mk 4 ZA606/BD Big Deal *(with cards illustrating the five GR.Mk 4/4A squadrons) from No. 14 Squadron.*

Tornado F.Mk 3

Equipped with the ASRAAM missiles and Towed Radar Decoy, Tornado F.Mk 3s from Nos 11, 25, 43 and 111 Squadrons were deployed to Al Kharj in Saudi Arabia. A number of the F.Mk 3s received art recognising previous aces (with their total 'kills'), while others received cartoon character and other less PC depictions. Unlike in Desert Storm the RAF F.Mk 3s were afforded a more active role in the campaign, escorting strike and bomber aircraft to their targets and protecting high-value assets such as tankers and AWACS. Known named F.Mk 3s appear below.

ZE731/YP	111 Sqn	*Desperate Dan/Bishop 72*
ZE158/UW	25 Sqn	*McCudden 57*
ZE161/UU	43 Sqn	*Op Telic 2003/Lacey 28*
ZE206/UI	43 Sqn	*Tracy Shaw Thing/Bader 22*
ZE737/YM	43 Sqn	*Dodger/Stanford-Tuck 28*
ZE162/UR	111 Sqn	*Proctor 54*
ZE692/XC	25 Sqn	*Dennis the Men'/Deere 27*
ZE968/XB	111 Sqn	*111F Have It!/Collishaw 60*
ZE758/YI	DARA	*Tremble/Caldwell 28*

Left:* ZE692/XC** Dennis The Menace/ Deere 27 ***from No. 25 Squadron.

Above:* ZE731/YP** Desperate Dan/ Bishop 72 ***from No. 111 Squadron.

Right:* ZE758/YI** Tremble/Caldwell 28 ***from DARA at St Athan.

New Tornado weapons options

MBDA Storm Shadow

Rushed into RAF service under an Urgent Operational Requirement (UOR) for its combat debut in Operation Iraqi Freedom, MBDA's Storm Shadow long-range cruise missile was devised to provide stand-off capability for the destruction of high-value surface targets, while minimising the risk of collateral damage. To be carried by both RAF GR.Mk 4s and Italian IDSs, the Storm Shadow can be launched as a fire-and-forget store by day or night and in all weathers, and has been designed with a minimal radar cross-section to bestow stealthy characteristics. MBDA claims pinpoint accuracy and low-level penetration for the missile thanks to the combination of INS, GPS and terrain reference navigation. The missile accelerates during the terminal phase high-speed dive before impact and detonation of a precursor charge, followed by the main large kinetic energy follow-through penetrator.

Above: Once released from the aircraft, the Storm Shadow's wings deploy and the missile 'flies itself' to the target under the power of a jet turbine engine. Dorsal and ventral stabilising fins manoeuvre the missile along the desired course and at the programmed altitude.

Left: A No. 617 Squadron Tornado, the unit that introduced the weapon into operational service, is seen with two Storm Shadows beneath the fuselage.

MBDA Brimstone

The Brimstone has been selected by the UK's Ministry of Defence as its new advanced air-launched anti-armour missile. Each Brimstone weapon comprises a 'smart' launcher and three missiles, of which any number can be launched at a given time. Brimstone can be targeted using Direct Mode (DM) where the pilot visually selects the target using on-board sighting systems (usually used against sudden targets of opportunity) or in Indirect Mode (IM) using third-party targeting information.

Once launched, Brimstone is fully fire-and-forget and is boosted to supersonic speed by a solid propellant rocket motor. A millimetric wave radar seeker ensures the missile can operate in all weathers and fly at a fixed height above the ground. The Tornado is capable of launching 18 missiles (aircraft at right carries 12) simultaneously, the missiles flying on separate paths spread out to cover the largest area. The onboard radar performs a comprehensive sweep for targets once in the programmed area of engagement, before destroying even the most durable main battle tanks with the tandem-shaped charge High-Explosive Anti-Tank (HEAT) warhead.

Raytheon AIM-120 AMRAAM

Widely regarded as the West's most capable beyond visual range (BVR) air-to-air missile, the AIM-120 Advanced Medium-Range Air-to-Air Missile (AMRAAM) was selected to replace the ageing Skyflash as the F.Mk 3's principle weapon. Following initial clearance trials in 1999-2000 and early operational experiences the MoD revealed that the aircraft/missile system was to undergo an improvement programme to incorporate mid-course guidance and increase BVR capabilities. F.Mk 3s have, in most cases, continued to operate with SkyFlash while the modification process was conducted.

The new equipment was tested successfully in Canada in May 2004 and AMRAAMs have now begun to replace the SkyFlash with operational squadrons. A Leuchars-based No. 43 Sqn F.Mk 3 is seen here in June 2004 carrying four of the weapons, recessed into the underside of the fuselage, in June 2004.

MBDA ASRAAM

Designed as an advanced competitor to the all-pervading AIM-9 Sidewinder, MBDA's Advanced Short-Range Air-to-Air Missile (ASRAAM) has been operational on the RAF's Tornado F.Mk 3s since September 2002, following earlier problems with the missile meeting the UK MoD's specifications.

ASRAAM can be targeted using the aircraft's radar or a helmet-mounted cueing system (available on Typhoon), but can also act in autonomous infrared search and track mode. The manufacturer claims the missile has unrivalled agility and intercept speed, powered by a low-IR signature rocket motor. Much work was done to provide ASRAAM with high levels of countermeasures resistance and it is fitted with a high lethality blast fragmentation warhead. The RAF is reported to be impressed with the missile's performance.

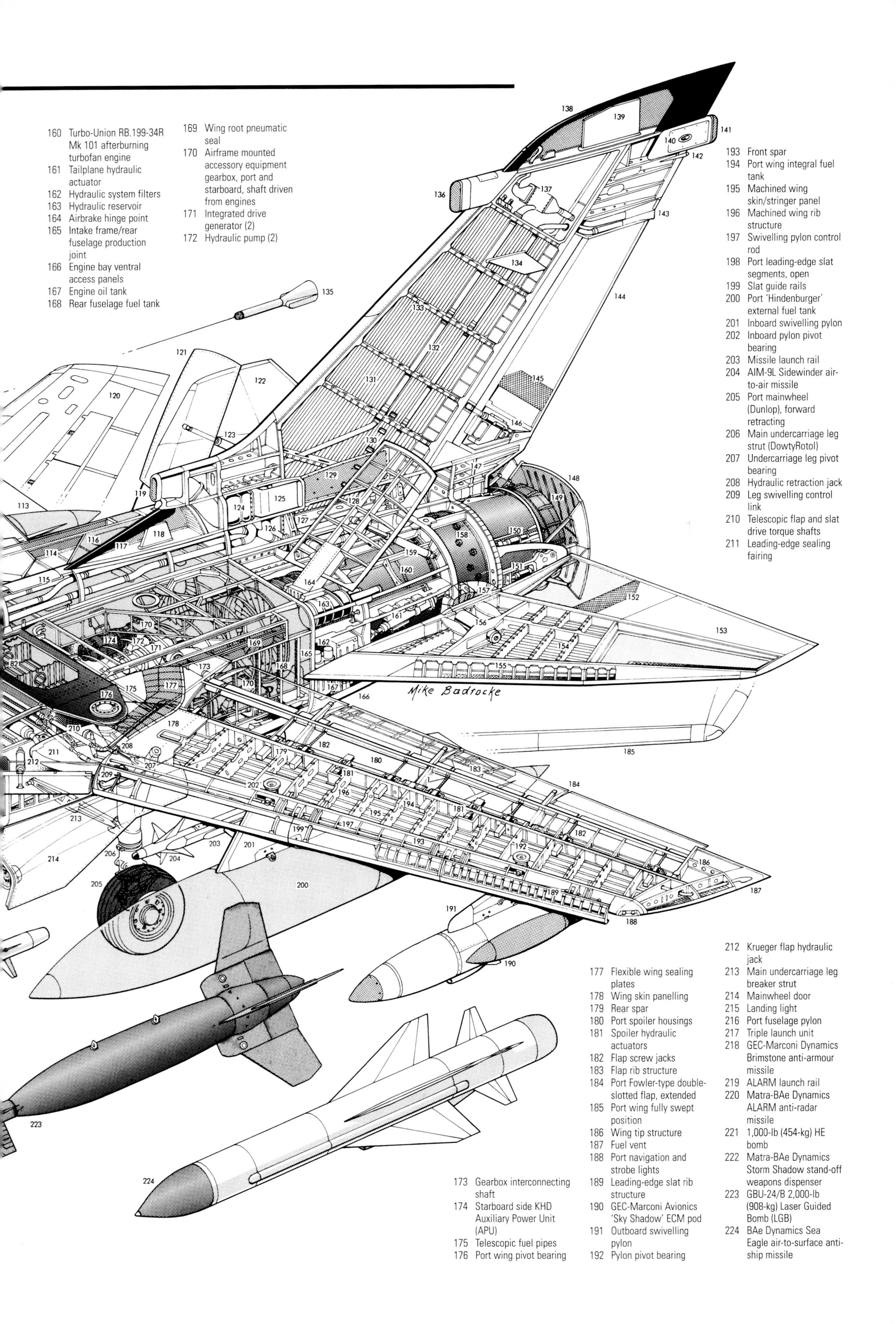

160 Turbo-Union RB.199-34R Mk 101 afterburning turbofan engine
161 Tailplane hydraulic actuator
162 Hydraulic system filters
163 Hydraulic reservoir
164 Airbrake hinge point
165 Intake frame/rear fuselage production joint
166 Engine bay ventral access panels
167 Engine oil tank
168 Rear fuselage fuel tank
169 Wing root pneumatic seal
170 Airframe mounted accessory equipment gearbox, port and starboard, shaft driven from engines
171 Integrated drive generator (2)
172 Hydraulic pump (2)
173 Gearbox interconnecting shaft
174 Starboard side KHD Auxiliary Power Unit (APU)
175 Telescopic fuel pipes
176 Port wing pivot bearing
177 Flexible wing sealing plates
178 Wing skin panelling
179 Rear spar
180 Port spoiler housings
181 Spoiler hydraulic actuators
182 Flap screw jacks
183 Flap rib structure
184 Port Fowler-type double-slotted flap, extended
185 Port wing fully swept position
186 Wing tip structure
187 Fuel vent
188 Port navigation and strobe lights
189 Leading-edge slat rib structure
190 GEC-Marconi Avionics 'Sky Shadow' ECM pod
191 Outboard swivelling pylon
192 Pylon pivot bearing
193 Front spar
194 Port wing integral fuel tank
195 Machined wing skin/stringer panel
196 Machined wing rib structure
197 Swivelling pylon control rod
198 Port leading-edge slat segments, open
199 Slat guide rails
200 Port 'Hindenburger' external fuel tank
201 Inboard swivelling pylon
202 Inboard pylon pivot bearing
203 Missile launch rail
204 AIM-9L Sidewinder air-to-air missile
205 Port mainwheel (Dunlop), forward retracting
206 Main undercarriage leg strut (DowtyRotol)
207 Undercarriage leg pivot bearing
208 Hydraulic retraction jack
209 Leg swivelling control link
210 Telescopic flap and slat drive torque shafts
211 Leading-edge sealing fairing
212 Krueger flap hydraulic jack
213 Main undercarriage leg breaker strut
214 Mainwheel door
215 Landing light
216 Port fuselage pylon
217 Triple launch unit
218 GEC-Marconi Dynamics Brimstone anti-armour missile
219 ALARM launch rail
220 Matra-BAe Dynamics ALARM anti-radar missile
221 1,000-lb (454-kg) HE bomb
222 Matra-BAe Dynamics Storm Shadow stand-off weapons dispenser
223 GBU-24/B 2,000-lb (908-kg) Laser Guided Bomb (LGB)
224 BAe Dynamics Sea Eagle air-to-surface anti-ship missile
Mike Badrocke

Inside the Tornado GR.Mk 4

Tornado GR.Mk 4 cutaway

1 Air data probe
2 Radome
3 Lightning conductor strips
4 Terrain-following radar antenna
5 Ground-mapping radar antenna
6 Radar equipment module, hinged position
7 Radome hinged position
8 IFF antenna
9 Radar antenna tracking mechanism
10 Radar equipment bay
11 UHF/TACAN antenna
12 Laser Ranger and Marked Target Seeker (LRMTS), starboard side
13 FLIR housing
14 Ventral Doppler antenna
15 Incidence probe
16 Canopy emergency release
17 Avionics equipment bay
18 Front pressure bulkhead
19 Windscreen rain dispersal air ducts
20 Armoured windscreen panel, gold film coated
21 Retractable, telescopic flight refuelling probe
22 Probe retraction link
23 Windscreen open position, instrument access
24 Pilot's wide-angle HUD
25 Instrument panel
26 Radar 'head-down' display
27 Instrument panel shroud
28 Control column
29 Rudder pedals
30 Battery
31 Cannon barrel housing, cannon deleted from port side
32 Nosewheel doors
33 Taxying light
34 Nose undercarriage leg strut
35 Torque scissor links
36 Forward-retracting twin nosewheels
37 Nosewheel steering unit
38 Nosewheel leg door
39 Electrical equipment bay
40 Ejection seat rocket pack
41 Engine throttle levers
42 Wing sweep control lever
43 Radar hand controller
44 Side console panel
45 Pilot's Martin-Baker Mk 10 'zero-zero' ejection seat
46 Safety harness
47 Ejection seat headrest/ drogue container
48 Upward hingeing cockpit canopy
49 Canopy centre arch
50 Navigator's radar display
51 Navigator's instrument console and weapons control panels
52 Foot rests
53 Canopy external latch
54 Pitot head
55 Additional avionics equipment in port cannon bay
56 Mauser 27-mm cannon
57 Cold air unit ram air intake
58 Transverse ammunition magazine, 180-rounds
59 Liquid oxygen converter
60 Cabin cold air unit
61 Stores management system computer
62 Port engine air intake
63 Intake lip
64 Cockpit framing
65 Navigator's Martin-Baker Mk 10 ejection seat
66 Starboard engine air intake
67 Intake spill duct
68 Canopy jack
69 Canopy hinge point
70 Rear pressure bulkhead
71 Intake ramp actuator linkage
72 Navigation light
73 Two-dimensional variable-area intake ramp doors
74 Suction relief doors
75 Wing glove Krueger flap
76 Intake by-pass air spill ducts
77 Intake ramp hydraulic actuator
78 Forward fuselage fuel tank, total internal capacity 1285 Imp gal (5842 litre; 1542 US gal)
79 Wing sweep control screw jack
80 Flap and slat control drive shafts
81 Wing sweep, flap and slat central control unit and motor
82 Wing pivot box integral fuel tank
83 Air system ducting
84 Anti-collision light
85 UHF antennas
86 Wing pivot box carry-through, electron beam welded titanium structure
87 Starboard wing pivot bearing
88 Flap and slat telescopic drive shafts
89 Starboard wing sweep control jack
90 Leading-edge sealing fairing
91 Wing root glove fairing
92 494-Imp gal (2250-litre; 694-US gal) 'Hindenburger' external fuel tank
93 AIM-9L Sidewinder air-to-air missile
94 Canopy open position
95 Canopy jettison unit
96 Pilot's rear view mirrors
97 Starboard three-segment leading-edge slat, open
98 Slat screw jacks
99 Slat drive torque shaft
100 Swivelling wing pylon angle control rod
101 Inboard pylon pivot bearing
102 Starboard wing integral fuel tank
103 Wing fuel system access panels
104 Outboard pylon pivot bearing
105 BOZ chaff/flare launcher pod
106 Outboard swivelling wing pylon
107 Starboard navigation and strobe lights
108 Wing tip fairing
109 Double-slotted Fowler-type flaps, extended
110 Flap guide rails
111 Starboard spoilers, open
112 Flap screw jacks
113 External tank tail fins
114 Wing swept position trailing-edge housing
115 Dorsal spine fairing
116 Aft fuselage fuel tank
117 Fin root antenna fairing
118 HF antenna
119 Heat exchanger ram air intake
120 Starboard wing fully swept position
121 Airbrake, open
122 Starboard all-moving tailplane (taileron)
123 Airbrake hydraulic jack
124 Primary heat exchanger
125 Heat exchange exhaust duct
126 Engine bleed air ducting
127 Fin attachment point
128 Port airbrake rib structure
129 Fin heat shields
130 Vortex generators
131 Fin integral fuel tank
132 Fuel system vent piping
133 Fin rib and machined skin panel structure
134 ILS antenna
135 Towed Radar Decoy (TRD)
136 Forward passive ECM housing
137 Fuel jettison and vent valve
138 Fin tip antenna fairing
139 VHF antenna
140 Tail navigation light
141 Aft passive ECM housing
142 Obstruction light
143 Fuel jettison
144 Rudder
145 Rudder honeycomb core structure
146 Rudder hydraulic actuator
147 Dorsal spine tail fairing
148 Thrust reverser bucket doors, open
149 Variable-area afterburner nozzle
150 Nozzle control jacks (4)
151 Thrust reverser door actuator
152 Honeycomb core trailing-edge structure
153 Port all-moving tailplane (taileron)
154 Tailplane rib structure
155 Leading-edge nose ribs
156 Tailplane pivot bearing
157 Tailplane bearing sealing plates
158 Afterburner duct
159 Airbrake hydraulic jack

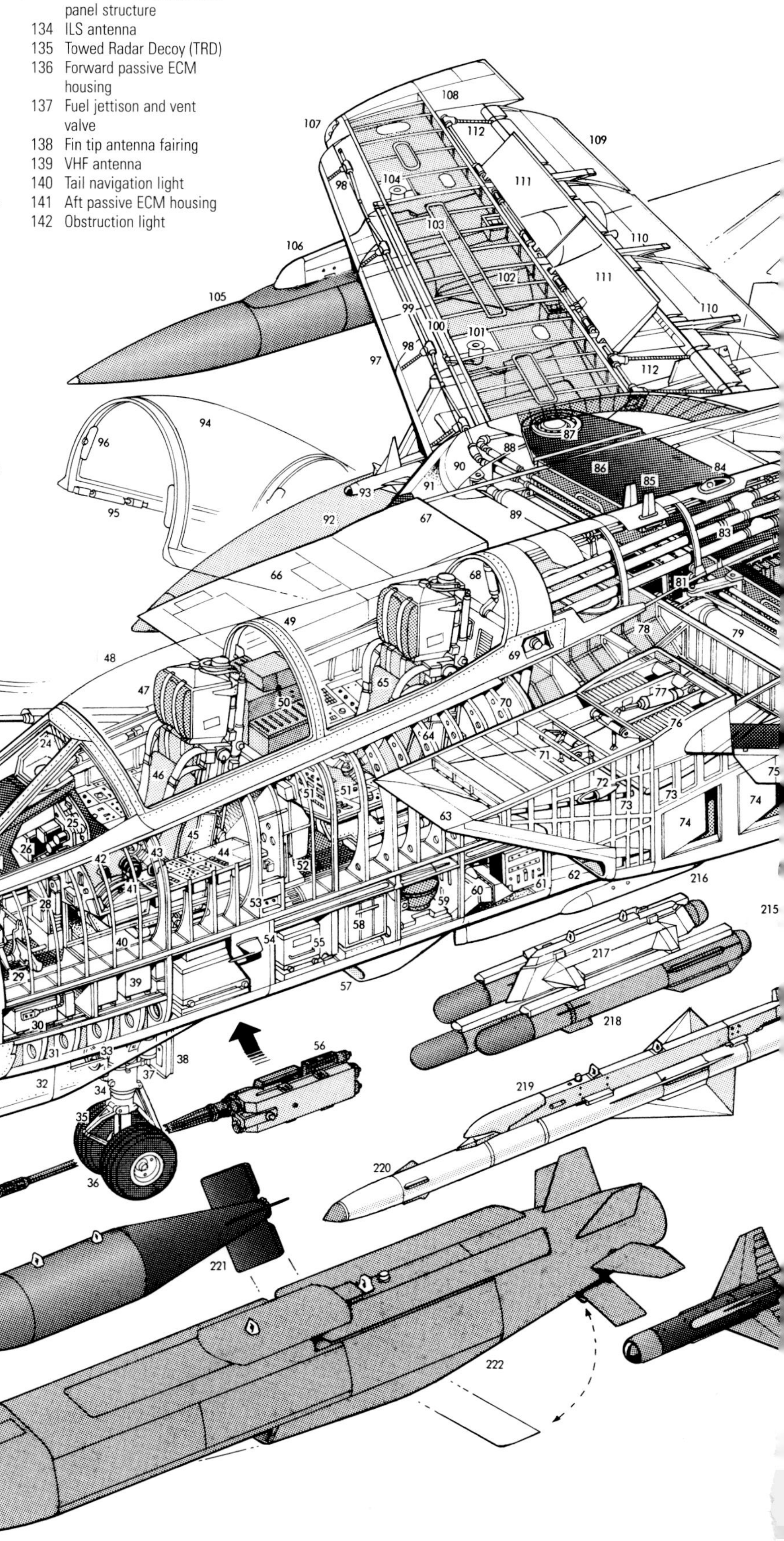

Paveway, Enhanced Paveway and Paveway IV

The RAF's Tornados have seen a massive improvement in precision-guided bomb delivery capabilities since the end of Desert Storm. The integration of the TIALD laser designating pod allowed the aircraft to self-designate targets for 1,000-lb (454-kg) Paveway II and 2,000-lb (907-kg) Paveway III LGBs. New software and GPS equipment in the GR.Mk 4 has allowed the integration of the all-weather Enhanced Paveway II and Enhanced Paveway III LGBs from 2001, both versions making a successful debut during Operation Telic.

In 2003 Raytheon was awarded a contract to produce a new 500-lb (227-kg) precision-guided bomb for the RAF, named Paveway IV. Planned for service entry in 2006, the Paveway IV will be deployed on Tornado GR.Mk 4s, Harrier GR.Mk 9s and Typhoons and is fitted with a 'jam-resistant' GPS guidance kit.

Taurus KEPD 350

After withdrawing from the French Apache programme, Germany selected the Taurus KEPD 350 conventionally-armed cruise missile to fulfil its stand-off precision-guided weapon needs. Developed by LFK of Germany and Saab-Bofors of Sweden, the weapon will be employed on the Saab Gripen as well as the Luftwaffe's Tornados.

With a range of some 400 km (249 miles), the Taurus, like the MBDA Storm Shadow, has wings which deploy in flight and is powered by a turbofan engine propelling the missile to speeds in excess of Mach 0.8. The tandem warhead, named Mephisto, comprises a precursor charge followed by the main explosive warhead. An intelligent programmable fuse calculates the optimum time for detonation of the charges to maximise their destructive ability.

Successful free-flight trial launches from a Tornado were carried out by the manufacturer and German test unit Wehrtechnische Dienststelle 61 in South Africa in 2002, and service entry is expected in 2005

Towed Radar Decoy

One of the most recent additions to the arsenal of active radar countermeasures is the Towed Radar Decoy or TRD. These systems deploy a unit externally from the aircraft on a cable (left). RAF Tornado F.Mk 3s have been operating with TRDs for a number of years and new versions of the TRD to be incorporated into modern defensive aids systems for the RAF and for German Tornados (below).

Rafael Litening and GBU-24

German Tornados are now fully operational with the Rafael Litening targeting pod, enabling the aircraft to deploy precision guided bombs for the first time. The weapon of choice is the GBU-24 LGB (as seen under the fuselage of the aircraft below). The pod and weapon integration programme was conducted at Eglin AFB, Florida, in 2001 and since that time a total of 20 Litening pods has been delivered to the Luftwaffe.

The Italian air force has also adopted a precision weapons capability with the integration of the Thomson-CSF Combined Laser Designator Pod (CLDP). The system was brought into the fleet in the late 1990s and made its combat debut in Operation Allied Force, delivering GBU-16 bombs. Additionally upgraded Italian Tornados will be equipped to deliver the Boeing GBU-31 GPS-guided JDAM and Enhanced Paveway III bombs.

AGM-88 HARM IV and HARM PNU

An important weapon for Luftwaffe and AMI Tornado ECRs and Marineflieger Tornado IDSs for many years, the AGM-88 High-speed Anti-Radiation Missile (HARM) has under gone a number of improvements. Seeker and mission software improvements have produced the HARM III (above) and HARM IV, adding accuracy and reliability to the original missile.

Both Italian and German MLU Tornados are slated to receive the new HARM Precision Navigation Upgrade (PNU) version, developed by BGT of Germany and Alenia of Italy alongside original manufacturer Raytheon. HARM PNU replaces the original mechanical gyroscopes with a laser inertial measuring system as part of a modern precision navigation system that will greatly increase the missile's accuracy.

Tornado cockpit upgrades

RAF Tornado GR.Mk 4

The Tornado GR.Mk 4's cockpit incorporated numerous changes to facilitate the aircraft's new capabilities. In the pilot's cockpit (below) the new FLIR sensor projects a wide-angle image onto a Ferranti holographic head-up display. Additionally, this image created by the FLIR can be displayed on a new Smiths Industries multi-function head-down display (MF-HDD) in the centre of the control panel and on the navigator's left TV display screen (right). The pilot's (MF-HDD) can be easily switched between modes and can be used to present a new digital moving map with the intended route of the Tornado and other vital information, such as threats and targets, superimposed onto the display. The top section of the pilot's control column has been altered to include hands-on throttle and stick (HOTAS) controls, allowing the pilot to control numerous functions without removing their hand from the control column. Unlike the GR.Mk 1, the GR.Mk 4 cockpit is fully Night Vision Goggle (NVG) compatible, incorporating equipment from Oxley Avionics.

A key feature of the next round of GR.Mk 4 incremental improvements is the integration of a new BAE Systems cockpit Pilot Multi-Function Display (PMFD). The display will allow the pilot to select a range of flight data, navigation and weapons systems information as required. The new equipment will be linked to the radar and the navigator's map processor as the Tornado Advanced Radar Display and Information System (TARDIS).

***Right:** Consigning the Heath-Robinson approach of using sticky tape to ensure various light sources are prevented from affecting operations with Night Vision Goggles (NVGs) to history, the Tornado GR.Mk 4's cockpit has full NVG compatibility. Here a GR.Mk 4 pilot demonstrates how the NVG system is integrated onto the crew's helmets. In case of an emergency ejection the NVG system is automatically released before the seat fires the crew member from the cockpit.*

Luftwaffe Tornado IDS

The current Tornado IDS pilot's (right) and navigator's (far right) instrument panel and cockpit displays are little changed from when the aircraft were delivered. The pilot's station has received a number of improved cockpit display systems, however, this is set to change with the introduction of all new displays for both the pilot and navigator.

The existing head-down displays will be replaced by new Astronautics units in each cockpit that will use the same Dornier digital moving map system as fitted to the Luftwaffe's Eurofighters. As well as displaying other important data derived from the aircraft's own sensors, the HDD will also be able to show tactical information provided by other aircraft or land-based units using the Link 16 datalink system. Both cockpits have already been modified for full NVG use with new interior lighting.

In the ECR version the pilot will receive a full-colour Electronic Warfare Indicator (EWI) and the Weapons Systems Operator (WSO) a programmable control and display unit – both linked to the new Defensive Aids Computer (DAC). This will provide both crew members with much improved situational awareness of any potential hostile threat to their aircraft.

Italian IDS aircraft will receive a similar cockpit upgrade with new HDD and EWI indicators for the ECR variant. In addition, Italian aircraft will also receive HOTAS flight controls and an NVG-compatible cockpit with internal and external lighting modifications.

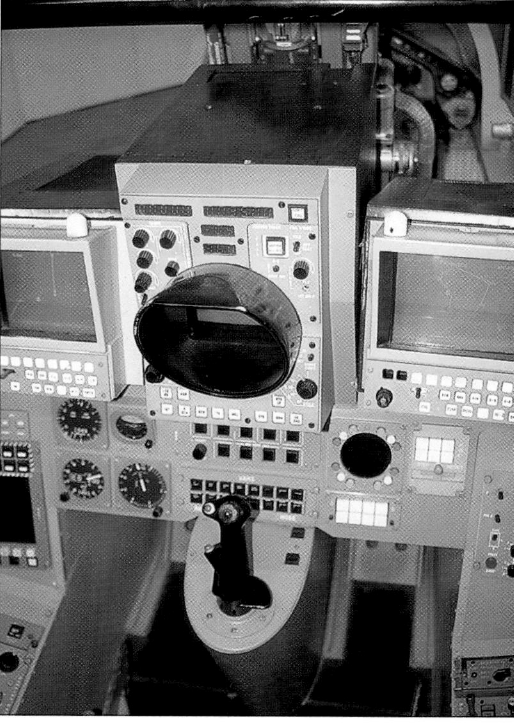

Above: A Tornado ECR, carrying its standard armament of two HARM missiles, takes on fuel via a Fletcher Sargent buddy refuelling pod. The aircraft wear the markings of Einsatzgeschwader 1 (EG 1) and JBG 32, the latter unit providing the bulk of the aircraft for the Piacenza-based detachment. EG 1's contribution to Operation Deliberate Force from 1995 marked the first use of Germon warplanes in a combat environment since the end of World War II. The ECRs carried live HARMs, but none was fired in anger until Operation Allied Force in 1999.

Below: Despite deliveries of the first batch of Eurofighters beginning to ramp up, the Tornado is destined to remain in Luftwaffe service until around 2025.

Upgrading the German fleet

The Luftwaffe and Marineflieger confirmed a requirement in January 1993 for the Mid-Life Improvement (MLI). An initial contract in April 1995 saw EADS (Daimler Benz Aerospace) take the lead role and trials for the stepped upgrade. The MLI got underway in earnest in 1998 with the conversion to Ada software under ASTT6 programme. This Phase One also included an improved mission computer, GPS/Laser Inertial Navigation System (LINS), improved cockpit display systems and video mission recording. 'Neue Avionikstruktur' (new avionics structure) includes: computer upgrade from 256 kbyte to 8 Mbyte with Avionics System Software Tornado in Ada (ASSTA 1) and a MIL-STD-1760 digital databus for precision-guided munition and stand-off missile capability. Rafael's Litening laser designator was selected in December 1995, with a purchase of 20 approved in January 1997.

Under 'Phase 1a' a number of software improvements to ASSTA 1 configuration were introduced, along with full Litening Laser Designation, GBU-24 bomb, Laser Inertial Navigation System/Global Positioning System (LINS/GPS) and AGM-88 HARM II integration.

In 2001, the Rafael Litening laser targeting/navigation pod and the GBU-24 Paveway III LGB were integrated on German Tornados following successful trials at Eglin AFB, Florida. The original Apache weapon programme was also subsituted the stand-off Taurus KEPD 350, which was test-launched in early 2002 in South Africa.

The German-Italian co-operative ASSTA 2 upgrade programme entered ground testing in mid-2002 with flight tests commencing in 2003 and scheduled for completion in 2005. This phase of the MLI will also modify cockpit displays and improve defensive aids systems, building on the previous upgrades. The programme, being developed for Panavia by EADS Military Aircraft and Alenia, started in Germany in December 2000 and Italy in July 2001, with development costs of about US$405 million.

A new Radar-Warning Receiver (RWR), produced by Saab Avionics, has been chosen for the German programme and it will be integrated into the Tornado Defensive Aids SubSystem (TDASS). This will give an increased frequency range and will introduce a Defensive Aids Management System (DMS) and a Defensive Aids Computer (DAC) that will co-ordinate threat-warning input and initiate jamming or dispense chaff and flares as required. The DMS will assist the crew in threat assess-

Right: A pair of reconnaissance-rolled IDSs, belonging to AKG 51 'Immelmann' transits at high altitude during an EG 1 deployment to Italy during the Allied Force operations of 1999.

Far right: EADS's new Radar Cross Section Test Facility for Military Aircraft in Manching has been used extensively during research to reduce the Tornado's radar signature.

***Left:* A trio of AKG 51 Tornado IDSs carries the new KS 153 reconnaissance pod. This replaces the previous MBB/Aeritalia recce pod, which had contained a pair of optical Zeiss wet-film cameras and an IRLS.**

***Below:* The Luftwaffe has selected the Rafael Litening targeting pod to provide its Tornado IDS fleet with a precision-guided weapons capability. The pod is mounted on the fuselage pylon, as seen below alongside a GBU-24 laser guided bomb.**

ment and selection of the best Electronic Counter Measures (ECM), and therefore will reduce the crew workload and improve the survivability even in a high-threat environment. It will be linked to a new Diehl/Thales colour Electronic Warfare Indicator for the pilot and a Canadian Marconi/Diehl Avionik programmable control and display unit for the weapons systems operator. Situational awareness will also be improved by the provisions of Astronautics colour head-down displays in both cockpits, which will provide a Dornier digital map system developed for the Eurofighter. Alternatively, these displays can be used to show situational awareness information from other units connected by NATO-standard Link 16 datalink or onboard systems status. Link 16 is currently planned as a follow-on item to the upgrade and by 2004 EADS had already performed prototyping activities, including flight demonstration, and proved the feasibility for a Link 16/MIDS integration into Tornado.

In conjunction with the new displays, the interior and the exterior lighting is adapted for the use of Night Vision Goggles. Other improvements include Missile Approach Warning System (MAWS), towed decoy, flare improvement, Telelens reconnaissance pod, Elint pod and SAR pod.

Phase 3 of the MLI will introduce software improvements to ASSTA 3 configuration, improved Identification Friend or Foe NATO Identification System (IFF/NIS), Infra-Red Imaging Sidewinder-Tail-controlled (IRIS-T), a differential GPS and a Command Stability Augmentation System (CSAS) upgrade. This will lead to the ultimate Phase 4, with improvements for the software to ASSTA 4 configu-

New Luftwaffe reconnaissance pod

To replace the ageing MBB/Aeritalia reconnaissance pods, inherited from the Marineflieger and Italian air force, German reconnaissance unit AKG 51 has now received a new system, developed to cover extended altitude ranges. The pod is available in two versions. The Reconnaissance Pod (right) contains a Honeywell infra-red linescanner system, a Zeiss KS153A Trilens 80 camera mounted in a vertical position (below right; centre of pod) and a second Trilens 80 mounted in the front section. A second version of the pod, called the Telelens, replaces the forward Trilens 80 camera with the Zeiss KS153A Telelens 610 camera, allowing area coverage from horizon to horizon (below). Both pod systems are controlled by a Honeywell Recce Management System and also incorporate Ampex digital recording equipment (below right; right end). Following trials conducted in 2001 the Luftwaffe has now received all 37 pods ordered adding full day/night/all-weather reconnaissance capabilities from low or medium altitudes. The modular design of the pod ensures that the internal equipment can be modified or upgraded with ease.

ration, Advanced Radar Missile (ARAMIS), high-precision weapons, measurements for incremental operational envelope improvements, test concept for crew workload reduction and reduced life cycle costs improvements.

An airframe upgrade package has also been developed by Panavia to extend airframe life from 4,000 to 8,000 hours. Testing of the structural upgrade was completed in 2002.

Italian AF

It was not just the RAF that participated in Operation Desert Storm in 1991. The Italian AF (AMI) deployed 12 Tornados to Al Dhafra (UAE) for 'Gulf War I' under Operation Locusta. The deployed AMI Tornados flew 226 sorties, dropping some 565 munitions.

Latterly, the AMI Tornado force was heavily involved in United Nations and NATO peacekeeping operations over the

Top: The sole remaining Marineflieger Tornado operator is Marinefliegergeschwader 2, based at Eggebek. Used in the attack, maritime strike and reconnaissance roles, MFG 2's Tornados still employ the Kormoran Mk 2 missile, however, the more usual weapon of choice today is the HARM missile (as seen on this example).

Above: Since the Tri-national Tornado Training Establishment closed in March 1999, all Luftwaffe Tornado training has been conducted by the Ausbildungsstaffel Tornado at Holloman AFB, New Mexico. Also home to the F-4F ICE aircraft of the 20th Fighter Squadron, 49th Fighter Wing, for training Luftwaffe Phantom pilots (until late 2004), Holloman was chosen thanks to its available areas for building new facilities and its average of 340 clear weather days per year.

ECR to the future

The Fast Emitter Location System (FELS) is being developed by EADS and will be first flown in a Tornado test aircraft in 2004. Like the current ELS, FELS is a passive system for radar emitter detection. Compared to the ELS the frequency range is extended to include the C-J Band.

The FELS can be introduced to 'any Tornado' by replacing the cannon and Krueger flaps with antenna arrays. Using passive ranging technology, the accuracy achieved is sufficient for SEAD (but not for Destruction of Enemy Air Defense (DEAD) tasks), radar sites instead being engaged using guided weapons. In this case the emitter positions ranged by the FELS will be the basis for an emitter/target refinement with the Laser Designator Pod in good weather, or the upgraded Tornado nose radar in all weather conditions. The Tornado Nose Radar (TNR) upgrade introduces enhanced radar modes with improved azimuth resolution (e.g. Doppler Beam Sharpening, Spot Synthetic Aperture Radar mode). The radar will also have an improved operating range and higher sensitivity giving an improved stand-off distance to the target, time for target identification and sorting. An EADS study has also showed that the adaptation of a conformal fuel tank (CFT) can double the HARM weapons loads per aircraft and increase the time on station.

Above: An AMI Tornado IDS, belonging to the Reparto Sperimentale di Volo, is seen conducting MBDA Storm Shadow carriage trials at Decimomannu, Sardinia, in September 2002. Upgraded AMI Tornados will be capable of delivering the weapon, as well the new AGM-88 HARM PNU, GBU-24 E/B LGBs and the IRIS-T air-to-air missile.

Left: The Gioia del Colle-based 156° Gruppo of 36° Stormo operates the IDS in the attack role, but is also the sole AMI Tornado unit trained for maritime strike operations. All but a handful of test aircraft, used by Alenia and the RSV, have now been repainted in this low-visibility grey camouflage. Aircraft codes and national insignia have also been toned-down to reduce conspicuity.

Former Yugoslavia, including maintaining the naval arms blockade of Yugoslavia by flying missions over the Adriatic. In September 1995, the Italian AF IDS deployment supporting Operation Deliberate Force dropped 'dumb' Mk 83 bombs, as NATO worked towards getting warring factions to the negotiating table. This attack underlined a shortfall in capability for the Italian Tornados, with the Thompson-CSF Combined Laser Designator Pod (CLDP) hurried into service to permit the use of precision laser-guided bombs (LGBs). The Italian Tornados continued to support operations in the region under Operation Joint Endeavor.

During Operation Allied Force in 1999, the Italian Tornado IDSs of 6° and 53° Stormo flew combat missions direct from Ghedi and Gioia del Colle, employing GBU-16 LGBs in tandem with the CLDP. The Italian ECRs were equally busy, with 50° Stormo from Piacenza flying SEAD missions alongside the Luftwaffe ECRs of JBG 32.

Italian enhancement

With the Italian IDS force due to stay operational alongside the Eurofighter Typhoon until around 2020, it became clear that an upgrade programme would be essential. Alenia Aerospazio's Aeronautics Division received the first of a series of contracts from the Italian Ministry of Defence in early June 2001. The subsequent Mid Life Update (MLU) concept was designed to extend the operational life of 90 Tornado Interdictor-Strike (IDS), Electronic Combat Reconnaissance (ECR) and trainer versions until at least 2020.

The first contract, known as Italian Mid-Life Upgrade (IT-MLU) '1st Upgrade' Standard (MLU Phase I), covered 18 Tornado IDSs to be upgraded to this standard and returned to AMI service in 2004, with avionics and weapon systems improvements. In parallel with the '1st Upgrade', Alenia Aeronautica is working on a complete upgrade package, known as IT-Full MLU. The development, evaluation and certification phase is due to be completed by February 2005 and will be incorporated on all Italian AF Tornado IDSs and ECRs, and even possibly trainer variants. In addition to the improvements introduced in Phase I, the Italian Tornado fleet is set to receive a revised cockpit with colour Multi-Function Displays (MFDs), HOTAS flight controls, upgraded RB.199 engines with digital control, an advanced Defensive Aids Sub-System (DASS) and other system improvements. Alongside the avionics and systems upgrade, Alenia is also analysing the AMI Tornado fleet's fatigue status.

The prototype Panavia Tornado IDS upgraded to Italian Mid-Life Upgrade (IT-MLU) '1st Upgrade' Standard (MM7048) first flew from Alenia's Casselle facility on 13 September 2002, before embarking on a comprehensive trials period. The first production IT-MLU '1st Upgrade' Tornado IDS (MM7063) was re-delivered to the Italian Air Force in July 2003 for trials. However, the full ASSTA 2 software will not be introduced until the 10th upgraded production example, scheduled to be returned to service in December 2004.

As the AMI has chosen to implement the IT-Full MLU package, including the new DASS, in a single depot maintenance period (at Alenia Aeronautica's Caselle plant), the programme is running behind the original delivery date of December 2005 for the first fully-upgraded Tornado.

The DASS package is still to be contracted and Alenia Aeronautica is proposing to deliver the first 17 AMI Tornados with provision for, but not equipped with, the DASS hardware and software. This will be provided with the 18th aircraft in March 2006, and subsequently installed on board the initial 17 aircraft.

As well as the Tornado IT-MLU '1st Upgrade' prototype, Alenia Aeronautica and the AMI's RSV Flight Test Wing (Reparto Sperimentale di Volo) are using two additional aircraft, one IDS and one ECR, to develop, test-fly and certify the full IT-MLU package. Additionally, the Tornado ECR will also support the integration certification of the MBDA Storm Shadow cruise missile and other weapon systems (including the AGM-88 HARM PNU, the IRIS-T air-to-air missile and the ACMI rangeless training pod). A further aircraft will be used for the DASS trials. The 87th (and last) AMI Tornado from the first batch of 18 IT-MLU '1st Upgrade' aircraft brought up to the full configuration will be redelivered in September 2010.

Into MLU

The IT-MLU '1st Upgrade' will provide the aircraft with the capability to deliver new all-weather, precision-guided weapons following the installation of dedicated new avionics. These include a GPIN laser-inertial navigation system with embedded LITEF GPS, advanced digital Have Quick I/II secure communications, improved Marconi Mobile M-425 NGIFF (Next-Generation IFF), a Night Vision Goggle-

Below: Based at Piacenza, 155° Gruppo of 50° Stormo is tasked with operating the AMI's fleet of 16 Tornado ECRs. The aircraft are dedicated to the SEAD role and will receive the new HARM PNU missile following the delivery of IT-MLU aircraft.

Right: The first prototype Italian MLU Tornado, MM7048 made its maiden flight on 13 September 2002 and is seen here 13 days later undergoing initial trials at Alenia's test facility at Decimomannu. The addition of the Mil Std 1760 databus, alongside the complementary computer and avionics upgrades, will allow Italian Tornados to carry a wide range of new munitions including Storm Shadow, GBU-31 JDAM and Enhanced Paveway GPS-guided bombs.

Above: Work on the first Tornado MLU prototype commenced in 2001 and the programme will eventually extend the operational life of all 87 Italian IDS and ECR aircraft until around 2020. The interim capability will continue until the 18th aircraft in March 2006, when the IT-Full MLU package becomes available. The first batch of 17 aircraft will then be brought up to this standard and will be the final aircraft redelivered to the AMI in 2010.

(NVG)-compatible cockpit and external NVG-compatible lighting modifications. The upgrade will also include the MIL-STD-1553B avionics databus and MIL-STD-1760B weapons system databus on weapons hardpoints (except the outer wing pylons) making them compatible with GPS/INS guided weapons, a new Thales TACAN, a Galileo Avionica radar altimeter and a Litton Italia SAHR (Secondary Altitude and Heading Reference). The Enhanced Main Computer will receive a new version of the operational software, based on the Ada computer language called ASSTA 2 (Advanced System Software Tornado Ada 2) WIP (Weapon Integration Package). This will provide functionality for the new avionics and systems. Also included is an improved EADS-LFK Navigation Hand Controller for the Weapons Systems Operator (WSO), similar to that used on the Tornado ADV.

New weapons options

Italian Tornados are now being equipped with the Boeing GBU-31 GPS-guided JDAM and Raytheon Enhanced Paveway III GPS/laser-guided kit for the Mk 84 and BLU-109 bombs. Alenia is also working on integration of the MBDA Storm Shadow ASM and improved AGM-88 HARM. ASSTA 2 software gave Initial Operational Clearance (IOC) for these weapons in early 2004.

The IT-Full MLU configuration is known to include a revised cockpit configuration with new multifunction colour displays (known as the Display Systems Upgrade or DSU) and new flight and mission control commands to an improved man-machine interface; an improved main computer with the full standard ASSTA 2 operational software; an advanced Defensive Aids SubSystem (DASS); and upgraded Turbo Union RB.199 Mk 103 engines with digital control.

The revised cockpit displays and main computer upgrade will be identical to the German Air Force Tornado MLI, thus maintaining some commonality among the fleets. The central head-down displays for the pilot (the Repeater and Projected Map Display (RPMD) for the IDS/trainer versions, and Combined Electronic Display and Map in the ECR) and the navigator's Combined Radar and Projected Map Display (CRPMD) will be replaced by multifunction Active Matrix Colour Liquid Displays (AMCLDs) provided by Astronautics of the US; while the TV-Tabs will have the same form-fit replacement AMCLD from BAE Systems as used on the UK's Tornado GR.Mk 4 MLU. Designated Pilot or Navigator Head-Down AMCLDs will present radar images; digital colour maps (with a new Digital Map Generator (DMG) replacing the current film-based system); the tactical situation picture; and the representation of aircraft status. The latter feature was not previously available to the pilot in the front cockpit.

Together with a revised cockpit layout for the WSO, incorporating the DSU and Dornier's DMG, the common Italian/German requirement will also include a Video Switching Unit; an Enhanced Digital Scan Converter for the Ground-Mapping Radar (GMR) video presentation; plus a new Electronic Warfare Indicator (EWI) for the pilot (provided by Diehl/Thales) and a programmable Control

The first IT-MLU 'Phase 1' prototype is now conducting trials with the Reparto Sperimentale di Volo (RSV) test unit at Practica di Mare. The aircraft is seen here during its UK public debut at the Farnborough air show in July 2004. It is one of three 'Phase 1' Tornados now involved in the development programme – another IDS and an ECR are shared by Alenia and the RSV for operational testing of the various new systems and weapons.

Top: In February 1995 the first Italian F.Mk 3 crews began conversion training with No. 56(R) Squadron at RAF Coningsby. Soon after the course began the first F.Mk 3 in Italian colours (ZE832/MM7202, coded 36-12) was delivered to Coningsby and is seen here in formation with a No. 56(R) Squadron example at the end of February. The Italian F.Mk 3s did not begin the transit to their new home with 12° Gruppo at Gioia del Colle until July of that year.

Above: 36° Stormo eventually incorporated both Tornado ADV Gruppi and continued to operate the type at full strength from Gioia del Colle until the rationalisation of the force into a single (12°) Gruppo in 2001. This example sits outside its hardened aircraft shelter (HAS) at Gioia in November 2003, shortly before transfer back to the UK for storage.

Above left: On 14 February 1997 the second F.Mk 3 Gruppo to form (21° Gruppo of 53° Stormo) received its first aircraft. It is seen here on arrival at Cameri, the Stormo commander Colonel Fabrizio Draghi greeting the crew with champagne.

Display Unit (CDU - from Canadian Marconi/Diehl Avionik) for the WSO. Both systems will support the DASS and, later, the Link 16/MIDS (Multifunctional Information Distribution System). In parallel with the MLU, Alenia is also upgrading the GMR under a separate programme.

The rear cockpit will receive pedal-activated control systems for radio communication, and laterally mounted hand grips (either side of the ejection seat). The latter will help the WSO keep track of threats and/or external awareness during violent aircraft manoeuvres, while allowing simultaneous release of chaff/flare decoys without looking into the cockpit. In order to improve engine response and extend engine life (reducing maintenance costs), the RB.199 engines will be equipped with a digital electronic control unit (DECU 2020). The DASS will be integrated by Alenia Aeronautica and will incorporate AR-3 Radar Warning Equipment (RWE), an internal Active Electronic CounterMeasures (AECM) system (Italian Tornados are already equipped with an internal jammer), chaff/flare-dispenser pods, new passive and active EW antennas and the Defensive Aids Computer (DAC) to manage the entire suite.

F.Mk 3s in Italy

In 1969 the Aeronautica Militare Italiana (Italian Air Force) commissioned into service the Lockheed F-104S, a new version of the Starfighter that had a more powerful engine, strengthened airframe, new hardpoints, upgraded radar and the ability to fire AIM-7E Sparrow radar-guided air-to-air missiles. A total of 205 aircraft was purchased and joined the fleet of F-104Gs already in service. The AMI gained an aircraft with acceptable operational, technical and logistical characteristics for the time, but the choice of this aircraft would have a less encouraging effect on future capability. In the following years, funds were directed to new programmes for attack aircraft (Tornado and AMX), and the air defence sector was allotted only minor upgrade programmes, leading to an extension of the service life of the Starfighters well beyond its logical limit. In the 1980s the European Fighter Aircraft (EFA) was launched, but the process of design, agreement, development, and introduction into service of this aircraft (finally designated EF2000 Typhoon) has been one of the longest and most difficult in aviation history. The Italian Air Force had no choice but to retain its Starfighters a good 10 years more than was expected in the beginning.

Against this backdrop, in 1993 the AMI was forced to face the problem of selecting an interim fighter that could bridge the gap between the progressive retirement of the F-104 and the arrival of the EF2000. The candidates at the time were the McDonnell Douglas F-15 Eagle, the General Dynamics F-16 and the Panavia Tornado ADV; the choice, once again dictated by economic, industrial and political reasons, fell on the Tornado. The Royal Air Force, which was the only NATO operator of the type, was committed to a transformation of its forces (the Options for Change programme) and was in a position to offer surplus aircraft to the AMI under a leasing contract. On 18 March 1994, the British and Italian defence ministries signed an MoU for the use of 24 aircraft (including four trainers); the contract also included aircrew training,

In their relatively short period of service AMI Tornado ADVs received a number of special schemes. In 1998 21° Gruppo marked its 80th anniversary by painting MM7234 '53-14' (top) in this special livery. The starboard side of the aircraft received AMI 75th anniversary markings, which occurred in the same year. Following the disbandment of 21° Gruppo, the 351ª Squadriglia was transferred to 12° Gruppo to maintain the unit's Tiger traditions. MM7234 '36-24' (above) received this appropriate livery for Tiger Meet 2002 at Beja, Portugal.

technician training, logistical support, 96 SkyFlash TEMP air-to-air missiles, and the exchange of personnel between the two air forces. At the time, the cost was indicated to be 700 billion lira, about £212 million.

With these aircraft, the AMI planned to re-equip two interceptor squadrons, the 12° Gruppo of the 36° Stormo at Gioia del Colle, and the 21° Gruppo of the 53° Stormo at Cameri.

The aircraft selected were all built in 1988-1989 and upgraded to the standard of Operation Granby. In February 1994 the aircraft's configuration was approved: they would have Stage One Plus radar, would have flown 1,300 to 1,600 hours each, and would be reconditioned completely by the RAF Maintenance Unit at St Athan under the 25 Fatigue Index programme. The aircraft also received the SRA-636 package which included ALE-40(V) chaff and flare units, Have Quick II secure voice radio and, later, the JTIDS (Joint Tactical Information Distribution System), Stage Two radar and compatibility with night-vision goggles. The aircraft were fitted with an Italian radar warning receiver library and an Elettronica ELT-553 Mk 3 electronic countermeasures system.

In mid-February 1995 the first group of crews from the 12° Gruppo began training with No. 56 (R) Squadron at RAF Coningsby. The conversion course included an intensive course in the English language and a programme of familiarisation with UK operations and procedures, which was carried out at RAF Valley on BAe Hawks. In mid-March, the first crew arrived at No. 56 Squadron and joined the 47 Long Course, B Flight. The flying phase of the training included some 70 hours for pilots and 55 for navigators, after which they were qualified as Limited Combat Ready. In the meantime, some 100 technicians attended courses at the Tornado Maintenance School at RAF Cottesmore and then completed their training at Coningsby.

First deliveries

The first Italian Tornado F.Mk 3 (ZE832/MM.7202, coded '36-12') was officially delivered at Coningsby to Lieutenant Colonel Paolo Falcone, commander of the 12° Gruppo, on 5 July 1995, and on the same day arrived at its new home base of Gioia del Colle for the official ceremony. In November 1995 the type flew for the first time in a national exercise, Mothia 95. The Gruppo accepted the last of its 12 aircraft on 15 March 1996, and was judged combat ready in June. The second unit to re-equip, the 21° Gruppo, started to send its crews to Coningsby in summer 1996; its first Tornado F.Mk 3 (ZE911/MM.7226) arrived at Cameri on 14 February 1997, and the squadron had completed its re-equipment by the following July.

In 1997 the Tornado F.Mk 3s of the 12° Gruppo were committed to their first operational task, a series of combat air patrol (CAP) missions over the Adriatic Sea in conjunction with Operation Alba, an air bridge from Italy to Albania that was part of air operations in the volatile Balkans area. The 21° Gruppo was then detached to Gioia del Colle for a period of on-the-job training, and during times of high tensions in the Balkans, detachments from the 21° Gruppo were always present at Gioia to reinforce the air defence of southern Italy.

In 1998 the 21° Gruppo also achieved full operational capability. It joined the 12° Gruppo in performing standard QRA service (at 15 minutes' readiness) and participating in various exercises in Italy and abroad, such as the Spring Flags, NATO TLP courses, and Exercises Dynamic Mix, Odax, NATO Air Meet, Central Enterprise, Tiger Meet, and others.

During this time it became evident that the distribution of the fleet between two bases 900 km (560 miles) apart was

Italian F.Mk 3s in Allied Force

Despite its short period of operational service, in 1999 the Italian Tornado F.Mk 3 fleet was involved in the first European war since 1945. Just before the outbreak of the conflict, the 21° Gruppo was deployed to Gioia del Colle and operated from that base for all of Operation Allied Force. War broke out on 24 March 1999, from when the Italian Tornado F.Mk 3s were tasked with defensive counter-air missions, flying CAPs in six patrol areas. Two aircraft were in the air 24 hours a day, a situation made possible in part by the RAF's delivery to Gioia del Colle of four spare engines. The F.Mk 3s initially operated over the Adriatic Sea, but later their areas of responsibility were moved to Albania and Macedonia, where they took advantage of the various KC-135, C-135FR, VC10 and TriStar tankers to stay on station for at least two hours. On several occasions the F.Mk 3s were directed by the AWACS to bogeys or they initiated an interception of enemy aircraft. At least once, an element of the 21° Gruppo arrived to lock onto a MiG-29, but the Serbian pilot immediately decided to head home and the interception ended. During the war, the two squadrons flew 295 sorties for a total of 760 flight hours. Standard configuration included four SkyFlash and four AIM-9L Sidewinder AAMs, two Hindenburger auxiliary fuel tanks, and rounds for the 27-mm (1.06-in) Mauser gun and the ALE-40(V) chaff and flare dispensers.

When four of 21° Stormo's Tornados moved to Gioia del Colle for participation in Operation Allied Force, they were unaware that they would never return to Cameri. '53-04' is seen here in May 1999 at the height of the conflict, fully-armed with underfuselage SkyFlash and wing-mounted AIM-9 Sidewinders, plus two 495-Imp gal (2250-litre) Hindenburger auxiliary fuel tanks. Although coming close on several occasions, Italy's F.Mk 3s were destined never to fire their weapons in combat.

Left: Seen in its final months of service with 53° Stormo, '53-06' is prepared for a sortie from Cameri in early 1999. The leasing of 34 F-16s from the USAF, pending delivery of sufficient Eurofighter Typhoons, spelled the end for the Tornado ADV in Italian service. The final example will be returned to the UK in November 2004, less than a decade after the first example was delivered

not ideal. In addition, many problems of serviceability arose (for instance, no spare engines had been acquired), and the level of efficiency of each squadron – only 50 per cent – did not allow realistic multi-aircraft training to be undertaken. Merging the squadrons on a single base increasingly looked like the best solution. Another difficulty that hampered operations was the incompatibility of the F.Mk 3's inflight-refuelling probe with the basket of the Sargent Fletcher pods used by the Italian B.707T/T: this led to a requalification programme carried out by the RSV, the Italian AF Test Wing.

After the end of Allied Force, it was decided to keep the 21° Gruppo at Gioia del Colle. Aircraft and crews remained in the south until the official departure from Cameri, whose ceremony was held on 28 July 1999. In 2000 the F.Mk 3s became fully operational with the JTIDS Link 16 system, and in October participated in Exercise JOTM (JTIDS Operator Tactical Meeting) at RAF Waddington, a NATO exercise intended to exploit the maximum capability of the system. In the same year, a force of six aircraft from the two Gruppi participated in Exercise Green Flag 04/00 at Nellis AFB, Nevada, from 31 July to 11 August, together with other Italian assets. The F.Mk 3s did not perform particularly well in this exercise, being tasked with sweep and escort missions (offensive counter-air) to which they are not really suited.

The end of 2000 brought the pending conclusion of the Tornado F.Mk 3 leasing period, more delays to EF2000 deliveries, and the possible leasing of F-16 fighters. This situation, coupled with a lack of personnel, led to the decision to disband one of the two F.Mk 3 Gruppi in order to merge the crews and aircraft into a single, more efficient unit. As such, on 1 March 2001, the 21° 'Tiger' Gruppo, which had been established on 25 May 1918, was disbanded; it was chosen because it was a few months younger than the 12° Gruppo. One of the three flights of the 21° Gruppo, the 351ª Squadriglia, was assigned to the 12° Gruppo in order to continue the NATO Tiger Club tradition.

The 12° Gruppo operated a fleet of 24 aircraft from the date of the disbandment until January 2003, when the squadron started returning its aircraft to the Royal Air Force. The Italian AF decided in February 2001 to lease 34 F-16A/B air defence fighters from the USAF, in order to bridge its fighter gap until operationally-capable EF2000s become available in 2010. This decision, coupled with Italy's refusal to enter the expensive RAF Capability Sustainment Programme for the Tornado, accelerated the return of the F.Mk 3. Funds earmarked for the Tornados will be diverted to cover the F-16 leasing. On 27 January 2003, the first F.Mk 3 (ZE837/MM.55057, coded '36-03') left Gioia del Colle for the RAF Maintenance Unit at St Athan. The aircraft will retire at a rate of approximately one per month until November 2004. On that date, the operational life of the Tornado F.Mk 3 with the Italian Air Force will come to an end. The 12° Gruppo should be able to perform QRA missions until mid-2004, when the three F-16 ADF squadrons will have completed their re-equipment.

Above: A trio of 12°Gruppo, 36° Stormo F.Mk 3s cruises over the Adriatic during a training sortie. Although possessing long-range and a sophisticated weapons suite, the F.Mk 3 could never be described as a dogfighter, and its performance at altitudes above 20,000 ft (6069 m) was woeful in comparison with contemporary fighters. The addition of the F-16s to the AMI's armoury will add a much improved offensive counter-air capability before the next-generation Eurofighter becomes Italy's primary air defence asset.

Even if the Tornado F.Mk 3 cannot be considered a true fighter, because it lacks agility and power, in Italian service it proved to be an excellent air defence aircraft, thanks to its radar (with a range of 100 nm/185 km), heavy armament, long endurance, two-man cockpit instrumentation and, above all, its JTIDS. The future combat tool of the Italian fighter force in the near term is represented by the F-16, but the aircraft that ultimately will form the backbone of Italian air defence is the EF2000 Tifone/Typhoon.

Saudi strikers

The Tornado export programme to the Royal Saudi AF (RSAF) under Al-Yamamah dates back to 1985 and initially encompassed 48 Tornado IDS and 24 ADVs. Six of the IDSs were delivered in GR.Mk 1A-type reconnaissance configuration, with deliveries commencing in March 1986. A second deal under Al-Yamamah II eventually led to the order and delivery of an additional 48 IDSs, with the first example heading to No. 75 Squadron at Dhahran on 3 October 1996. The last of this batch was delivered to Nos 75 and 83 Squadrons at Khamis Mushait in 1998 – marking the end of the Tornado production line.

Persistent rumours suggest that the RSAF is interested in an upgrade for its Tornado IDS fleet. This upgrade could take a similar form to the RAF's GR.Mk 4 Mid-Life Upgrade, with BAE Systems taking the lead role as prime contractor.

Jamie Hunter; additional material by Dr R. Niccoli and Daniel March

Below: The Tornado ADV will eventually be replaced in both RAF and Italian service by the Eurofighter Typhoon. Here a Tornado GR.Mk 4 from No. 31 Squadron joins a No. 29(R) Squadron Typhoon T.Mk 1 on the first leg of a sales tour to Singapore in July 2004.

Tornado operators

Featuring **Danger Mouse** *nose art, Tornado GR.Mk 4 ZA542 from No. 31 Squadron returns to the UK after seeing combat in Operation Telic during the spring of 2003.*

Royal Air Force

Since the end of Operation Desert Storm a series of defence cuts and the withdrawal of RAF assets from RAF Germany has seen a reduction in the overall strength in both the GR.Mk 4 and F.Mk 3 communities.

The closure of the Tri-national Tornado Training Establishment (TTTE) in 1998 allowed the RAF to concentrate its GR.Mk 4s at two bases (Marham and Lossiemouth), each with four squadrons. This includes the Tornado OCU, No. XV(R) Sqn, located at Lossiemouth.

Only four operational F.Mk 3 squadrons remain, based at RAF Leeming and RAF Leuchars. The training unit, No. 56(R) Squadron, is based at the latter.

Additionally, the RAF's Fast Jet Operational Evaluation Unit (FJOEU) operates a handful of GR.Mk 4s and F.Mk 3s from RAF Coningsby.

Tornado GR.Mk 4

No. IX(B) Sqn, RAF Marham

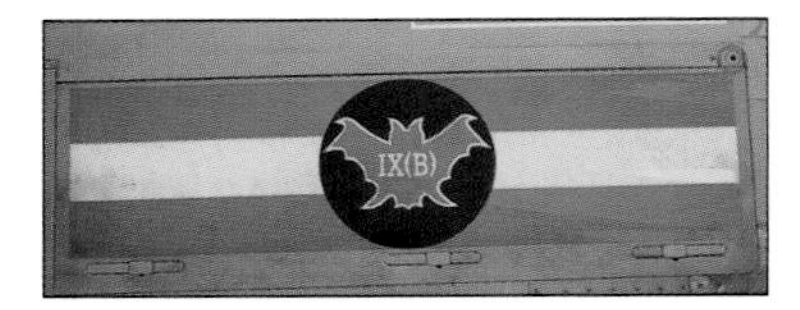

On 1 June 1982 No. IX(B) Squadron became the world's first front-line Tornado unit at RAF Honington. In 1986 the unit moved to RAF Brüggen, Germany, and took its GR.Mk 1s to the Gulf for Desert Storm in 1991, taking the ALARM missile into combat for the first time. In 1998 No. IX became the first operational GR.Mk 4 squadron before moving to its current base at RAF Marham . Along with attack duties the squadron remains the RAF's premier SEAD unit armed with ALARMs.

No. 12(B) Sqn, RAF Lossiemouth

Famous as a Buccaneer operator, No. 12(B) Squadron became the last front-line Tornado GR.Mk 1 unit to form, receiving its initial equipment in October 1993. The 'Foxes' flew the GR.Mk 1B variant, equipped with the Sea Eagle missile for the additional maritime strike role from RAF Lossiemouth, where the squadron remains today. The unit participated in Desert Fox in 1998 before the arrival of the GR.Mk 4 spelled the end of its secondary maritime strike role.

Wearing the unit's famous 'Fox head' badge on the tail fin a No. 12 Squadron GR.Mk 4 gets airborne carrying a Sky Shadow countermeasures pod beneath the port wing.

No. 14 Sqn, RAF Lossiemouth

No. 14 Squadron received its first Tornado GR.Mk 1s at RAF Brüggen in 1985 and its crews played an active role in Desert Shield/Storm in 1990-91. In 1993 TIALD pods were received and the unit became highly proficient in LGB delivery. In January 2001 the squadron returned to the UK and, having converted to the GR.Mk 4, remains active in the attack role.

No. XV(R) Sqn, RAF Lossiemouth

As the first operational RAF Germany Tornado squadron, No. XV received its first GR.Mk 1s in July 1983. Home-based at RAF Laarbruch, No. XV Squadron aircraft and crew played a major role in Operation Desert Storm and were granted the Battle Honour 'Gulf 1991'. The unit disbanded shortly after the conflict to reappear in 1992 as No. XV(R) Squadron and operating as the Tornado Weapons Conversion Unit at RAF Lossiemouth. The squadron today remains responsible for all RAF conversion training for the GR.Mk 4/4A.

No. 31 Sqn, RAF Marham

The first of the RAF Brüggen-based Tornado units, No. 31 Sqn received GR.Mk 1s from June 1984. The unit participated in Desert Storm in 1991 and Allied Force in 1999, by which time it had become the second ALARM-capable squadron. Now flying the Mk 4, No. 31 currently forms part of the Marham Tornado Strike Wing alongside No. IX Sqn.

No. 617 Sqn, RAF Lossiemouth

The second front-line Tornado squadron, No. 617 Sqn, reformed at Marham in 1982. Involved in Southern Watch duties in the early 1990s, it converted to the GR.Mk 1B at Lossiemouth in 1993. The squadron now operates the Mk 4 and flew the first combat Storm Shadow missions during Operation Telic.

Tornado GR.Mk 4A

No. II(AC) Sqn, RAF Marham

One of the original Royal Flying Corps squadrons, No. II(AC) has operated in the reconnaissance role for many years. The first GR.Mk 1As were delivered to No. II(AC) at RAF Laarbruch in 1988 and conducted 'Scud hunting' and other recce sorties in Desert Storm, before moving to RAF Marham in December 1991. Mk 4As (far right) were received in time for Operation Telic, where the new RAPTOR recce pod was used for the first time.

No. XIII Sqn, RAF Marham

No. XIII was the second RAF squadron to receive the GR.Mk 1A in 1990, just in time for participation in Desert Storm. No. XIII Sqn continues in the role today, having become the first operational GR.Mk 4A unit in 1998.

Tornado F.Mk 3

No. 11(F) Sqn, RAF Leeming

As the first RAF Leeming-based F.Mk 3 operator, No. 11 Squadron reformed at Coningsby in April 1988 before moving to its current base three months later. The squadron participated in Deny Flight operations over Bosnia and conducted numerous Jural deployments before participating in Operation Telic in 2003.

A No. 111 Sqn F.Mk 3 displays an impressive weapons load of four AMRAAM and four ASRAAM air-to-air missiles. The unit's badge consists of three black seaxes in the centre of a Jerusalem cross.

No. 25 Sqn, RAF Leeming

Established at Leeming as the third and final F.Mk 3 squadron at the base in October 1989, No. 25 Squadron continues to operate the type in the air defence role today. The squadron has seen regular deployments to the Gulf region throughout its existence as an F.Mk 3 operator, culminating in participation in Operation Telic, flying escort and defensive patrols.

No. 43(F) Sqn, RAF Leuchars

As the most important RAF fighter station during the Cold War, Leuchars's F.Mk 3 units were tasked with Northern QRA duties to repel any Soviet intrusions. Gaining F.Mk 3s in 1989 No. 43 Sqn crews combined QRA duties with deployments to the Gulf and for Operation Deny Flight in 1994.

No. 56(R) Sqn, RAF Leuchars

Formerly 229 OCU, No. 56(R) Squadron is responsible for all F.Mk 3 conversion and weapons training (badge at right). The unit was based at RAF Coningsby alongside Nos 29 and 5 Sqns until the disbandment of these led to No. 56(R) relocating to RAF Leuchars in 2003, allowing the RAF to concentrate its air defence assets at just two bases.

No. 111(F) Sqn, RAF Leuchars

The second of the Leuchars-based F.Mk 3 units to form, No. 111 Squadron conducted Northern QRA and Southern Watch deployments following its re-equipment with the type in May 1990. The squadron committed several aircraft to Operation Telic in 2003.

LUFTWAFFE

The Luftwaffe recently has scaled back its Tornado force as it prepares for total conversion to the MLU version of the IDS. Jagdbombergeschwader (JBG) 34 was disbanded in 2003, leaving three attack, one reconnaissance and one SEAD wings, plus the training squadron based at Holloman AFB.

TORNADO IDS

AKG 51 'Immelmann', Jagel

AKG 51 currently operates two squadrons (511 and 512 Staffel) of reconnaissance-rolled Tornado IDSs from Jagel. Its aircraft participated in Allied Force in 1999.

JBG 31 'Boelcke', Nörvenich

JBG 31 has operated the IDS in the attack role since 1983. The wing comprises 312 and 312 Staffel. The unit has now received the Rafael Litening targeting pod for LGB delivery.

JBG 33, Büchel

JBG 33 has been an attack-roled IDS operator since 1985. The unit's two squadrons (331 and 332 Staffel) have been tasked in the nuclear strike role.

In 2004 this JBG 33 Tornado IDS received this special paint scheme with an enlarged version of the unit's diving eagle badge on the tail fin.

Far left: JBG 31 badge.

Left: JBG 38 badge.

Above: AKG 51 badge

JBG 38 'Friesland', Jever

JBG 38 was the first Luftwaffe IDS wing to form (in 1983) tasked with weapons training for Luftwaffe and Marineflieger air crew. A single staffel (381) is operated.

Ausbildungsstaffel Tornado, Holloman AFB, NM

After the closure of the TTTE in 1998 Luftwaffe Tornado conversion training has been conducted in the USA.

TORNADO ECR

JBG 32, Lechfeld

JBG 32 is the Luftwaffe's premier SEAD and ECM unit, tasked with both training and operational duties using the Tornado ECR. Armed with HARM missiles, the unit flew combat missions during Allied Force and will imminently have its capabilities enhanced with MLU aircraft and the addition of the HARM PNU.

MARINEFLIEGER

The Marineflieger acquired the Tornado IDS to replace the Starfighter in the attack, recce and anti-shipping roles. In 1994 strength was reduced from two to one Marinefliegergeschwader (MFG 2), with MFG 2 currently operating around 48 Tornados.

MFG 2, Eggebek

Receiving its first Tornado IDS in 1986, MFG 2 Tornados are truly multi-role – the most common stores being HARM and Kormoran 2 missiles. The aircraft can also carry the buddy-buddy refuelling system.

Left: An MFG 2 IDS in special markings.

Right: MFG 2's swooping eagle unit badge.

AERONAUTICA MILITAIRE ITALIANA (AMI)

The AMI has budgeted to put 87 IDSs and ECRs through the MLU process. Tornados are currently assigned to three Stormi and the RSV. The F.Mk 3 will retire from AMI service with the return of the last example to the UK in late 2004.

TORNADO IDS

6° Stormo, Ghedi

102° and 154° Gruppi are assigned to 6° Stormo based at Ghedi. The aircraft have been tasked with conventional and nuclear attack and delivered GBU-16 LGBs during Allied Force in 1999.

36° Stormo, Gioia del Colle

36° Stormo operates both ADV and IDS versions of the Tornado. 156° Gruppo is the IDS operator and is likely to be the first operational unit to receive the MLU aircraft. The unit is equipped with the CLDP pod for PGM delivery.

50° Stormo, Piacenza

155° Gruppo became part of 50° Stormo in 1990 and operates primarily in the SEAD role. Although the unit mainly operates the ECR version it also employs Tornado IDSs in the attack role.

TORNADO F.MK 3

36° Stormo, Gioia de Colle

The nine-year operational career of the Tornado F.Mk 3 is drawing to a close, 12° Gruppo returning its final aircraft to the UK in November 2004. The AMI operated two full Gruppi of F.Mk 3s, until the decision was made to lease 36 F-16s, pending the arrival of the Eurofighter. The sole combat deployment of the F.Mk 3 occurred during Allied Force in 1999, when 21° Gruppo of 53° Stormo conducted combat air patrols in Bosnian airspace.

A 50° Stormo Tornado ECR is seen with its usual weapons load of two HARM anti-radiation missiles. The unit operated alongside Luftwaffe ECRs during Allied Force.

TORNADO ECR

50° Stormo, Piacenza

155° Gruppo operate 16 Tornado ECRs in the SEAD role. The aircraft will undergo full MLU upgrade and will be equipped with the HARM PNU missile for improved anti-radar capabilities following service clearance trials.

ROYAL SAUDI AIR FORCE

The Royal Saudi Air Force is the sole Tornado IDS operator yet to commit its fleet to a mid-life upgrade programme. The RSAF eventually received 96 Tornado IDSs (including the last aircraft to emerge from the production line) and 24 Tornado ADVs. Both versions saw combat during Operation Desert Storm in 1991.

TORNADO IDS

No. 7 Sqn, Dhahran

No. 7 Squadron conducts operational training and has 10 'twin stick' IDSs.

No. 66 Sqn, Khamis Mushait

No. 66 Sqn's assets include the six recce-configured IDSs delivered in Batch 1.

No. 75 Sqn, Khamis Mushait

No. 75 Sqn formed with 24 of 48 Batch 2 IDSs delivered in 1998.

No. 83 Sqn, Khamis Mushait

No. 83 Sqn IDSs may have a SEAD role.

TORNADO ADV

No. 29 Sqn, Dhahran

No. 29 Squadron is now the sole Tornado ADV Squadron in the RSAF following its absorption of No. 34 Sqn in the early 1990s. Its ADVs conducted combat air patrols during Desert Storm alongside RSAF F-15 Eagles.

Russian Federation

Part 2: 37th VA VGK (SN), 61st VA VGK (VTA), KSN

37th Vozdushnaya Armiya Verkhovnogo Glavnogo Komandovaniya (Strategicheskogo Naznacheniya)

37th Air Army of the High Supreme Command (Strategic Purpose) HQ: Moscow

Before the outbreak of World War II, and in the war's first phase, Soviet long-range bombers were subordinated to commanders of military districts (fronts) and were often used for tasks inappropriate to their range performance and ability. On 5 March 1942, the Aviatsiya Dalnego Deystviya (ADD), or Long-range Aviation, was separated from the VVS and subordinated directly to the High Supreme Command. In 1944, it was incorporated again into the VVS as the 18th Air Army, but this move was quickly recognised to be an error and in April 1946 the 18th Air Army was transformed into the Dalniaya Aviatsiya Vooruzhonnykh Sil (DA VS) or Long-range Aviation of the Armed Forces.

The DA VS existed in the Soviet Union as a separate air service until 1980. In January of that year, Dalniaya Aviatsiya Command was disbanded and the heavy bomber units were organised into three Air Armies: the 37th Vozdushnaya Armiya Verkhovnogo Glavnogo Komandovaniya (Strategicheskogo Naznacheniya – VA VGK (SN)) or Air Army of the High Supreme Command (Strategic Purpose) – with headquarters in Moscow; the 46th VA VGK (SN) with headquarters in Smolensk; and the 30th VA VGK (SN) with headquarters in Irkutsk. The pendulum swung back in 1988 when the Dalniaya Aviatsiya Command was restored again, a situation that continued until 1997.

In 1998, when the VVS and PVO were joined, the Long-range Aviation was transformed into the 37th VA VGK (SN). The commanding officer of the 37th VA VGK (SN) is Major General Igor Khvorov; born 8 March 1953, he was appointed to this post on 15 November 2002. He has more than 4,000 hours of flying time on all types of long-range aircraft. His second in command is Major General Pavel Androsov.

Training for the strategic bomber force is handled by the 43rd TsBPiPLS, headquartered at Ryazan-Dyagilevo. It has on strength a few examples of the front-line types, plus a squadron of Tu-134UBL pilot trainers, with elongated Tu-22M-style noses. These operate mainly at Tambov under the auspices of the 3119th AB.

Long-range Aviation celebrates on 23 December every year. On this day in 1914, the Tsar approved the decision to create the first squadron of Ilya Muromets heavy bombers (this four-engined bomber was designed by Igor Sikorsky, who later became one of the most celebrated aircraft designers in the USA and is credited with inventing the helicopter).

43rd Tsentr Boyevogo Primeneniya i Pereuchivaniya Lyotnogo Sostava

43rd Centre for Combat Application and Training of Air Crew – Ryazan (Dyagilevo)

The 1st Vysshaya Ryazanskaya Shkola Shturmanov (1st Navigator High School) was established on 18 September 1940 at Dyagilevo airfield near Ryazan. During the war between 1941 and 1945, the school prepared and sent to the front 480 aircrews (in this period the school was evacuated to Karshi in Uzbekistan). In 1944, the school became the 1st Vysshaya Shkola Shturmanov i Lyotchikov Aviatsii Dalnego Deystviya (1st High School of Navigators and Pilots for Long-range Aviation). In May 1958, it was renamed as the 43rd Tsentralnye Kursy Podgotovki Lyotnogo Sostava (Central Courses of Flight Crew Training); in 1968, the school received its current designation as the 43rd TsBPiPLS. The centre includes two squadrons stationed in Ryazan for the Tsentralnye Offitserskiye Kursy (Central Officer Courses), which trains pilots and ground personnel for new aircraft types, as well as a squadron stationed at Tambov airfield (within the 3119th AB).

In the past, the Ryazan Centre trained about 1,000 specialists from other countries that bought long-range bombers from the USSR, such as China, Indonesia, Iraq, Libya and Egypt (one of the students was Egyptian President Hosni Mubarak).

The centre has eight Tu-22M3 medium bombers (as of 2002), four Tu-95MS strategic bombers (six in 2003), Tu-134UBL training aircraft and An-26 transport aircraft. Its commanding officer is Major General Igor Krasilnikov.

22nd Gvardeyskaya Krasnoznamionnaya Donbasskaya Tiazholaya Bombardirovochnaya Aviatsionnaya Diviziya

22nd Donbass Heavy Bomber Air Division of Guards awarded with Red Banner – Engels

Before the disintegration of the Soviet Union, the 22nd TBAD was stationed in Bobruysk (now in Belarus) and included three heavy bomber regiments (200th TBAP, 121st TBAP and 203rd TBAP) equipped with Tu-22 bombers. In 1992 the division was transferred to Russia: division headquarters plus the 121st and 203rd regiments transferred to Engels, and the 200th regiment to Belaya air base near Irkutsk. In 1994 the 22nd TBAD was disbanded, but in 1997 its number, colours and titles were given to the 201st TBAD at Engels. That unit had been formed at Engels as early as 1954 and was equipped with Myasishchev M-4 and 3M strategic bombers. Major General Anatoliy Zhikhariev has been the commanding officer of the 22nd TBAD since 27 July 2000.

121st Gvardeyskiy Krasnoznamionnyi Sevastopolskiy Tiazholyi Bombardirovochnyi Aviatsionnyi Polk

121st Sevastopol Heavy Bomber Air Regiment of Guards awarded with Red Banner – Engels

The regiment came into being in 1940 at Khotunok airfield near Novocherkassk as the 81st Dalniy BAP, with TB-3 and then DB-3F bombers. On 27 June 1941, the regiment became operational with 61 DB-3F bombers. During World War II it took part in all important operations, attacking strategic targets in Romania, Finland, Poland, Czechoslovakia,

Spearhead of the bomber fleet is the Tu-160 'Blackjack', which has been in limited service since 1987 and with the 121st TBAP since 1994. The aircraft are named: above is Bort 03* Pavel Taran, *while at right is Bort 02* Vasiliy Reshetnikov, *seen being towed past a line of Tu-95MS 'Bear-Hs' at the principal strategic bomber base at Engels.

Left: Whereas the Tu-160s are named after people, the Tu-95MS fleet has acquired city names.* Ryazan *is named for the city near to the 43rd TSBPiPLS's airfield at Dyagilevo.

Hungary and Germany. On 18 August 1942, the regiment was rated among the Guards and renamed as the 5th BAP DD of Guards. On 24 May 1944 it was awarded the honorary title 'Sevastopolskiy' for its participation in operations to liberate Sevastopol. Its last combat action during World War II was an air raid on Swinemünde (Swinoujscie) harbour on 30 April 1945. On 18 August 1945 the regiment was awarded the Red Banner, and on 20 September 1946 it was renamed as the 121st Bomber Air Regiment.

In 1951, the unit received the reconnaissance version of the B-25 and was renamed as the 121st ODRAP long-range reconnaissance regiment. In 1953, the regiment was equipped with Tu-4 reconnaissance aircraft, followed by Tu-16s in 1958 and Tu-22s in 1964. In 1969, the regiment received Tu-22K missile-carriers and Tu-22P electronic warfare aircraft, and was renamed again as a TBAP bomber regiment. From 1952 until 1994, the regiment was stationed at Machulishchi air base in Belarus.

Another regiment, the 1096th TBAP, came into being on 20 September 1954 at Engels air base. The unit was equipped with Myasishchev M-4 strategic bombers, which in 1957 were replaced by improved 3M aircraft. In 1970s, 3M bombers were converted into air tankers and in the late 1980s, they were withdrawn from service. The first three Tu-160 bombers were incorporated into the 1096th TBAP in 1992. On 1 June 1994, the 1096th TBAP was renamed as the 121st TBAP, adopting its colours and tradition.

In 1998, Tu-95MS strategic bombers from Mozdok air base were temporarily assigned to the 121st TBAP. On 1 September 2000, the Tu-95MS bombers were taken over by the newly re-established 184th TBAP stationed at the same air base in Engels, leaving the 121st with just Tu-160s (two squadrons). The commanding officer of the 121st TBAP is (2003) Colonel Alexei Serebryakov. The regiment day is 27 June.

184th Gvardeyskiy ordena Lenina Krasnoznamionnyi Poltavsko-Berlinskiy Tiazholyi Bombardirovochnyi Aviatsionnyi Polk

184th Poltava-Berlin Heavy Bomber Air Regiment of Guards awarded with Lenin Order and Red Banner – Engels

The regiment was formed on 15 July 1938 in Kursk as the 51st Skorostnoy BAP (51st High-speed Bomber Air Regiment) with SB aircraft. In 1940 it took part in the war against Finland. In the first days of war with Germany during June 1941, the regiment undertook day operations without fighter escort and suffered heavy losses: 40 aircraft and 128 casualties in 10 days. On 6 March 1942, the regiment was renamed as the 749th AP DD. In summer 1942, nine crews of the regiment flew air raids against Berlin, Danzig (Gdansk) and Stettin (Szczecin). Other crews operated at Stalingrad (Volgograd).

On 26 March 1943, the regiment was rated among the Guards and renumbered as the 9th Long-range Regiment of Guards. On 24 August 1943 the regiment was awarded the honorary title 'Poltavskiy' in recognition of its active operations in the Poltava region, and on 11 June 1945 with 'Berlinskiy' for its Berlin operations. The last combat action of the regiment was the air raid on Swinemünde (Swinoujscie) on 28 April 1945.

In April 1946, the regiment received a new number – 184 – and was stationed at Pryluky, Ukraine. At the beginning of 1950 it re-equipped with Tu-4 bombers, in 1955 with Tu-16s (versions A, K-26 and P), in 1984 with Tu-22M3s, and in 1987 with Tu-160s. After the disintegration of the USSR, on 8 May 1992 the regiment swore the oath of allegiance to Ukraine and for several years it served in that country's armed forces, operating Tu-22M3 bombers (Tu-160 bombers were also stationed in Pryluky, but did not fly). Several years later, Ukraine disbanded the 184th TBAP.

37 VA VGK (SN) HQ: Moscow

UNIT (1ST LEVEL)	UNIT (2ND LEVEL)	BASE	INVENTORY/NOTES
43rd TsBPiPLS		Ryazan (Dyagilevo)	Tu-22M3, Tu-95MS, Tu-134UBL, An-26
22nd TBAD		Engels	
	121st TBAP	Engels	Tu-160
	184th TBAP	Engels	Tu-95MS
	52nd TBAP	Shaykovka	Tu-22M
	840th TBAP	Soltsy	Tu-22M
326th TBAD		Ukrainka (Seryshevo)	
	182nd TBAP	Ukrainka (Seryshevo)	Tu-95MS
	79th TBAP	Ukrainka (Seryshevo)	Tu-95MS
	200th TBAP	Belaya near Irkutsk	Tu-22M3, Tu-22MR
	444th TBAP	Vozdvizhenka (Ussuriysk)	Tu-22M3
203rd OAP SZ		Ryazan (Dyagilevo)	Il-78, Il-78M
3119th AB		Tambov	Tu-134UBL, Tu-134Sh, L-39, An-26, MiG-23 (in storage), MiG-27 (in storage)
1425th CSFP DA		Irkutsk region	

Three regiments are equipped with the Tu-95MS, which is principally used as a stand-off missile-carrier. As with the Tu-160, an update programme is in the process of adding considerable conventional attack capability.

A new bomber regiment was formed at Engels on 1 September 2000 with the Tu-95MS aircraft taken from Mozdok (formerly belonging to the 182nd TBAP, in 1998-2000 to the 121st TBAP); The new regiment adopted the name and colours of the 184th TBAP, and its regiment day is 15 July.

52nd Gvardeyskiy Tiazholyi Bombardirovochnyi Aviatsionnyi Polk
52nd Heavy Bomber Air Regiment of Guards – Shaykovka

The 52nd BAP was formed at Migalovo airfield in December 1944 with 18 crews and TB-3 bombers transferred from the Russian far east. On 22 February 1945 the regiment was raised to Guards status. In August 1945 it took part in an air show at Tushino, when TB-3 bombers paraded above Moscow for the last time. In 1946, the regiment was transferred to Balbasovo airfield (Orsha) and re-equipped with B-25 aircraft. In 1949 it was one of the first units to receive Tu-4 bombers. In 1954 the regiment was equipped with the first Tu-16 bombers and on 1 May 1995 it took part in an air parade over Moscow's Red Square with these aircraft. In 1958-59 the regiment moved to Shaykovka airfield, where it is still based. The 52nd TBAP mastered instrument landing and inflight-refuelling as one of the first units of Soviet Long-range Aviation. In 1958, its Tu-16s were the first jet-propelled bombers to land on the drifting ice near polar research station SP-6 (Severnyi Polyus, North Pole). In September 1961, a Tu-16 (crew of Lieutenant Colonel A. I. Salnikov) dropped an H-bomb on the Novaya Zemlya testing ground. On 12 March 1982 the regiment commenced operations with Tu-22M3s, which it flies today. In June 1983, the crew of Major Rumiantsev made the first launch of a Kh-22 missile from a Tu-22M3. In 1988, four Tu-22M3s of the 52nd TBAP flew air raids against Afghanistan. The regiment day is 4 October.

840th Krasnoznamionnyi Tiazholyi Bombardirovochnyi Aviatsionnyi Polk
840th Heavy Bomber Air Regiment awarded with Red Banner – Soltsy

The 840th BAP ADD was formed on 4 April 1942 at the Dyagilevo airfield near Ryazan with Il-4s and began combat operations on 24 June at Kursk. In spring 1944 the regiment was re-equipped with Tu-2 bombers. During the final stage of World War II, the regiment operated from Kurland, in Pommern and flew air raids on Berlin. After the end of war operations in Europe, the regiment was transferred to the far east and attacked Japanese troops from 9 to 11 August 1945. In 1951 the regiment came back to Europe, to Soltsy air base near Novgorod, where it was equipped with Tu-4 bombers. In 1955 it converted to Tu-16 bombers and in 1976 to Tu-22M2s. The 840th TBAP made 71 combat sorties in Afghanistan, dropping 122 OFAB-3000 bombs; later, the regiment took part in operations in Tajikistan and Chechnya. Three Tu-22M3s of the 840th TBAP paraded over Moscow during the 50th anniversary of victory over Germany, on 9 May 1995. The regiment day is 24 June.

326th Tarnopolskaya ordena Kutuzova Tiazholaya Bombardirovochnaya Aviatsionnaya Diviziya
326th Tarnopol Heavy Bomber Air Division awarded with Kutuzov Order – Ukrainka

The division was formed in 1943 and until 11 June 1992 its headquarters was stationed in Tartu (now the Estonian Republic). It consisted of three bomber regiments with Tu-22M aircraft: the 132nd TBAP at Tartu, the 402nd TBAP at Balbasovo, and the 840th TBAP at Soltsy. After the collapse of the Soviet Union, the division headquarters was moved to Soltsy and then to Ukrainka air base in the Russian far east. Between 1993 and 1997, the commanding officer of the 326th TBAD was Igor Khvorov, who now commands the 37th Army. The 326th's current commanding officer is Major General Konstantin Dementyev (2002).

79th Tiazholyi Bombardirovochnyi ordena Krasnoy Zvezdy Aviatsionnyi Polk
79th Heavy Bomber Air Regiment awarded with Red Star Order – Ukrainka

The 79th Air Regiment was founded on 13 July 1938 at the airfields of Kirovskoye and Zonalnoye on Sakhalin. The regiment consisted of four squadrons with SB bombers as well as I-15bis and I-16 fighters. From 8 August to 2 September 1945, the regiment fought in the war against Japan and for these operations was awarded the Red Star Order. From 1945-1953, the regiment was stationed at Burevestnik airfield on Iturup Island in the Kuril Islands. In 1953 it moved to Ukrainka airfield, where it remains. In October 1956 the regiment was transferred from the tactical command to the long-range bomber command, renamed as the 79th TBAP and given Tu-16 bombers. In 1959 it re-equipped with M-4 strategic bombers (latterly 3Ms and 3MDs), and from 1985 to 1986 operated Tu-95K-22 bombers. Currently, the regiment is armed with Tu-95MS bombers received by Russia from Kazakhstan. The regiment day is 28 August.

182nd Gvardeyskiy Krasnoznamionnyi Sevastopolsko-Berlinskiy Tiazholyi Bombardirovochnyi Aviatsionnyi Polk
182nd Sevastopol-Berlin Heavy Bomber Air Regiment of Guards awarded with Red Banner – Ukrainka

The regiment was founded on 10 May 1943 at Drakino airfield (near the town of Serpukhov) as the 18th BAP DD. During World War II, the regiment became part of the Guards. It was awarded the honorary titles 'Sevastopolskiy' for actions during the liberation of Sevastopol and 'Berlinskiy' after the taking of Berlin. In 1946 the regiment was incorporated into Long-range Aviation and renamed as the 182nd BAP, stationed at Nezhin, Belarus. It was armed with Tu-4s, and from 1956 with Tu-16 bombers. In 1961 the regiment moved to Mozdok airfield and re-equipped with Tu-95K strategic bombers. In 1985, the 182nd received its first Tu-95MSs. In the first half of 1998, when Mozdok turned out to be too near the trouble spot of the North Caucasus, the aircraft were

With shorter range than the Tu-160, the Tu-22M3 'Backfire' is not considered a true intercontinental bomber, but nevertheless is an impressive performer, capable of high speed at low level. Four regiments operate the type.

Above: The Tu-22M3 shoulders the main burden of the penetration mission, despite the appearance of the Tu-160. 'Backfire' bases are located in the northwest, Moscow district, southern Siberia, and in the Far East.

moved to Engels and organised into a new TBAP, the 184th. At the same time, the headquarters of the 182nd TBAP was moved to Ukrainka where it took over the aircraft that came from Kazakhstan. The regiment day is 10 May.

200th Gvardeyskiy Krasnoznamionnyi Brestskiy Tiazholyi Bombardirovochnyi Aviatsionnyi Polk

200th Brest Heavy Bomber Air Regiment of Guards awarded with Red Banner – Belaya

The regiment was formed during World War II as the 112th Night Blocking BAP, tasked with suppressing enemy anti-aircraft defence of objects that were to be attacked by Soviet bombers. On 10 January 1944, the regiment was transformed into the 26th Long-range Regiment of Guards, armed with B-25 bombers. In the final stage of the war, the regiment was awarded the honorary title of 'Brestskiy' and, in April 1946, its number was changed to 200.

After the war, the regiment was stationed at Bobruysk airfield, Belarus. In 1949 it re-equipped with Tu-4 bombers, in 1955 with Tu-16 bombers, and in 1964 with Tu-16K missile-carriers with KSR-2 stand-off missiles. On 23 February 1968, the regiment was awarded the Red Banner Order. In 1986 it converted to Tu-22M3 bombers. In November-December 1994, after the disintegration of the USSR, the regiment moved from Bobruysk to Belaya airfield near Irkutsk. Belaya has always been a large air base: in the 1970s and 1980s, two heavy bomber regiments – the 1125th TBAP and 1129th TBAP with Tu-16s and then Tu-22Ms – and the 350th IAP fighter regiment with Tu-128 heavy interceptors were based there. In 1998, the 1129th TBAP – which had been stationed at Belaya since 1955 – was disbanded and its Tu-22M 'Backfire' aircraft were taken over by the 200th TBAP. The commanding officer is Colonel Wellington Cirengarmayev (2003). The regiment day is 13 January.

The Tu-22MR is a specialist 'Backfire' version for reconnaissance. The small number are operated by the 200th TBAP (which also has standard bombers) from its base near Irkutsk. This location is close to the Mongolian border – and China.

444th Krasnoznamionnyi Tiazholyi Bombardirovochnyi Aviatsionnyi Polk

444th Heavy Bomber Air Regiment awarded with Red Banner – Vozdvizhenka (Ussuriysk)

The regiment was formed in August 1941 at Khabarovsk as a composite unit with DB-3 bombers and I-16 fighters. Its main task was training pilots for the German-Soviet front. On 15 August 1943, the regiment was awarded the Red Banner Order. In August 1945 the regiment attacked Japanese positions with bombs. It was stationed in North Korea immediately after the war but returned to the USSR in 1948, first to Pozdeyevka airfield and then in 1953 to Vozdvizhenka, where it was stationed until disbandment. From 1951-1956, the regiment operated Tu-4 bombers, from October 1956 Tu-16s, and from 1975 Tu-16K (KSR-5M) missile-carriers. In 1992, the Tu-16 bombers were withdrawn from service and the regiment was disbanded.

Its history did not end there, however. Several years later (possibly 1997) the name and colours of the 444th TBAP were assumed by another unit, the 132nd TBAP. It was established in 1940 and was equipped with Tu-4s (from 1953), Tu-16s (from 1957) and Tu-22M3s (from 1987). From November 1951 until the collapse of the USSR, the regiment was stationed in Tartu, Estonia, from where it was moved in 1992 to Zavitinsk air base in the Russian far east, and on 1 September 1997 to Vozdvizhenka. It is likely that the regiment was renamed the 444th TBAP just after its arrival at Vozdvizhenka. The regiment day is 15 August.

203rd Gvardeyskiy Otdelnyi Aviatsionnyi Orlovskiy Polk (Samolyotov Zapravshchikov)

203rd Independent Orel Air Regiment of Guards (Air Tankers) – Ryazan

This unit came into being on 6 July 1941 at Monino air base near Moscow as the 412th AP air regiment, with TB-7 (Pe-8) bombers; several weeks later the regiment was renamed as the 432nd AP. On 10 August 1941, 10 crews took part in an air raid on Berlin and then on Danzig (Gdansk), Königsberg (Kaliningrad) and Bucharest. On 3 December 1941, the regiment was renamed as the 746th AP DD. In April 1942, one of its TB-7s carried Soviet Foreign Minister Vyacheslav Molotov to the United States on a route across Germany to Great Britain and then over the Atlantic Ocean. On 19 September 1943, the regiment achieved Guards status and was renamed as the 25th AP DD of Guards. On 27 May 1944, it was awarded the honorary title of 'Orlovskiy' for participation in the liberation of the town of Orel.

After the war and until November 1951, the regiment was stationed at Balbasovo airfield and then at Baranovichi in Belarus. In June 1948, it began training with new Tu-4 bombers, in February 1954 with Tu-16s, and in 1962 with Tu-22 supersonic bombers (Tu-22K in 1966 and

Although seen here during an airshow flypast while refuelling two versions of the 'Flanker' tactical fighter, the Il-78 'Mainstay' is first and foremost a strategic asset, and is a key part of the 37th VA roster. The tanker fleet is operated by the 203rd OAPSZ at Ryazan, having moved from Engels in 2000.

Tu-22P in 1969). In 1994, following the disintegration of the USSR, the Tu-22s were moved to the 6213th BKhUAT storage facility at Engels air base in Russia.

On 1 December 1994, the name and colours of the 203rd TBAP were taken up by another unit then stationed at Engels, the 1230th Air Tanker Regiment which became the 203rd OAP SZ. The 1230th was established around 1960 using M-4-2 air tankers (converted from M-4 bombers) and later also new -3MS2 and -3MN2 versions. Between 1989 and 1993, they were replaced by about 20 Il-78 and Il-78M tankers. The new 203rd OAP SZ was stationed in Engels for only a short time: on 1 September 2000, it moved to Dyagilevo air base near Ryazan. Its commanding officer is Colonel Vadim Prozorov.

3119th Aviatsionnaya Baza
3119th Air Base – Tambov

In the past, Tambov had housed a pilot school for bomber aviation with a training regiment consisting of two squadrons of Tu-134UBLs and one squadron of An-26s. In 1998 the school was disbanded so that Tambov could become the 3119th Air Base. The base's 'flying' component is an independent squadron of Tu-134UBL and An-26 training aircraft belonging to the 43rd TsBPiPLS in Ryazan; the squadron also has at least one Tu-134Sh navigator trainer. On the base is a small storage and scrapping facility for old aircraft (L-39, Tu-134UBL, MiG-23 and MiG-27), and also a museum with An-12, An-26, Tu-134, MiG-23, MiG-27, Su-25 and other aircraft.

Springboard airfields

Russian strategic aviation uses Engels and Ukrainka air bases situated near the southern border of the country. Therefore, so-called 'springboard airfields' are necessary for mounting attacks in a northerly direction (for example, against the US via the North Pole). They are usually inactive airfields located in the far north of Russia, where a small staff and fuel stores are kept and the airstrip is maintained in working condition. If necessary, Tu-95 or Tu-160 aircraft from the south land at these springboard airfields for refuelling, take off and continue their mission. The springboard airfields of the 37th VA VGK (SN) include Anadyr, Chekurovka, Magadan, Nagurskoye, Norilsk, Rogachevo, Vorkuta (Sovetskiy airfield in Vorkuta is the most northerly point in Russia with a railway terminal) and Tiksi. In the past, the springboard airfields fell under the command of the Opperativnaya Gruppa v Arktikie (OGA) or Operational Group in the Arctic with headquarters in Tiksi, having at its disposal the 24th OTAE transport squadron with light aircraft and helicopters. It is not clear whether this group still exists.

Until recently, the 37th Air Army also parented the 6213th BKhUAT storage and utilisation facility at Engels, tasked with demolition of old Myasishchev 3M 'Bison' and Tupolev Tu-22 'Blinder' bombers. The facility was disbanded in 2001 after the destruction of the last bombers had been completed.

61st Vozdushnaya Armiya Verkhovnogo Glavnogo Komandovaniya (Voyenno-Transportnaya Aviatsiya)

61st Air Army of the High Supreme Command (Military Transport Aviation) HQ: Moscow

During World War II, the units of transport aviation were allocated to individual air armies. A separate airlanding and transport aviation command was created only in 1946 (within airlanding forces) and in summer 1955 all transport air units were consolidated into the Voyenno-Transportnaya Aviatsiya (VTA) or Military Transport Aviation, subordinated to the Air Forces. As the result of the great reorganisation of 1998, the VTA was transformed into the 61st VA VGK (VTA) Air Army with two air divisions – the 3rd VTAD at Krechevitsy and the 12th VTAD at Tver' – with a total of nine air regiments and the 610th TsBPiPLS training centre. A few years later the 3rd VTAD was disbanded, as was the 235th VTAP regiment in Ulyanovsk. It should be noted that all military units of transport aviation are stationed in the European part of Russia and there are none between the Ural Mountains and the Pacific coast.

Personnel strength of the 61st VA VGK (VTA) is currently around 10,000. The main military transport aircraft in Russia is the Il-76, complemented by a few dozen heavy An-124s and An-22s, as well as An-12 medium transports (other An-12s and smaller transport aircraft such as the An-24, An-26 and An-72 come under the control of the VVS and PVO armies).

The commanding officer of the 61st VA VGK (VTA) is Lieutenant General Viktor Denisov. Born on 14 June 1948, he commanded a transport air regiment from 1983 to 1987, and between 1987 and 1991 was commanding officer of an air division. In 1993, he graduated from the General Staff Academy and was appointed to the post of second in command of the VTA; from 1997, he was the commanding officer of the VTA and then commanding officer of the 61st Army.

The official celebration day of the 61st VA VGK (VTA) is 1 June.

A true Cold War relic, the An-22 is used sparingly for outsize military cargo duties, plus the occasional civilian charter with the 224th LO, to augment the hard-pressed An-124 fleet. The survivors fly from Tver'-Migalovo with an independent squadron within the 12th VTAD organisation.

610th Tsentr Boyevogo Primeneniya i Pereuchivaniya Lyotnogo Sostava imeni Marshala Aviatsii N. S. Skripko
610th Centre for Combat Application and Training of Aircrew named after Air Marshal N. S. Skripko – Ivanovo

The centre has its origins in the 25th Officer Courses of the Air Forces, which was established in December 1967 at Seshcha and transferred in 1968 to Ivanovo. In December 1974, the Courses were renamed the 610th TsBPiPLS training centre of transport aviation. On 6 May 1995, the centre was named after Marshal Nikolai Skripko, former commander of the VTA. The centre was established to improve the skills of transport aviation crews and ground personnel, to train aircraft captains, and to conduct research into the operation of transport aviation. So far, the centre has trained 12,000 specialists. The commanding officer of the centre is Major General Aleksandr Akhlyustin (2001). The day of the 610th Centre is 1 September.

Some units of transport aviation have always functioned within the 610th Centre, the first being the 374th VTAP with An-12s. In 1979 the regiment was divided into two independent instructor squadrons, one with An-12s and one with Il-76s, and later one more research squadron was created. On 11 October 1993, the three squadrons were joined together into the 517th VTAP instructor transport air regiment with Il-76 aircraft. At the same time, the centre's branch in Fergana was disbanded and a new instructor squadron was formed at Ivanovo with An-12s that had belonged to that branch. Aircraft from the centre took part in the Afghan War (1980-1982) and the Chechnya conflict (1994-1996), and now are active in Afghanistan alongside American aircraft. Around 2002, the 517th Regiment was reduced to squadron size; the number of this squadron remains unknown. The squadron day is 1 September, the same as that of the 610th Centre.

61 VA VGK (VTA) HQ: Moscow

UNIT (1ST LEVEL)	UNIT (2ND LEVEL)	BASE	INVENTORY/NOTES
610th TsBPiPLS		Ivanovo-Severnyi	
	unknown IVTAE	Ivanovo-Severnyi	Il-76
12th VTAD		Tver' (Migalovo)	
	196th VTAP	Tver' (Migalovo)	Il-76
	566th VTAP	Seshcha	Il-76, An-124
	unknown OVTAE	Tver' (Migalovo)	An-22
	unknown BKhAT	Tver' (Migalovo)	An-22
103rd OVTAP		Smolensk	Il-76
110th OVTAP		Krechevitsy	Il-76
117th OVTAP		Orenburg	Il-76, An-12
334th OVTAP		Pskov	Il-76
708th OVTAP		Taganrog	Il-76
76th OVTAE		Klin-5	Il-76
224th LO VTA		Tver' (Migalovo)	An-124, Il-76MD

12th Mginskaya Krasnoznamionnaya Voyenno-Transportnaya Aviatsionnaya Diviziya

12th Mga Military-Transport Air Division awarded with Red Banner – Tver' (Migalovo)

The division was created on 5 May 1943 at Monino air base as the 12th Aviatsionnaya Diviziya Dalnego Deystviya (AD DD) long-range air division with Li-2 (Douglas DC-3) aircraft. In May 1944 the division was awarded the honorary title 'Mginskaya' and in November that year received the Red Banner Order. On 1 July 1946, the division was reorganised into the 12th Transportnaya Aviatsionnaya Diviziya of airlanding forces and in 1956 it was incorporated into the VTA. After World War II the division headquarters was at Tula, but in 1975 it moved to Seshcha and in 1983 to Tver' (then Kalinin), where it is still stationed. The commanding officer is Major General Genrikh Levkovich (2001). The division day is 4 June.

196th Gvardeyskiy Minskiy Voyenno-Transportnyi Aviatsionnyi Polk

196th Minsk Military-Transport Air Regiment of Guards – Tver' (Migalovo)

The organisation of this regiment, as part of the 12th Division, dates to 25 May 1943 at Monino air base, where it initially was known as the 110th AP DD and equipped with Li-2 (DC-3) aircraft. For its participation in the liberation of Minsk in August 1944, the regiment was awarded the honorary title 'Minskiy'. In November 1944 it was given Guards status and renamed as the 33rd Minsk Long-range Air Regiment of Guards. On 1 February 1946, the regiment was transferred for a brief time to bomber aviation command under the new name 196th Long-range Bomber Air Minsk Regiment of Guards; several months later, on 1 June, it was moved to transport command as the 196th Independent Transport Air Regiment of Guards. In March 1957, the regiment began training with Tu-4D heavy transport aircraft, in May 1961 with An-12s, and in February 1980 with Il-76Ms. After World War II the regiment was stationed in Tartu, Estonia, until the disintegration of the USSR, and in autumn 1992 it moved to Migalovo air base near Tver' where it remains. The regiment day is 25 May.

566th Solnechnogorskiy Krasnoznamionnyi ordena Kutuzova Voyenno-Transportnyi Aviatsionnyi Polk

566th Solnechnogorsk Military-Transport Air Regiment awarded with Red Banner and Kutuzov Order – Seshcha

This is the only regiment of military transport aviation to operate the world's heaviest transport aircraft, the An-124. The regiment was created on 12 September 1941 as the 566th Shturmovoy Aviatsionnyi Polk (Attack Air Regiment) with Il-2 aircraft. For its participation in the fighting at Solnechnogorsk in May 1943, the regiment was awarded the honorary title 'Solnechnogorskiy'. In December 1944, it was distinguished with the Red Banner Order and in May 1945 with the Kutuzov Order (third degree). After the war the regiment remained at Rakvere air base in Estonia, where on 1 August 1946 it was transformed into the 566th Aviatsionno-Transportnyi Polk (Transport Air Regiment) with Li-2 aircraft. Subsequently, the regiment was equipped with Il-12 aircraft and Tsibin-25 airlanding gliders (from 1948), Tu-4Ds (1956) and An-12s (1959).

In 1970 the unit accepted the An-22, the then-largest transport aircraft in the world, which in 1987 was replaced by the even larger An-124. The aircraft of the 566th Regiment took part in many combat operations, including the invasions of Czechoslovakia (1968) and Afghanistan (1979-1984), as well as the fighting in Chechnya (1994-1995). It is not certain when the regiment moved from Rakvere to Seshcha, but the most probable year is 1970, after it received the An-22s. The regiment day is 7 September.

unknown Otdelnaya Voyenno-Transportnaya Aviatsionnaya Eskadrilya

unknown Independent Military-Transport Air Squadron – Tver' (Migalovo)

At the beginning of the 1990s, An-22 aircraft were operated by two transport air regiments: the 8th VTAP in Tver'-Migalovo and the 81st VTAP in Ivanovo. After about 30 years of operation with this type, in November 1997 the 81st regiment was disbanded. On 21 January 1998 the serviceable An-22s were moved to the 8th regiment at Migalovo airfield, where a storage facility (BKhAT) was established for those aircraft withdrawn from service. At the end of 2001, the 8th VTAP was disbanded and the remaining An-22s joined an independent squadron commanded by Lieutenant Colonel V. Borisenko.

103rd Gvardeyskiy Krasnoselskiy Krasnoznamionnyi Otdelnyi Voyenno-Transportnyi Aviatsionnyi Polk imeni Geroya Sovietskogo Soyuza V.S. Grizodubovoy

103rd Krasnoye Selo Independent Military-Transport Air Regiment of Guards awarded with Red Banner named after Hero of Soviet Union V.S. Grizodubova – Smolensk

The regiment was formed in April 1942 at Chkalovskiy air base near Moscow as the 101st Transport Air Regiment under the command of the famous Soviet woman pilot Valentina Grizodubova. In May 1944, for its participation in the liberation of Krasnoye Selo, the regiment was awarded the honorary title 'Krasnoselskiy'. In August of that year it was awarded the Red Banner Order and in November was ranged among the Guards and renamed as the 31st Long Range Air Regiment of Guards. In December 1945, the regiment received a new

Outsize cargo transport is the role for the An-124, the world's largest production aircraft. The type flies on military duties with the 566th VTAP at Seshcha, and on civilian charters with the Migalovo-based 224th LO. The fleet wears a mixture of VVS and Aeroflot colours.

Antonov's An-12 – Russia's answer to the Hercules – operates in small numbers with a variety of transport units, including the VTA's 117th OVTAP at Orenburg. Other 'Cubs' are assigned at theatre level to the numbered air armies. Despite its age, the An-12 usefully fills the capacity gap between the An-26 and Il-76.

number – 186. In June 1946, it relocated to the airfield at Smolensk, where it remains. In August 1953, the regiment was reduced to squadron size (as the 20th Independent Squadron of Transport Aviation) and returned to regiment size only in April 1988. Also in 1988, it received its current name of the 103rd VTAP. The regiment was equipped with An-12s and An-26s until 1995, when it re-equipped with the Il-76. That same year, it was given the name of Grizodubova. The regiment day is 25 May.

110th Komsomolskiy Transil'vanskiy Krasnoznamionnyi Otdelnyi Voyenno-Transportnyi Aviatsionnyi Polk

110th Komsomol Transylvania Independent Military-Transport Air Regiment awarded with Red Banner – Krechevitsy

The regiment was formed on 3 July 1942 on the initiative of Komsomol (the Communist youth organisation) at Alatyr' airfield in Chuvashya as the 930th Light Bomber Air Regiment with Polikarpov U-2 aircraft. In July 1944, the regiment was awarded the Red Banner Order and in November the honorary title 'Transil'vanskiy'. In May 1946, it was reorganised as the 374th Transport Air Regiment, but four months later the number 930 was restored. It is not clear when the regiment received its present number of 110 or when it transferred to its present base of Krechevitsy.

In 1957 the regiment converted to Il-14s and, two years later, to An-8s. In August 1967 it began training with An-12s, and in the 1980s it received Il-76s. The regiment's crews took part in the invasion of Hungary in 1956. During the Afghan War (1979-1989) the 110th VTAP made the greatest number of transport missions of all Soviet air transport units. From 1982 to 1989, one of the regiment's squadrons was always based at Bagram air base in Afghanistan. The regiment day is 3 July.

117th Berlinskiy ordena Kutuzova Otdelnyi Voyenno-Transportnyi Aviatsionnyi Polk

117th Berlin Independent Military-Transport Air Regiment awarded with Kutuzov Order – Orenburg

The regiment came into being on 1 July 1938 as a bomber unit; in 1939 it took part in fighting against the Japanese in Mongolia and China, and then in winter 1939/40 in the war against Finland. In April 1945 the regiment was awarded the honorary title 'Berlinskiy' and the Kutuzov Order (third degree); the regiment number then was 345. After the war, as the 345th Heavy Bomber Air Regiment, it was equipped with Tu-4 and then Tu-16 bombers, and was stationed at Kirovabad (now Ganca) in Azerbaijan. In November 1959 it became the 179th Independent Bomber-Tanker Squadron with Tu-16s. The next dramatic change came 10 years later, in November 1969, when the squadron was transformed into the 117th Independent Radio Air Regiment of Special Purpose, equipped with An-12PP and An-12PPS electronic warfare aircraft and moved to Siauliai airfield in Lithuania. As an electronic warfare unit the 117th Regiment took part in all important exercises, as well as in real combat operations in Syria. In November 1994 the regiment again changed its designation and location: it moved to Orenburg and, equipped with Il-76MD transport aircraft (although one of the squadrons uses An-12s), became the 117th VTAP. The commanding officer is Colonel Sergey Zakharov (2001). The regiment day is 1 July.

334th Berlinskiy Krasnoznamionnyi Otdelnyi Voyenno-Transportnyi Aviatsionnyi Polk

334th Berlin Independent Military-Transport Air Regiment awarded with Red Banner – Pskov

The regiment was formed on 13 March 1944 at Vorotinsk air base near Kaluga as the 334th AP DD, with Li-2 (DC-3) aircraft. For participation in Berlin operations the regiment was awarded the honorary title 'Berlinskiy' in June 1945. Since 1948 the regiment has been continuously stationed at Kresty airfield in Pskov. Immediately after World War II, the regiment flew Il-12s, then Il-14s and Tu-4Ds; in May 1963 it received An-12s, and in 1979 accepted Il-76s, which the 334th VTAP still uses. The commanding officer is Colonel Sergey Nikiforov (2001).

708th Gvardeyskiy Kerchenskiy Krasnoznamionnyi Otdelnyi Voyenno-Transportnyi Aviatsionnyi Polk

708th Kerch Independent Military-Transport Air Regiment of Guards awarded with Red Banner – Taganrog

The regiment was created on 7 October 1942 as the 2nd Transportnyi Aviatsionnyi Polk Osobogo Naznacheniya (Transport Air Regiment for Special Purpose) equipped with Li-2 and C-47 aircraft. In February 1949, it was incorporated into transport aviation and renamed as the 708th VTAP. In 1952, the regiment was equipped with Il-12s, in 1957 with Il-14s, in 1963 with An-8s and in 1965 with An-12s. In 1987, it received Il-76MDs, which it still uses. In 1962 the regiment performed a special mission that involved transporting and airlanding troops and equipment from North Vietnam to Laos. The unit was stationed at Ganca (formerly Kirovabad) in Azerbaijan until 1992, when it moved to Taganrog in Russia. In 1995 the 708th Regiment took over the honorary titles 'Gvardeyskiy' and 'Kerchenskiy' and the Red Banner Order from the disbanded 192nd VTAP. The commanding officer is Colonel Aleksandr Burlachenko (2002) and the regiment day is 7 October.

Unquestionably the backbone of the 61st VA VGK, the Ilyushin Il-76 serves with eight transport regiments, plus the training centre at Ivanovo and the VTA's 'airline' – the 224th LO. As with the other transport types, the fleet is split between full militarily-marked aircraft (above) and those carrying Aeroflot schemes (right). The latter are more commonly used for flights outside Russia.

76th Otdelnaya Voyenno-Transportnaya Aviatsionnaya Eskadrilya

76th Independent Military-Transport Air Squadron – Klin-5

This unit is the former 978th VTAP regiment of PVO troops, and since 1960 has been based at Klin. Among other operations, the regiment took part in transporting Soviet troops to Czechoslovakia in 1968. In 1998, when the VVS and PVO were merged, the regiment was incorporated into the 61st Army of Air Forces. Around 2000, the 978th VTAP regiment was reduced to squadron size and received a new number – 76. Currently, the 76th VTAE operates about 10 Il-76s.

224th Lyotnyi Otriad Voyenno-Transportnoy Aviatsii

224th Air Detachment of Military-Transport Aviation – Migalovo (Tver')

Separate from the 61st Army structure and registered as an airline company, the detachment is equipped with Il-76MD and An-124 heavy transport aircraft and mainly conducts charter flights to China. Currently, two Il-76MDs belonging to the 224th LO are being modernised to meet new requirements introduced in western Europe, including the installation of TCAS systems and new PS-90A-76 engines.

Komandovaniye Specialnogo Naznacheniya

Special Purpose Command HQ: Moscow

Special Purpose Command, tasked with protecting the capital city of Moscow, is the strongest group in tactical aviation. Unlike its status in the past, it has become a border district. Under the Air Forces reform of 1 June 1998, the Moskovskiy Okrug PVO (Moscow Air Defence District) and the 16th Vozdushnaya Armiya (Air Army) of the VVS were consolidated into a single Moskovskiy Okrug VVS i PVO (Moscow District of the VVS and PVO). In October 2002, the District was transformed into the Komandovaniye Specialnogo Naznacheniya (KSN) or Special Purpose Command. The commanding officer of the KSN is Lieutenant General Yuriy Solovyov.

In addition to air units, the Special Purpose Command has the 1st Korpus PVO (Air Defence Corps) with four brigades of anti-aircraft missiles protecting Moscow (this corps has no aircraft), as well as regiments of anti-aircraft missiles and radio technical units. The zone of responsibility of the Special Purpose Command equates to the territory of Moscow's Military District.

In the wake of the VVS and PVO merger of 1998, reductions continue: for example, two regiments equipped with MiG-31 interceptors – the 736th IAP in Pravdinsk and the 153rd IAP in Morshansk – were disbanded recently.

16th Krasnoznamionnaya Vozdushnaya Armiya

16th Air Army awarded with Red Banner – Kubinka

The 16th Army was formed in August 1942 at Stalingrad. Some 204 Heroes of the Soviet Union fought in this army during World War II, including triple Hero of the Soviet Union Ivan Kozhedub, the highest-scoring Allied pilot. The combat history of the 16th Army ended in Berlin. After this, the army remained in Germany for several decades (from 1949-1968 as the 24th Air Army) with headquarters in Zossen. On 27 May 1994, the 16th VA was withdrawn to Kubinka, Russia. On 1 June 1998 it was disbanded and its units were incorporated in bulk into Moscow's District of the VVS and PVO. This move was soon revealed to be an error, and as early as 25 November the same year, the 16th Smeshannyi Aviatsionnyi Korpus (SAK) or Composite Air Corps was formed within the district. On 1 February 2002, the 16th SAK was again transformed, into the 16th Air Army.

The commanding officer of the 16th VA is Lieutenant General Valeriy Redunskiy (born on 6 March 1952), former commanding officer of the 16th SAK. He has more than 2,000 flying hours with 10 aircraft types.

105th Smeshannaya Aviatsionnaya Diviziya

105th Composite Air Division – Voronezh

The division was formed in July 1950. In the 1980s, as the 105th Aviatsionnaya Diviziya Istrebiteley-Bombardirovshchikov (ADIB) or Fighter-Bomber Air Division, it was stationed in East Germany with its HQ at Grossenhain, and operated MiG-27s and MiG-23BNs. In 1993, the division was withdrawn from Germany to Voronezh and then reorganised into the 105th Bombardirovochnaya Aviatsionnaya Diviziya (BAD) or Bomber Air Division with three regiments of Su-24s (the 20th, 164th and 455th BAP). At the end of the 1990s, two regiments of Su-24s were disbanded and replaced in the division structure by the 899th attack air regiment with Su-25 aircraft; the 105th division was then given its current name.

455th Bombardirovochnyi Aviatsionnyi Polk

455th Bomber Air Regiment – Voronezh (Baltimore)

The regiment came into being on 28 April 1953 at Tambov, and soon after, in July 1954, it moved to Voronezh. This instructor regiment was subordinated to Lipetsk's 4th TsBPiPLS combat training centre. From 1953-1990, the 455th regiment was equipped with bomber and reconnaissance types including the Il-28, Yak-25, Yak-27, Yak-28, MiG-21 and Su-7. In the 1990s, the regiment acquired MiG-25RB, Su-17M and Su-24 aircraft. More than 1,500 pilots were trained by this regiment, including future commanding officers of air units of the Soviet Air Forces and other countries of the Warsaw Pact. In November 1993, the regiment was transformed into a combat unit (the 445th BAP) with Su-24s and was included in the 105th Composite Air Division. The commanding officer of the 455th BAP is Colonel Vyacheslav Durasov (2003). The regiment day is 28 April.

899th Gvardeyskiy Orshanskiy dvazhdy Krasnoznamionnyi ordena Suvorova Shturmovoy Aviatsionnyi Polk imeni F.E. Dzerzhinskogo

899th Orsha Attack Air Regiment of Guards awarded with Red Banner (twice) and Suvorov order named after F.E. Dzerzhinsky – Buturlinovka

This regiment originates from an air detachment of the Tsarist Russian armed forces, formed on 18 June 1914. The regiment fought in World War I and during the Civil War in Russia, in Spain, in China and Mongolia, then against Finland and during World War II. From 1979, as the 899th Aviatsionnyi Polk Istrebiteley-Bombardirovshchikov (APIB) or Fighter-Bomber Air Regiment with MiG-21bis and MiG-21SMT aircraft, it was stationed at Lielvarde in Latvia. At the end of the 1980s it re-equipped with MiG-27Ds. In 1992 the regiment was withdrawn from Latvia to Baturlinovka in Russia and transformed into the 899th OShAP with Su-25s (inherited from the 80th OShAP withdrawn from Azerbaijan and the 357th OShAP withdrawn from Germany). The regiment took part in the Chechnya war from 1994-1996. Several Su-25 aircraft from this regiment are stationed in Tajikistan as a part of the 670th Air Group at Dushanbe.

One of three internationally renowned fast-jet display teams parented by the 237th TsPAT at Kubinka, the 'Strizhi' ('Swifts') fly the MiG-29. They were the first team to regularly use flares in their routine, although this was swiftly adopted by the 'Russian Knights'.

This Su-27UB wears the patriotic colours of the 'Russkiye Vityazi' ('Russian Knights') display team, which has performed in several overseas countries. In addition to its Su-27/27UBs, the team also flies Su-27Ms. The 237th TsPAT also has an L-39 team.

14th Gvardeyskiy Leningradskiy dvazhdy Krasnoznamionnyi ordena Suvorova Istrebitelnyi Aviatsionnyi Polk imeni Zhdanova

14th Leningrad Fighter Air Regiment of Guards awarded with Red Banner (twice) and Suvorov order and named after Zhdanov – Kursk (Khalino)

The regiment was formed in 1938 as the 7th IAP, and on 7 March 1942 it was ranged among the Guards and given the new number of 14. During World War II it flew Yak-1 and Yak-9D fighters. After the war, the regiment was stationed in Ungru, Estonia. In 1957, after the suppression of the Hungarian uprising, the unit was transferred to Hungary where it remained until 1989 – first at Kolosca (until 1960) and then at Kiskunlachaza air base. By the end of the 1980s, the regiment was flying MiG-23MLD fighters. After its withdrawal to the USSR it was stationed at Zherdievka, where it was rearmed with MiG-29 fighters. In 1998, the 14th IAP and its aircraft moved 360 km (224 miles) west to Khalino airfield near Kursk (Khalino had formerly housed the MiG-23P-equipped 472nd IAP, which was disbanded on 1 May 1998). The commanding officer of the 14th IAP is Colonel Aleksandr Isakov (2003).

28th Gvardeyskiy Leningradskiy ordena Kutuzova Istrebitelnyi Aviatsionnyi Polk

28th Leningrad Fighter Air Regiment of Guards awarded with Kutuzov order – Andreapol

The unit was founded as the 153rd IAP shortly before World War II. For services at the front, on 22 November 1942 it was ranged among the Guards and received the new number of 28. At this time, it flew Yak-9 fighters. From November 1950 until March 1951, the regiment fought in Korea. In the 1960s, it moved to Andreapol and was equipped with Su-9 interceptors. In 1994, the regiment was disbanded and the MiG-23P fighters it was then flying were sent to a storage facility.

At the same time, the number and traditions of the 28th IAP were taken by another unit, the 33rd IAP, which had withdrawn to Andreapol from Germany. Prior to 7 April 1994 the 33rd IAP was stationed at Wittstock, East Germany, and had operated MiG-29s since January 1986. The 28th IAP of Guards from Andreapol should not be confused with another unit, the 28th IAP PVO (not ranged among the Guards), which until the beginning of the 1990s was stationed in Krichev, Belarus, and was armed with MiG-25P interceptors.

47th Gvardeyskiy Borisovsko-Pomeranskiy dvazhdy Krasnoznamionnyi ordena Suvorova Razvedyvatelnyi Aviatsionnyi Polk

47th Borisov-Pomeraniya Reconnaissance Air Regiment of Guards awarded with Red Banner (twice) and Suvorov order – Shatalovo

The unit was formed in July-August 1941 as the 2nd Aviatsionnyi Polk Razvedki (Reconnaissance Air Regiment); on 8 February 1943 it was ranged among the Guards with the new number of 47. During World War II, the regiment used Pe-2, Pe-3, Il-4, Tu-2 and other aircraft. Right after the war, the unit remained in Poland, at Torun and Inowroclaw; in April 1946 it moved to the USSR and, from June 1946, was based at Klokovo near Tula. In 1954 the regiment received reconnaissance Il-28Rs and moved to the larger airfield at Migalovo near Tver'. A move in March 1959 took the regiment to Shatalovo airfield, where it remains.

In 1957 the regiment received Yak-25R reconnaissance aircraft, then Yak-27Rs (1959) and Yak-28Rs (1973). A new chapter in its history began in 1970, when it accepted MiG-25Rs; in 1971 and 1973, several of its MiG-25s were stationed in Egypt and flew reconnaissance missions over Israeli territory. The first Su-24MR came to the regiment in 1973.

In 1987, the Air Forces brought together all the aircraft from different reconnaissance regiments, causing the 47th ORAP to exchange its MiG-25s for more Su-24MRs. However, this move was not considered a success and, one year later, the MiG-25s came back to Shatalovo. From 1988-1993, the regiment was transformed into the 1046th TsBPiPLS combat training centre of reconnaissance aviation and gained a squadron of Su-17M3R and M4R aircraft. The unit reassumed the 47th RAP designation on 1 November 1993, and since 1994 a group of its aircraft has taken part in Chechnya operations.

In 1998 the unit adopted the titles of the disbanded 871st IAP fighter air regiment, including the honorary title 'Pomeranskiy' and the second Red Banner order. Presently, the regiment has two squadrons, one with MiG-25RBs and one with Su-24MRs.

237th Gvardeyskiy Proskurovskiy Krasnoznamionnyi ordenov Kutuzova i Aleksandra Nevskogo Tsentr Pokaza Aviatsionnoy Tekhniki imeni Marshala Aviatsii I.N. Kozheduba

237th Proskurov Air Technology Demonstration Centre of Guards awarded with Kutuzov and Aleksandr Nevskiy orders named after Air Marshal I.N. Kozhedub – Kubinka

This is one of the most elite units of the Russian Air Forces. It originates from the 19th Independent Fighter Air Regiment (OIAP) formed on 22 March 1938 at Gorelovo near Leningrad. The unit, formed from pilots who had fought in defence of the Spanish Republic, was equipped with I-15bis, I-153 and I-16 fighters. The regiment then saw action against Poland in late September 1939, in the Winter War against Finland and, from June 1941, against German aggression. The regiment completed its World War II combat over Berlin flying La-5 and La-7 fighters. On 19 August 1944, for courage in battle, the 19th IAP was renamed the 176th IAP of Guards.

The regiment returned to Russia in 1946 and was deployed at Tyoplyi Stan air base, near Moscow. In November 1950 some of its aircraft went to the Korean War; the remainder were formed into a new unit, the 234th IAP, which took over the colours, awards and tradition of the 176th IAP. In its final move, at the beginning of 1952, the regiment was sent to Kubinka air base where it remains. The next change to the unit's name came on 15 January 1989, when it was renamed as the 237th Smeshannyi Aviatsionnyi Polk (Pokaznoy) or Composite Air Regiment (Demonstration). On 13 February 1992, the regiment received its present name of the 237th TsPAT, although on 10 August 1993 that was extended further with "named after Air Marshal I.N. Kozhedub".

The 237th TsPAT has three aerobatics teams: 'Russkiye Vityazi' with Su-27s (since 2003, also five Su-27Ms), 'Strizhi' with MiG-29s, and 'Nebesnye Gusary' with L-39Cs (formerly Su-25s). The regiment day is 22 March and the commanding officer is Colonel Anatoliy Omelchenko (2003).

The 47th RAP is the VVS's primary reconnaissance unit, flying both the MiG-25RB 'Foxbat' (illustrated) and the Su-24MR 'Fencer'. The regiment is based at Shatalovo, close to the border with Belarus.

Kubinka is home to the 226th OSAP, which flies electronic warfare missions with a variety of converted types. Above is an Mi-8PPA, while below is an An-12PPS.

KSN HQ: Moscow

UNIT (1ST LEVEL)	UNIT (2ND LEVEL)	BASE	INVENTORY/NOTES
1st KPVO		Balashikha	surface-to-air missiles only
16th VA		Kubinka	
	105th SAD	Voronezh	
	455th BAP	Voronezh	Su-24
	899th ShAP	Buturlinovka	Su-25
	14th IAP	Kursk (Khalino)	MiG-29
	28th IAP	Andreapol	MiG-29
	47th RAP	Shatalovo	MiG-25RB/RU, Su-24MR
	237th TsPAT	Kubinka	MiG-29, Su-27, Su-27M, L-39C
32nd KPVO		Rzhev	
	611th IAP	Dorokhovo	Su-27, MiG-31
	790th IAP	Khotilovo	MiG-31, MiG-25U
226th OSAP		Kubinka	Mi-8, Mi-9, An-12, An-24, An-26, An-30
5th ODRAO		Voronezh	An-30
Army Aviation component			
45th OVP		Oreshkovo (Vorotinsk) near Kaluga	Mi-24
440th OVP BU		Vyaz'ma	Mi-24, Mi-8
490th OVP BU		Klokovo near Tula	Mi-24, Mi-8
865th BRV		Protasovo near Ryazan	Mi-24, Mi-8

611th Peremyshlskiy Krasnoznamionnyi ordena Suvorova Istrebitelnyi Aviatsionnyi Polk

611th Peremyshl Fighter Air Regiment awarded with Red Banner and Suvorov order – Dorokhovo near Bezhetsk

During World War II the regiment flew I-153, Yak-1, Yak-9 and Yak-3 fighters. In the 1950s it was equipped with Yak-25 interceptors and then for a number of years with Su-15s (even in 1992, the regiment had 39 of these aircraft). In 1993 the regiment received Su-27 fighters and, in 2001, a few MiG-31s which were formed into a separate squadron. The commanding officer of the 611th IAP is Colonel Aleksandr Gorbaniov (2004).

790th ordena Kutuzova Istrebitelnyi Aviatsionnyi Polk

790th Fighter Air Regiment awarded with Kutuzov order – Khotilovo

The regiment was formed in August 1942. During World War II it operated in the Caucasus with LaGG-3 and La-5 fighters. It was the first combat unit to receive Su-15 interceptors, in 1965; it later converted to the MiG-25 (in 1992 it still had 38 MiG-25s). In 1994 it received MiG-31 interceptors, and after 2000, the regiment also took over at least some of the MiG-31 aircraft from the disbanded 153rd IAP regiment in Morshansk. The regiment's commander (2004) is Valeriy Knysh.

226th Otdelnyi Smeshannyi Aviatsionnyi Polk

226th Independent Composite Air Regiment – Kubinka

The 226th OSAP moved to Kubinka air base from Sperenburg, Germany, in 1991, together with the 16th Air Army. The regiment operates various auxiliary aircraft, including transport types An-12, An-24 and An-26 (including An-26RT radio relay aircraft), Mi-8 and Mi-9 executive helicopters, An-30 reconnaissance (aerial photography) aircraft, and others. In 1998/99, the 226th OSAP absorbed the former 297th squadron with its electronic warfare Mi-8PPAs, Mi-8SMVs, Mi-8MTPBs and others.

5th Otdelnyi Dalniy Razvedyvatelnyi Aviatsionnyi Otriad

5th Independent Long-range Reconnaissance Air Detachment – Voronezh

This unit was stationed at Belaya near Irkutsk, and used one or two An-30 reconnaissance/ aerial photography aircraft in Afghanistan operations during the 1979-1989 war. Since 1992 it has been stationed at Voronezh. During the first Chechnya war of 1994-1996, and probably after, the 5th ODRAO's aircraft operated over that territory.

45th Otdelnyi Vertoliotnyi Polk

45th Independent Helicopter Regiment – Oreshkovo (Vorotinsk) near Kaluga

This regiment was stationed at Weimar-Nohra, Germany, as the 336th OVP with Mi-8s and Mi-24s. In August 1992 the regiment moved to Oreshkovo in Russia and was incorporated into the 22nd Army of the Ground Forces of the Moscow Military District. In 2000 or 2001, the 336th OVP was redesignated as the 45th OVP. From January 2003, together with the whole of army aviation, the regiment was incorporated into the VVS.

440th Otdelnyi Vertoliotnyi Polk Boyevogo Upravleniya

440th Independent Helicopter Regiment for Battle Control – Vyaz'ma

From July 1992 the 440th OVP BU was stationed at Stendal-Borsel in Germany with Mi-8, Mi-9 and Mi-24 helicopters, including special version Mi-24Ks and Rs. After the move to Russia, it was integrated into the 20th Army of the Ground Forces of the Moscow Military District.

490th Otdelnyi Vertoliotnyi Polk Boyevogo Upravleniya

490th Independent Helicopter Regiment for Battle Control – Klokovo near Tula

Before being incorporated into the Air Forces, the 490th OVP BU was subordinated to the 20th Army Corps of the Moscow Military District. The regiment has been stationed in Klokovo for several decades.

865th Baza Rezerva Vertoliotov

865th Helicopter Reserve Base – Protasovo near Ryazan

This unit is the former 225th OVP helicopter regiment which was stationed at Allstedt, Germany, with Mi-24s and Mi-8s. In April 1991 the regiment moved to Protasovo, Russia, where it had existed for several years. About 1998, the 225th regiment was disbanded and replaced by the 865th BRV reserve base, which took over the regiment's helicopters.

Above: Among the 226th OSAP's varied fleet of specialist aircraft is the An-30, a version of the An-26 used for photo-survey. This type is also by Russia as its 'Open Skies' platform.

Another Antonov 'special' on the 226th OSAP roster is the An-26RT, easily distinguished by its large aerials. This version has a radio relay role, extending communications, and was first assigned to the regiment when it was based at Sperenburg in East Germany.

PHOTO FEATURE

Grupo 6 de Caza

Base Aérea Militar Tandil

Photographed by Chris Knott and Tim Spearman

Situated 300 km (186 miles) to the south of Buenos Aires, BAM Tandil was opened in 1948 and initially housed South America's first operational jet fighters in the shape of the Meteor F.Mk 4. After the Meteors moved, the base housed an anti-aircraft artillery unit, before being reactivated in 1979 to operate IAI Daggers. Today it is home to two squadrons of Mirages, Daggers and Fingers, forming the operational element of VI Brigada Aérea.

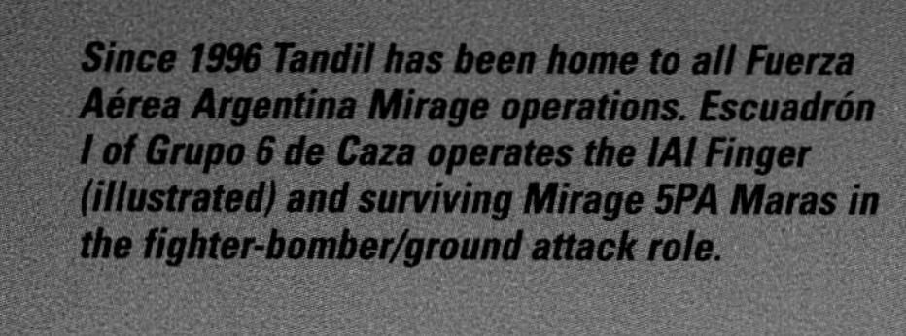

Since 1996 Tandil has been home to all Fuerza Aérea Argentina Mirage operations. Escuadrón I of Grupo 6 de Caza operates the IAI Finger (illustrated) and surviving Mirage 5PA Maras in the fighter-bomber/ground attack role.

Preserved at Tandil

Meteor F.Mk 4 I-057 is preserved as a memorial to the type's association with the base. Meteors entered service at Tandil with Regimiento 4 de Caza Interceptora from March 1948. In June 1949 Regimiento 6 also began flying the Meteor, both units forming VI Brigada Aérea. In January 1951 the two units were renamed as Grupo 2 and 3 de Caza, shortly before moving to VII Brigada at Morón. Some spare airframes remained in storage at Tandil for a while after the move.

Considerably modified and painted to represent a Dagger, Mirage IIICJ C-711 is displayed outside the main entrance to the base. After the 1982 war 19 IIICJs and three IIIBJs were acquired from Israel for service with Grupo 4's Escuadrón 55 at Mendoza and Escuadrón 10 of Grupo 10 at Rio Gallegos. They were withdrawn from service in 1991.

Dassault Mirage IIIEA
The first true supersonic fighter for the Fuerza Aérea Argentina was the Dassault Mirage IIIEA, the first 10 of which were delivered in 1972 along with two IIIDA two-seaters. A further seven IIIEAs were added in 1979, plus two IIIDAs in December 1982, and the type remains in use with II Escuadrón of Grupo 6 in the air defence role. Initially operating from BAM Mariano Moreno on the outskirts of Buenos Aires with Grupo 8, the Mirages relocated to BAM Tandil in March 1988, partly as a result of budgetary constraints and partly because of the encroachment of development around their former base. Formerly I and II Escuadrones de Caza Interceptora of Grupo 8, the Mirage IIIEA squadrons initially became III and IV Escuadrones of Grupo 6, before further consolidation saw them reduced to one squadron. I-018 (above) still wears the special markings applied in 2002 to celebrate 30 years of Mirage operations.

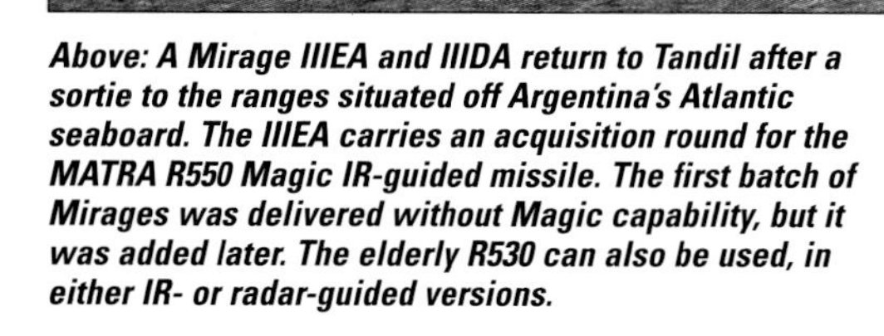

Above: A Mirage IIIEA and IIIDA return to Tandil after a sortie to the ranges situated off Argentina's Atlantic seaboard. The IIIEA carries an acquisition round for the MATRA R550 Magic IR-guided missile. The first batch of Mirages was delivered without Magic capability, but it was added later. The elderly R530 can also be used, in either IR- or radar-guided versions.

Right: A freshly painted Mirage IIIEA taxis from the Tandil flightline. Grupo 6 has 11 Mirage IIIEAs remaining in service from 17 delivered.

IAI Finger A

Under a programme code-named Dagger, 26 ex-Israeli IAI Neshers (Mirage 5s) were acquired in 1978 to counter a perceived territorial threat from Chile to islands in the south of the country. They were joined in 1981 by a further 13 aircraft for a total of 39. Of these, 35 were single-seat Dagger As (Nesher S) and four were two-seat Dagger Bs (Nesher T). The Nesher was an unlicensed copy of the Mirage 5J/DJ and its Atar 09C engine, and was originally intended as a day fighter with Shafrir missiles. Its seven hardpoints made it a useful fighter-bomber, and it was in this role that it primarily served during the 1982 South Atlantic War. Following the conflict, the surviving Dagger As were upgraded to Finger standard. This involved the installation of Kfir C7 avionics, including a Kfir-style nose housing Elta EL-2001 ranging radar. A HUD, INS and RWR were also installed in a three-phase update programme. Around 20 are still in use, flying alongside the handful of Mirage 5PA Maras in Escuadrón II. All of the Mirage force now wears a toned-down overall grey scheme with low-visibility national insignia and unit marks.

The Daggers of Grupo 6 were heavily involved in the Falklands conflict, flying their missions from San Julian and Rio Grande in Patagonia. Eleven aircraft and six pilots were lost in the conflict, the latter being honoured by the six stars in the unit's badge. Many of the unit's aircraft still carry mission marks to commemorate their involvement, C-415 (above) carrying a ship silhouette to signify the mission in which it damaged HMS Brilliant on 21 May, while C-412 (below) also wears markings for the attack on HMS Brilliant as well as one for an attack on HMS Arrow on 1 May. Both ships sustained 30-mm cannon damage during the attacks. Grupo 6's war veterans also carry a silhouette of the Islas Malvinas/Falkland Islands on the engine intake sides.

Dassault Mirage IIIDA (above)
The long-serving Mirage IIIDA has been in service since the first pair was delivered in 1971, initially flying with VII Brigada Aérea at Mariano Moreno. Two more were delivered in December 1982, but one of the original aircraft was lost in 1979. Today the three survivors remain in the training role with II Escuadrón. The IIIDA was based on the French Mirage IIIBE – which partnered the IIIE multi-role single-seater – and introduced a simplified nose with camera gun port. It was built with a number of designations in the IIIDx and 5Dx range, but all were essentially similar.

IAI Dagger B (left and below)
Training for I Escuadrón's Finger/Mara pilots is performed using the Dagger B, of which four were delivered and three remain in use. As the aircraft did not undergo the Finger upgrade, they retain the Dagger name. They can be distinguished from the Mirage IIIDAs by having a black-tipped nose.

Dassault Mirage 5PA Mara
Although it was widely reported that the 10 Mirage 5Ps acquired from Peru were acquired as a 'gift' to make good war losses, they were actually purchased in December 1981 for US$50 million, but not delivered until just after the 1982 war. They were initially used by Grupo 6 as replacement for lost Daggers (even adopting the serials of 10 of the 11 lost aircraft), but following rerfurbishment to 5PA Mara standard (not to be confused with the Pakistani Mirage 5PA) they were reassigned to Escuadrón 10 at Rio Gallegos as part of Grupo 10. This was named the 'Cruz y Fiero' (fire and cross) squadron. Operations at Rio Gallegos came to an end in 1996 and the Mirage 5PAs rejoined Grupo 6 at Tandil, forming III Escuadrón. Further consolidation saw all the fighter-bomber Mirages pooled in I Escuadrón. The Mara upgrade was performed locally by Aerocuat and brought the aircraft in line with the Daggers. RWR, VLF Omega and Dagger noses were fitted, and the aircraft were wired to carry the Shafrir missile.

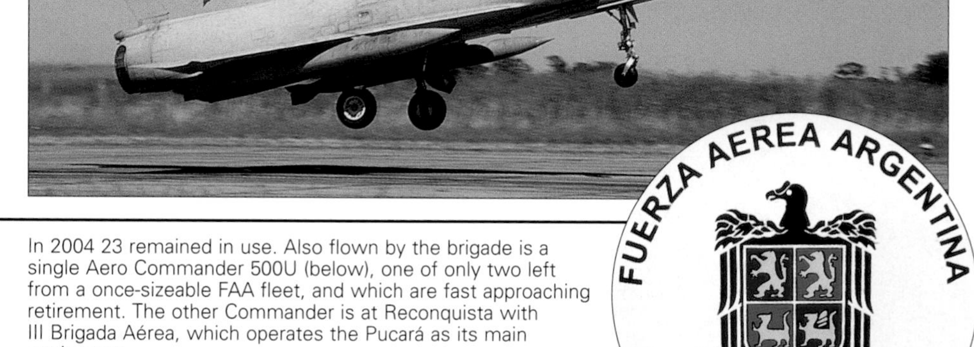

FUERZA AEREA ARGENTINA
VI Brigada Aérea

VI Brigada Aérea

In common with other FAA brigades, VI Brigada Aérea has a base flight (Escuadrilla de Servicios) which operates the FMA-built Cessna A182 (below) on general liaison duties. The type was introduced to FAA service in 1966 and 40 were delivered. In 2004 23 remained in use. Also flown by the brigade is a single Aero Commander 500U (below), one of only two left from a once-sizeable FAA fleet, and which are fast approaching retirement. The other Commander is at Reconquista with III Brigada Aérea, which operates the Pucará as its main equipment.

Bristol Blenheim
Wartime workhorse

Designed as a high-speed bomber whose speed would confer protection against (slower) fighters, the Blenheim's performance was completely inadequate by the time it went to war, and the aircraft proved horrifyingly vulnerable. Large numbers had been procured to form the backbone of the RAF's expansion and, as a result, the type bore the brunt of early operations, and suffered correspondingly heavy losses. The Blenheim's poor performance and vulnerability placed its crews in 'harm's way' as a matter of routine, and there were numerous instances of great heroism as young aircrew 'pressed on' with their missions. In doing so, often in the face of impossible odds, no fewer than three Blenheim pilots won the Victoria Cross – Britain's highest award for gallantry. Bristol and various licencees produced an eventual total of 4,422 Blenheims, and the type saw service in every theatre.

By the time the improved Blenheim Mk IV entered service the aircraft had lost its decisive speed advantage over contemporary fighters. Insufficient defensive armament, and the mistaken tactic of sending RAF Blenheim squadrons on unescorted daylight raids, led to mounting losses. A massacre of the Blenheim force was only prevented by the skill of the aircrew and the aircraft's ability to absorb a great deal of punishment. The RAF rapidly withdrew the type from UK-based frontline light-bomber units, but the type continued to serve with home-based operational training units such as this trio of Mk IVs from No. 13 OTU, based at Bicester.

The Blenheim was developed from a twin-engined, six-passenger transport monoplane which was designed as a rival to the Douglas DC-1, the first of which was ordered by the younger of the Harmsworth brothers, Lord Rothermere, the intensely air-minded proprietor of the *Daily Mail*, who had served as the first Secretary of State for the Royal Air Force under Lloyd George.

Although Rothermere wanted the aircraft to encourage air-mindedness, and to stimulate the use of commercial aviation, he also used it to demonstrate to the Air Ministry the inadequacy of its fighters against modern light bombers.

Legend often has it that the Air Staff beat a path to Bristol's door, desperate to procure a medium bomber from the company after seeing the performance demonstrated by Rothermere's Bristol 142 transport. In fact, the RAF was interested rather than enthusiastic. The Chief of the Air Staff wrote to Bristol pointing out that the type 'could be considered' as a medium bomber, if Bristol could supply it in reasonable numbers, and offered to test the aircraft free of charge at Martlesham Heath.

The RAF was not really interested in medium bombers, whose usefulness was felt to be limited to the Colonial Policing and Army Co-operation roles, or in tactical bombing, preferring to concentrate on the type of strategic bombing pioneered by the Great War Independent Force. The RAF felt that its role was to bomb 'root industries' including manufacturing plants, electrical power stations and coking plants, though many expected that accuracy would not be sufficient to allow this, and that industrial cities would be attacked as area targets, with the principal aim of provoking a collapse in civilian morale. This was what Trenchard famously called bombing for 'Moral effect' (mainly because he could not spell 'psychological').

The Bristol 142 was clearly unsuitable for such a bombing campaign, though in the short term the use of medium bombers did allow a rapid build up of aircrew and squadrons for the planned four-engined heavy bombers. Because the Air Ministry regarded the 142M as an interim type, Bristol were given permission to export the type to friendly governments after the RAF's urgent requirements had been met.

RAF interest

Rothermere loaned his new aircraft to the RAF for testing at Martlesham Heath, even before it could take up its allocated civil registration. The Bristol 142 impressed RAF test pilots, and the RAF asked to borrow the machine, and then to purchase it. Rothermere responded by presenting the £18,500 aircraft to the nation without charge. With a top speed of 307 mph (494 km/h) the 142 was faster than the prototype Gladiator, though no production Blenheim would ever achieve such a speed, and none would fly at all for another two years.

Bristol had already offered a convertible Model 143F model with interchangeable nose and tail sections to allow it to fulfill the passenger, freight, air ambulance and fighter-bomber roles but now proposed a dedicated 142M military bomber derivative to the Air Ministry.

The 142M bomber differed from Rothermere's *Spirit of Britain First* in having its wing raised by about 16 in (40.6 cm) to make space for bomb bays below the wing spars. The structure was beefed up, and all passenger doors and windows were removed. A bomb-aimer's station was added in the nose and a Vickers K 0.303-in (7.7-mm) machine-gun was added in the port outer wing, while a semi-retractable rear gunner's turret, containing a single 0.303-in (7.7-mm) machine-gun, was fitted just aft of the wing trailing edge. The Mercury VIS 2 640-hp (477-kW) radial engines used by the original Bristol 142 were replaced by 840-hp (626-kW) Mercury VIIIs, though these were not sufficient to outweigh the increased weight of the bomber, whose performance was slightly poorer than Rothermere's transport!

The Air Ministry placed an order for 150 of the 142Ms in August 1935, straight off the draw-

Attempts to restore a Blenheim to airworthy status for display on the airshow circuit, using various Bolingbroke Mk IVT airframes imported to the UK from Canada, have been beset by difficulties. Following the June 1987 crash of the first example returned to flying condition, only a month after its first post-restoration flight, a second example flew in June 1993. It is seen here in the markings of No. 82 Sqn shortly before it too suffered a landing accident in 2003.

By the time the Blenheim Mk I was entering frontline service in spring 1937 its major asset – speed – had been negated by the introduction of the new generation of monoplane fighters, such as the Messerschmitt Bf 109 and Hawker Hurricane. This trio of aircraft from No. 44 Squadron, RAF Waddington, is seen on a pre-war practice bombing sortie.

ing board, to meet Specification 28/35. In doing so, the Air Ministry ordered what was the first modern, all-metal, stressed-skin, cantilever-wing monoplane to be placed in production for the RAF. The aircraft was certainly modern by British standards, but claims that the Blenheim was 'at the leading edge of the technology of the day' were wide of the mark. The aircraft lacked adequate heaters for the crew, de-icing systems and modern navigation and radio equipment, while it was too cramped and cluttered to allow efficient operation, and control surfaces were still fabric-covered.

By May 1936, the new bomber had been allocated the name Blenheim, and the first of the new aircraft (which was not, strictly speaking, a prototype) made its maiden flight on 25 June 1936.

Deliveries commence

Initial deliveries were made to No. 114 (B) Squadron at Wyton in March 1937, though the first Blenheim to land was written off when the pilot braked too fiercely and flipped the machine onto its back. Further squadrons rapidly re-equipped with the new medium bomber monoplane, trading in their biplane light bomber Hawker Hinds, Harts and Audaxes. Five squadrons had re-equipped by the end of the first year. Nor was the new Blenheim restricted to home-based units. No. 30 Squadron in Iraq re-equipped with Blenheim Mk Is from January 1938, and No. 60 Squadron in India followed in March 1939. By August 1939, the Mk Is of No. 11 Squadron were operating from Singapore.

The first export customers also received their first Blenheims during 1937. Bristol built an initial batch of 18 Blenheim Mk Is for the Finnish Air Force (Ilmavoimat) in 1937-38, and the Yugoslav government purchased two more Blenheim Mk Is and a licence for Ikarus AD at Zemun to build 50. The first of Turkey's 30 Blenheim Mk Is were delivered the same year.

No-one seemed to care over-much that the production Blenheim Mk I was significantly slower than *The Spirit of Britain First* had been, nor did anyone really notice that claims that the aircraft was 'faster than contemporary fighters' had been overtaken by the emergence of aircraft like the Hawker Hurricane. Presumably the Air Ministry had failed to notice when the Messerschmitt Me 209 beat the world speed record, or assumed that the Bf 109 was no faster than a Gladiator. In any event, the belief was actively encouraged that the Royal Air Force was equipped with the best bomber in the world, and no-one mentioned the inconvenient fact that the new high-speed, single-seat, multi-gun monoplane fighters had effectively rendered the fast medium bomber obsolete.

Massive demand

Blenheim Mk I orders soon poured in, with the initial batch of 150 soon being joined by 434 more, and then by a further 134. With such massive orders, Filton's output (which reached 24 aircraft per month by December 1937), was augmented by production by Avro at Chadderton, and by Rootes Securities at Speke.

By September 1938, the Blenheim Mk I equipped 17 Bomber Command squadrons, though No. 1 Group discarded its Blenheims before going to France as part of the Advanced Air Striking Force, and No. 5 Group soon began to convert to the Hampden. Despite this loss of Blenheim units, 16 of Bomber Command's 33 squadrons flew Blenheims when war broke out, though the force was less impressive in reality than it appeared on paper.

Long-serving Bomber Command squadrons were split up to form the basis of newly forming units, with one flight from an existing squadron forming the cadre of the new squadron. The long-serving peacetime pilots were soon widely scattered, and flights would often be commanded by a Flying Officer, or even a senior Pilot Officer, who might himself have only two years of service behind him. The rest of the new squadron's officer aircrew would often be very young, and very green Volunteer Reservists. The situation was exacerbated by the fact that the two-man Hinds were being replaced by an aircraft whose crew complement required an observer/bomb-aimer, in addition to the usual light bomber crew of pilot and wireless operator/air gunner. In addition to the pilot, observer and wireless operator/gunner the Blenheim Mk I could carry one passenger or light freight in the well in the wing centre-section carry-through, and in emergency, it was possible to carry at least six passengers.

Many believed that the auxiliary squadrons were actually more combat ready than the regular squadrons, since they retained their experienced COs and flight commanders, and many of their long-serving pilots.

But while pilot quality was variable, the new squadrons did at least have improved equipment. Within Bomber Command, the B.Mk I

Progenitor of the Blenheim, the Bristol Type 142 was the aircraft which shocked the complacent UK Air Ministry of the mid-1930s into action. Outperforming any RAF fighter in service at the time of its first flight, the Type 142 was lent to the RAF for testing by its owner, Lord Rothermere. The result of the tests was a substantial order for a militarised version known as the Type 142M.

was rapidly giving way to the Blenheim Mk IV, allowing the Mk Is to be 'cascaded' to units overseas, or to be converted to the fighter role. It has frequently (and erroneously) been stated that all Bristol Blenheim Mk I bombers were out of front-line service with UK-based squadrons by September 1939. In fact 301 Blenheim Mk Is remained in home-based Bomber Command service at the outbreak of war, 168 of them in frontline squadrons. Some 243 more Blenheim Mk Is served with overseas squadrons, which used no other Blenheim variant. The bulk of the overseas Blenheims served in the Middle East, 99 equipping four squadrons in Egypt, 70 with two units in Iraq, and 21 with a single squadron in Aden. Some 53 more served with two Squadrons in India. The move of Blenheim Mk Is to overseas commands, and to Fighter Command, accelerated after war broke out.

Although the Blenheim Mk IV had originally been designed to meet a Coastal Command requirement, it marked enough of an improvement over the Mk I that it simply replaced the earlier version on the Bristol, and later the Avro and Rootes, production lines, and was delivered to Bomber Command, Army Co-operation Command and overseas units, as well as to Coastal Command squadrons, and a few were even delivered as radar-equipped night-fighters to Fighter Command.

Long-nosed Blenheims

The new Blenheim Mk IV was originally designed as an interim general reconnaissance version of the Blenheim for Coastal Command, with increased fuel tankage and a larger navigator's position. Bristol was directed to deliver as many of the 134 aircraft already ordered to the new standard without disrupting the pace of deliveries. In the event, the long nose and extra fuel tanks were phased in with the 67th aircraft of the batch.

The Blenheim B.Mk IV was fitted with Mercury XV engines, which could produce 920-hp (686-kW) in 'combat boost' (9 psi) using 100 octane fuel. This allowed Bristol to claim anew that the Blenheim was the world's fastest bomber, though it was, in fact, slower than the Junkers Ju 88, or the Douglas DB-7.

When the Blenheim Mk IV entered service in January 1939, it did so with No. 53 Squadron – then a night reconnaissance unit based at Odiham and reporting to No. 22 Group, Bomber Command. From 1 December 1940 No. 22 Group formed the backbone of the new Army Co-operation Command formed that day under the command of Air Marshal Sir Arthur Barrett. The first Blenheim Mk IV bombers were delivered to No. 90 Squadron, and by September 1939 No. 2 Group had seven Mk IV squadrons.

The introduction of the Blenheim Mk IV soon freed up Blenheim Mk Is for service overseas. Moreover, with the replacement of the Hawker Hart, Hind and Audax in the bomber and army co-operation roles, the biplane Demon fighter began to look increasingly anachronistic, and the opportunity was taken to convert some 200 redundant Blenheim Mk I bombers into long-range fighters.

No. 11 Group had wanted an extensively modified single-seat Blenheim fighter, without the heavy, high drag power-operated gun turret. Others pushed for a two-seater, with the observer in a low-drag 'Beaufighter-type' dorsal cupola. Early plans called for Blenheim fighter

RAF Blenheim Mk I colour schemes

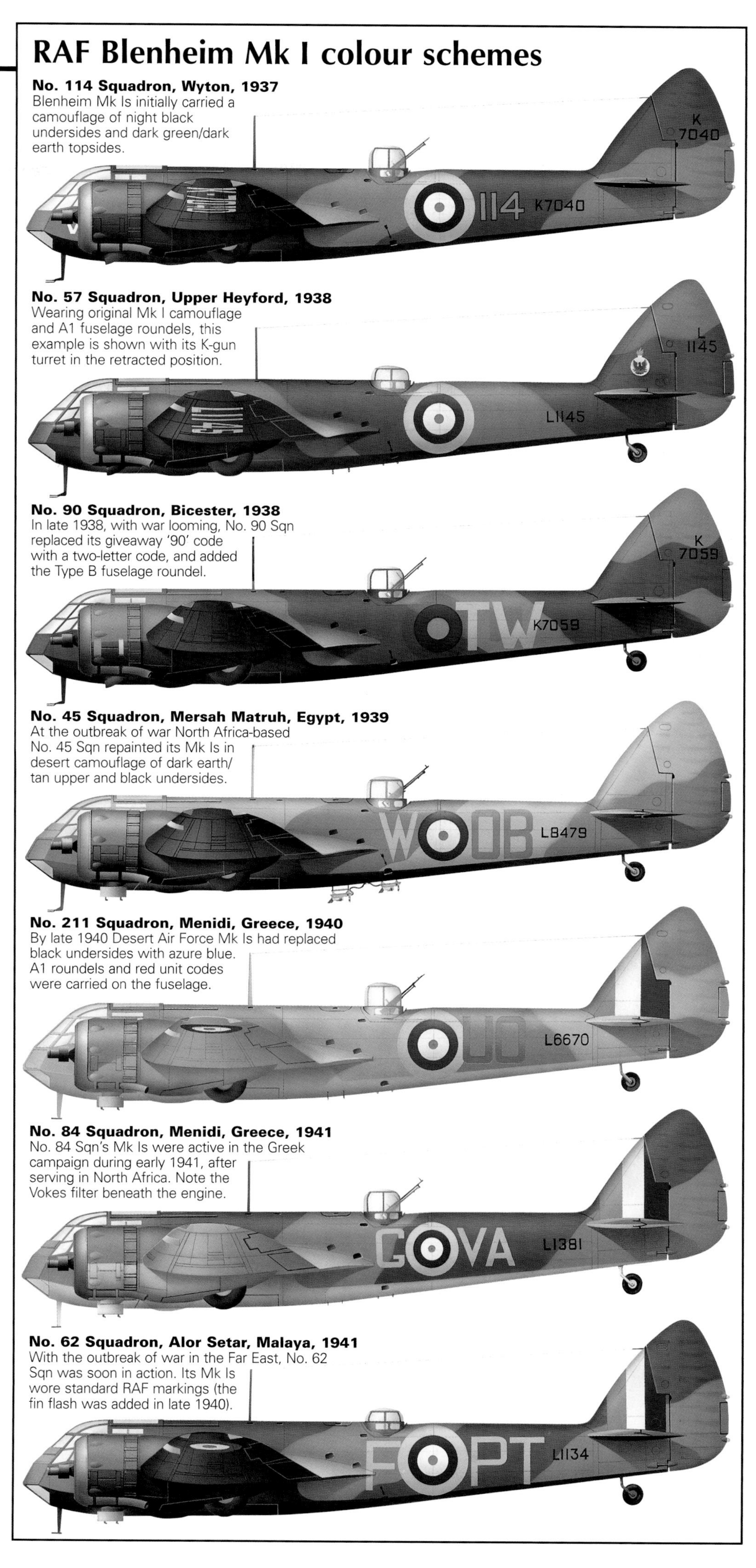

No. 114 Squadron, Wyton, 1937
Blenheim Mk Is initially carried a camouflage of night black undersides and dark green/dark earth topsides.

No. 57 Squadron, Upper Heyford, 1938
Wearing original Mk I camouflage and A1 fuselage roundels, this example is shown with its K-gun turret in the retracted position.

No. 90 Squadron, Bicester, 1938
In late 1938, with war looming, No. 90 Sqn replaced its giveaway '90' code with a two-letter code, and added the Type B fuselage roundel.

No. 45 Squadron, Mersah Matruh, Egypt, 1939
At the outbreak of war North Africa-based No. 45 Sqn repainted its Mk Is in desert camouflage of dark earth/tan upper and black undersides.

No. 211 Squadron, Menidi, Greece, 1940
By late 1940 Desert Air Force Mk Is had replaced black undersides with azure blue. A1 roundels and red unit codes were carried on the fuselage.

No. 84 Squadron, Menidi, Greece, 1941
No. 84 Sqn's Mk Is were active in the Greek campaign during early 1941, after serving in North Africa. Note the Vokes filter beneath the engine.

No. 62 Squadron, Alor Setar, Malaya, 1941
With the outbreak of war in the Far East, No. 62 Sqn was soon in action. Its Mk Is wore standard RAF markings (the fin flash was added in late 1940).

squadrons to be equipped with 14 fully-converted Blenheim Mk IEs (with cupolas) and five interim Blenheim Mk IRs. In the end, though, the fighter Blenheim conversion was more modest, with a reflector gunsight and a slab-sided ventral gun pack (built in Southern Railways' workshops) containing four 0.303-in (7.7-mm) Browning machine-guns fitted over the closed bomb bay doors. This limited the fighter Blenheim to a maximum speed of about 263 mph (423 km/h) due to the drag and weight.

With five forward-firing 0.303-in (7.7-mm) guns (the wing-mounted machine-gun was retained) the Blenheim fighter packed a more powerful punch than the Demon, Fury, Gauntlet or Gladiator but was lightly armed by comparison with the eight-gun Spitfire and Hurricane. Moreover, the 0.303-in (7.7-mm) machine-gun was inadequate against modern bombers, which could survive tens or even hundreds of hits by such lightweight rounds. Ironically, Bristol did convert a number of surplus Mk Is as fighters for Yugoslavia with a pair of forward-firing 20-mm cannon, a much more viable armament.

The first Blenheim fighters were delivered to No. 600 Sqn at Hendon in September 1938, while No. 25 Sqn re-equipped in December. Seven squadrons of Blenheim fighters were operational by the time war broke out, and these included small numbers of similarly converted Mk IVFs.

On 3 September 1939, just hours after the declaration of war, a Blenheim Mk IV of No. 139 Squadron flew the first operational sortie of the

Showing the upper fuselage entrance hatches for both pilots and gunner, this pre-war image shows a Bristol-built Blenheim Mk I, belonging to No. 90 Squadron, in early 1938. At this time the RAF used the squadron number as an identifying code (90) with individual aircraft within the squadron allocated its own letter (B). This system was abandoned later in the year as the threat of war increased. To prevent easy enemy identification of which unit aircraft belonged to, each unit was issued with a two-letter code identifier. No. 90 Squadron was initially allocated 'TW' codes, but this was rapidly replaced by the 'WP' prefix that the unit wore as war broke out in 1939. The high visibility A1 roundels (with a yellow outer segment) were also replaced by the less obvious Type B roundel.

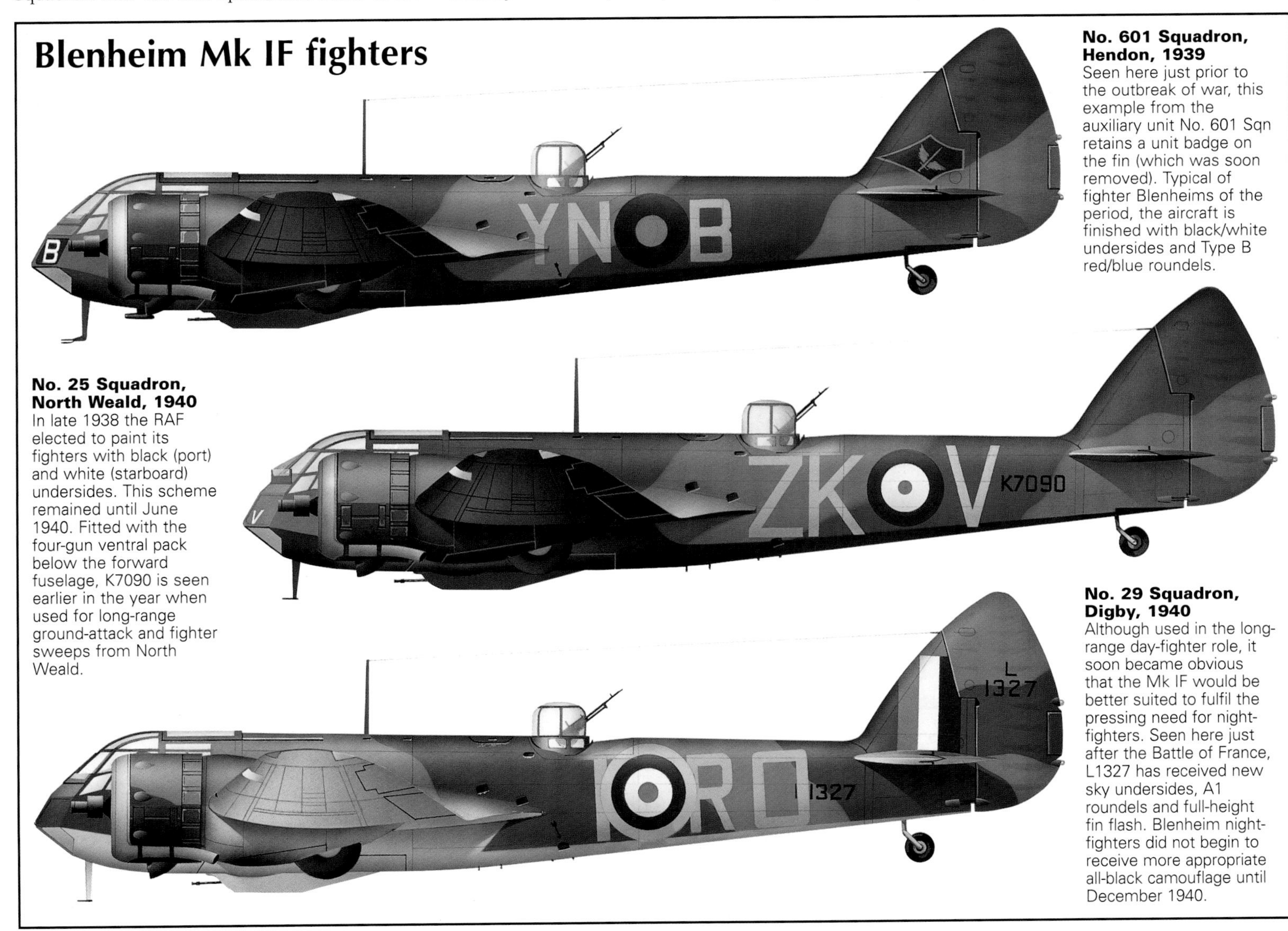

Blenheim Mk IF fighters

No. 601 Squadron, Hendon, 1939
Seen here just prior to the outbreak of war, this example from the auxiliary unit No. 601 Sqn retains a unit badge on the fin (which was soon removed). Typical of fighter Blenheims of the period, the aircraft is finished with black/white undersides and Type B red/blue roundels.

No. 25 Squadron, North Weald, 1940
In late 1938 the RAF elected to paint its fighters with black (port) and white (starboard) undersides. This scheme remained until June 1940. Fitted with the four-gun ventral pack below the forward fuselage, K7090 is seen earlier in the year when used for long-range ground-attack and fighter sweeps from North Weald.

No. 29 Squadron, Digby, 1940
Although used in the long-range day-fighter role, it soon became obvious that the Mk IF would be better suited to fulfil the pressing need for night-fighters. Seen here just after the Battle of France, L1327 has received new sky undersides, A1 roundels and full-height fin flash. Blenheim night-fighters did not begin to receive more appropriate all-black camouflage until December 1940.

Mk IF radar-equipped night-fighter

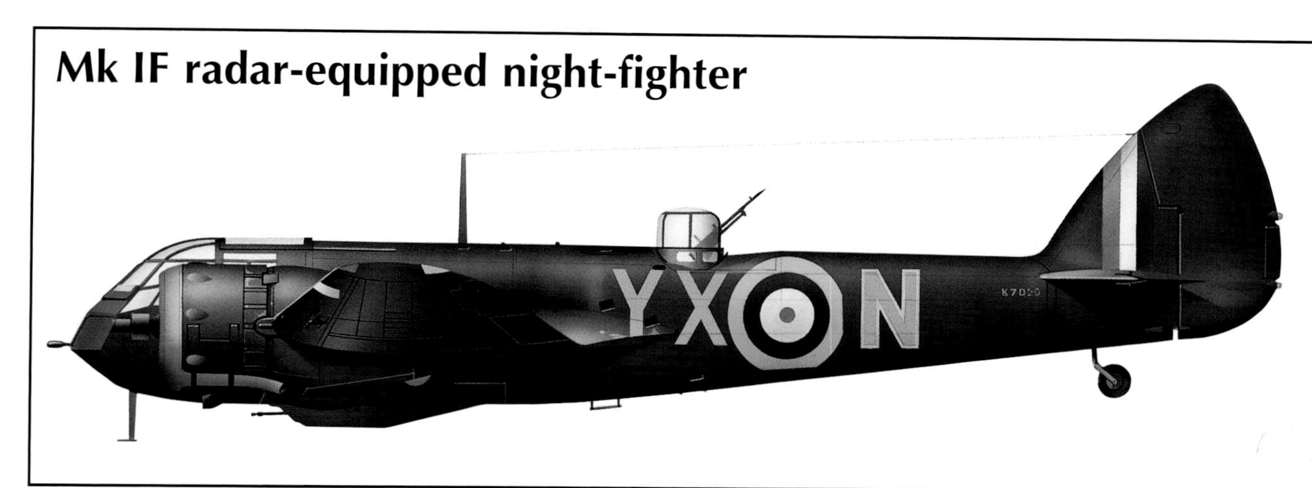

No. 54 OTU, Church Fenton, 1941
Although the Blenheim Mk IF had been operating with some successes as a night-fighter, the advent of the AI.Mk III radar (antennas on the nose and under the port wing) created much-improved interception rates. This Mk IF is seen serving with the night-fighter training unit, No. 54 OTU, in September 1941. The all-over black colour scheme was not entirely successful and was abandoned before the end of the year.

war. Flying Officer MacPherson and his crew made a solo reconnaissance of the German fleet at Wilhelmshaven, making sketches and radio reports when their cameras failed due to the freezing temperatures. The same crew returned at low level the next day, when 15 Blenheim Mk IVs of Nos 107, 110 and 139 Sqns mounted the RAF's first bombing raid of the conflict. These aircraft attacked the German fleet as it lay anchored in the Schilling Roads, hitting several enemy ships but losing six aircraft in the process.

Blenheim at war

Early operations included a daring raid by the Blenheim Mk IF fighters of Nos 25 and 601 Squadrons on the enemy seaplane base at Borkum on 28 November 1939.

During the winter months Blenheim squadrons joined the bomber element of the Advanced Air Striking Force in France, and flew regular missions into enemy territory during the so-called Phoney War. Frequently encountering enemy fighters, the Blenheims suffered heavy losses, in a grim foretaste of what was to come.

The Blenheim scored another RAF first on 11 March 1940, when Squadron Leader Delap of No. 82 Squadron sank an enemy submarine.

The Blenheims performed surprisingly well during the initial German attack on France and the Low Countries which began on 10 May 1940, taking a toll on the attacking enemy forces, though losses mounted steadily, until 200 had been lost or damaged in 20 days, and on a number of occasions only one or two aircraft returned from whole squadrons.

After Dunkirk Bomber Command's Blenheim squadrons were the only units capable of undertaking offensive operations, and were given the task of attacking enemy-held ports and rivers where invasion barges were being

***Top:** Blenheim Mk IF night-fighter K7159 (as depicted in the profile artwork) is seen during a training flight from Church Fenton in 1941 and is equipped with the standard four-gun pack. As visual bomb-aiming was not necessary for night-fighter versions, the lower part of the nose glazing was painted to prevent reflections in the cockpit.*

***Above:** No. 601 Squadron was an early operator of the Blenheim Mk IF, flying the variant between January 1939 and February 1940. This example, seen with the dorsal turret retracted, was flown by Squadron Leader Roger Bushell, the mastermind of the 'Great Escape' from Stalag Luft 3 prison in 1944. Known by the codename 'Big X', Bushell was re-captured and executed by the Gestapo.*

***Left:** A formation of No. 90 Squadron Blenheim Mk Is floats serenely over the English countryside during a pre-war exercise. By the time war erupted the squadron had re-equipped with the improved Mk IV.*

Mk Is in foreign service

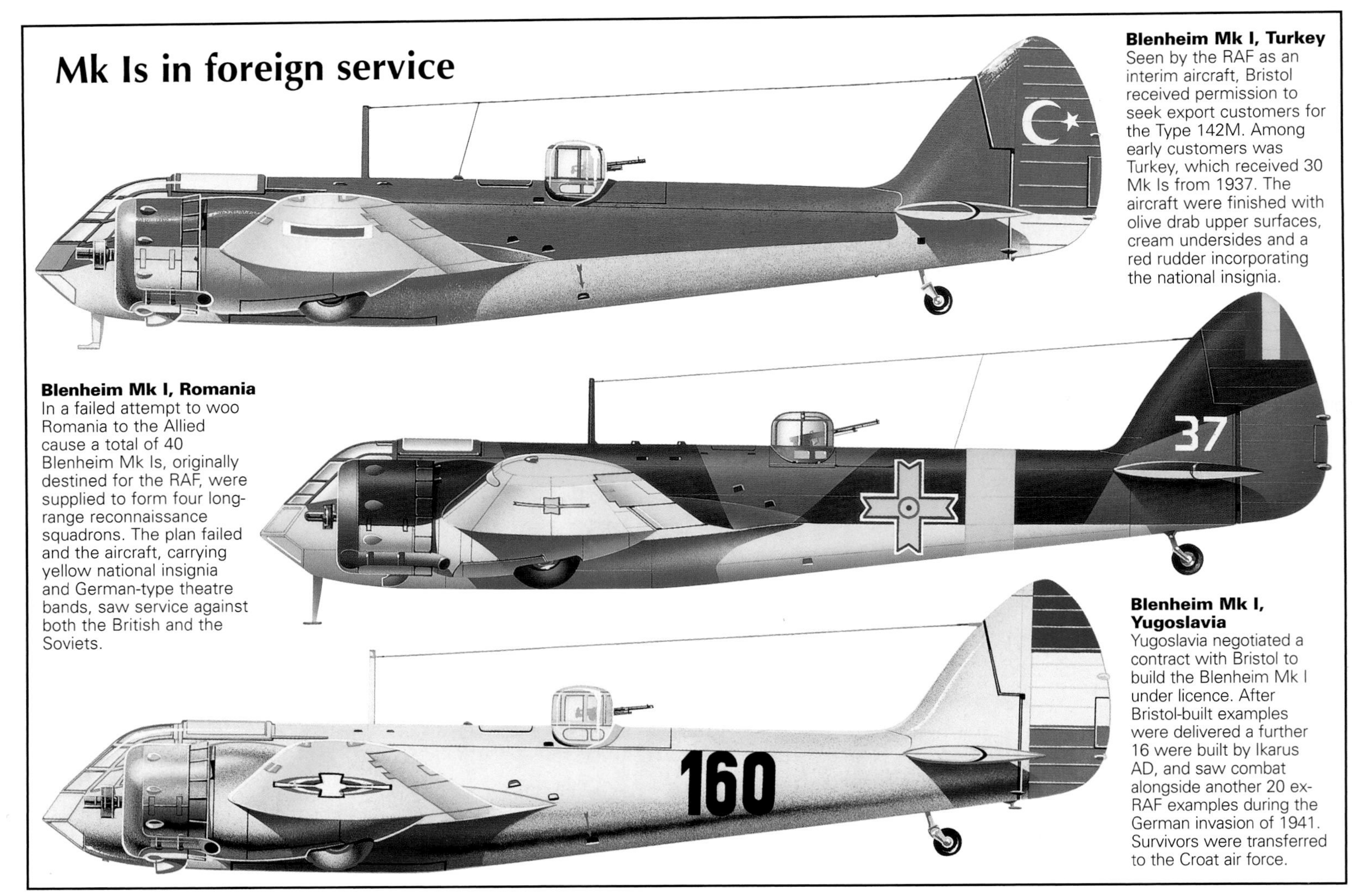

Blenheim Mk I, Turkey
Seen by the RAF as an interim aircraft, Bristol received permission to seek export customers for the Type 142M. Among early customers was Turkey, which received 30 Mk Is from 1937. The aircraft were finished with olive drab upper surfaces, cream undersides and a red rudder incorporating the national insignia.

Blenheim Mk I, Romania
In a failed attempt to woo Romania to the Allied cause a total of 40 Blenheim Mk Is, originally destined for the RAF, were supplied to form four long-range reconnaissance squadrons. The plan failed and the aircraft, carrying yellow national insignia and German-type theatre bands, saw service against both the British and the Soviets.

Blenheim Mk I, Yugoslavia
Yugoslavia negotiated a contract with Bristol to build the Blenheim Mk I under licence. After Bristol-built examples were delivered a further 16 were built by Ikarus AD, and saw combat alongside another 20 ex-RAF examples during the German invasion of 1941. Survivors were transferred to the Croat air force.

gathered for Operation Sea Lion, the planned invasion of the British Isles, and against enemy airfields. Operating at low level and often without fighter escort, the Blenheims suffered heavy losses, but maintained an offensive of sorts.

Battle of Britain

During the Battle of Britain the Blenheim fighter proved to be unequal to the Luftwaffe's fighters by day, but proved effective and useful when properly used. On 15 August, for example, three Blenheim Mk IVFs intercepted an unescorted formation of Heinkel 111 bombers and shot four of them down, and the type proved even more useful at night.

Non-radar equipped Blenheims scored a number of successes against German bombers, and the first radar-directed kill was made on 5 February 1940, when Wing Commander Farnes at Bawdsey vectored Flight Lieutenant Christopher 'Blood Orange' Smith to shoot down an He 111. Smith's aircraft was actually based at Martlesham Heath, for calibrating the Chain Home radar station at Bawdsey.

The Blenheim enjoyed greater endurance than the single-engined fighters, and was easier and safer to operate by night. No. 23 Squadron's Blenheims accounted for five Luftwaffe bombers during the German night bombing raid on London on 18 June, and from July, some of No. 600 Squadron's Blenheim Mk IF's at Manston were fitted with AI.Mk III radar. Early AI radar sets weighed around 600 lb (272 kg), and required an operator, and this in turn necessitated a twin engined aircraft to carry them. The Blenheim could carry radar, yet was agile enough to manoeuvre onto the tail of enemy night bombers. The first success came on the night of 22/23 July, when Flying Officer Ashfield and his crew from the Fighter Interception Unit made the first ever successful interception using airborne radar, and shot down a Dornier 17Z into the sea off Brighton. As radar development continued, and as high-speed Beaufighters began to arrive, Blenheim Mk IFs with earlier, obsolete AI sets accompa-

Above: One of the 30 Blenheim Mk Is supplied to Turkey is seen prior to delivery in 1937. The Turkish Blenheim force was augmented by small numbers of Blenheim Mk IVs and Mk Vs in 1943, with all three variants continuing to serve for a number of years following the end of the war.

Left: The crew of a Yugoslav Blenheim Mk I poses in front of its aircraft in 1937. Allocated to two bomber regiments and one reconnaissance group, the Blenheims suffered heavy losses to Luftwaffe fighters during 1941.

nied Blenheim bombers on offensive night intruder operations over occupied Europe.

Though costly, offensive operations over occupied Europe continued – and the Blenheims, as the least vulnerable bombers available, bore a disproportionate share of the burden. The first 'Circus' operations were mounted in January 1941, using very heavily escorted Blenheim raids to draw enemy fighters into the air to be destroyed by the escorting Spitfires. The tactic proved disappointing and costly, accounting for only 103 enemy aircraft (and not the 731 claimed!) while 123 RAF fighter pilots were lost in one six-week period alone.

Painful sacrifice

From March until November 1941 the Blenheim squadrons of No. 2 Group mounted even more costly 'Roadstead' and 'Channel Stop' operations against enemy ships, surface raiders and U-boats. No. 2 Group's commander had to be directly ordered to conserve his aircraft and crews, after failing to take the hint when Churchill remarked that the famously wasteful and unproductive Charge of the Light Brigade was 'eclipsed in brightness by the almost daily deeds of fame' by No.2 Group's Blenheim units.

Over land and sea, the Blenheims operated at very low-level, and one aircraft returned to base having killed an enemy soldier during a low level turn, knocking him off his bicycle with an ominously dented wingtip. During anti-ship attacks, enemy anti-aircraft gunners frequently had to hold their fire to avoid hitting other ships, and more than one Blenheim limped home with damaged propeller tips after glancing off the sea surface.

When Blenheim Mk Is were replaced in Bomber Command squadrons by newer Blenheim Mk IVs, many of those not converted to fighter configuration were cascaded to overseas units. RAF Blenheims fought a fierce rear guard action against advancing Italian forces after Mussolini declared war on 10 June 1940, as General Wavell's British forces retreated to Egypt. Three Blenheim squadrons were deployed to Greece to help counter the Italian invasion in October 1940 augmenting 12 Greek air force Blenheims. The Greek campaign slipped away from the Allies after Germany joined the campaign in April 1941, and the Blenheims covered the British withdrawal to Crete, and then back to North Africa. From April 1941, deployed Blenheim squadrons flew missions from the vital island fortress of Malta, attacking enemy shipping, as well as targets in North Africa and Italy.

There were four Blenheim squadrons (one operating in the night-fighter role) within the RAF's Far East Air Force when Japan launched its attack against the Malayan peninsula and Singapore on 8 December 1941, the day after the attack on Pearl Harbor. The Blenheims were quickly depleted by enemy airfield attacks, and more were lost to prowling enemy fighters.

In the North Nos 27, 34 and 62 Sqns were based at Sungei Patani, Tengah and Alor Setar respectively. To cover the southeast region No. 60 Sqn was based at Kuantan, equipped to serve in the night-fighter role.

The Japanese assault began with a feint attack at Kota Bahru to divert the defenders attention from the main landings at Singara in Siam.

RAF Blenheim Mk IVs

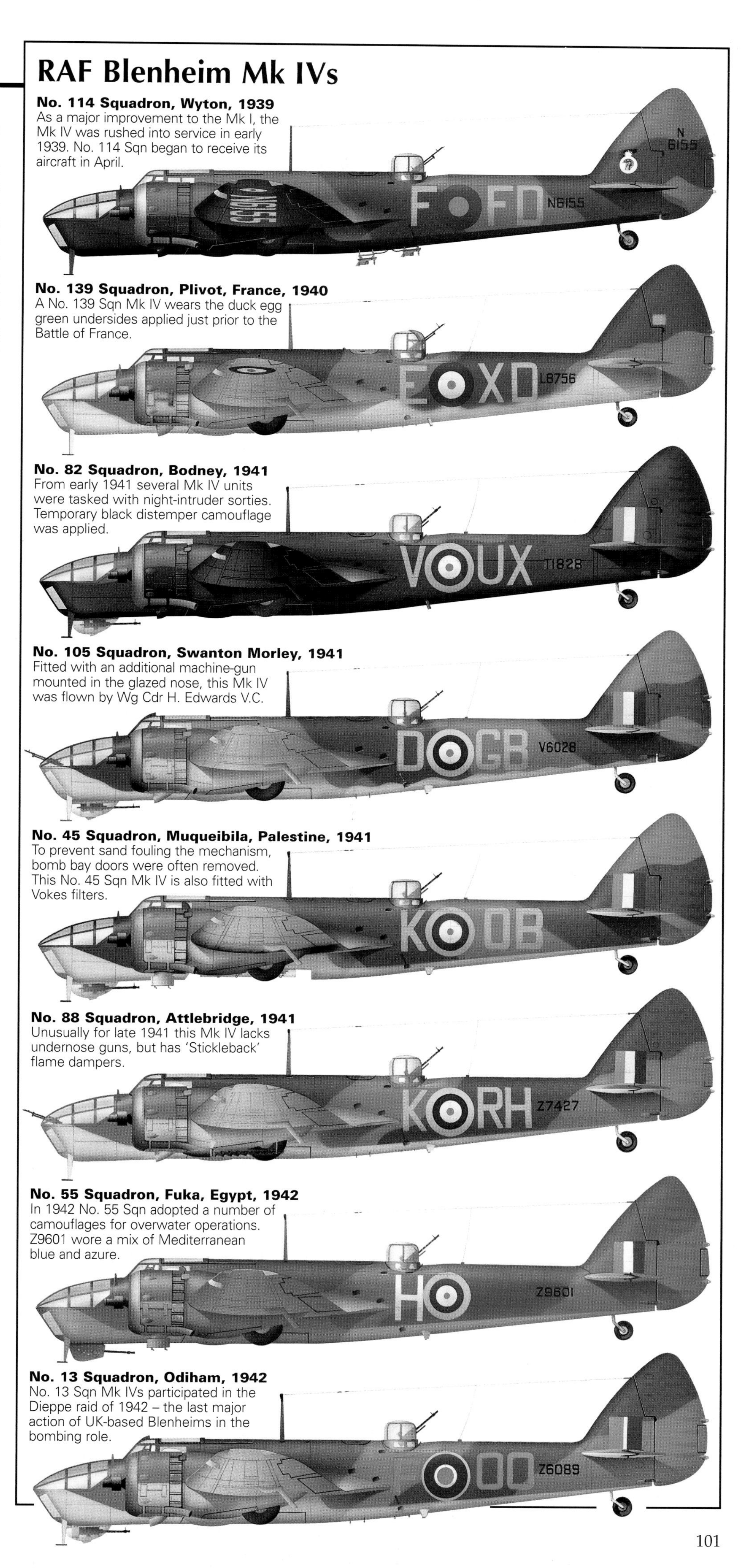

No. 114 Squadron, Wyton, 1939
As a major improvement to the Mk I, the Mk IV was rushed into service in early 1939. No. 114 Sqn began to receive its aircraft in April.

No. 139 Squadron, Plivot, France, 1940
A No. 139 Sqn Mk IV wears the duck egg green undersides applied just prior to the Battle of France.

No. 82 Squadron, Bodney, 1941
From early 1941 several Mk IV units were tasked with night-intruder sorties. Temporary black distemper camouflage was applied.

No. 105 Squadron, Swanton Morley, 1941
Fitted with an additional machine-gun mounted in the glazed nose, this Mk IV was flown by Wg Cdr H. Edwards V.C.

No. 45 Squadron, Muqueibila, Palestine, 1941
To prevent sand fouling the mechanism, bomb bay doors were often removed. This No. 45 Sqn Mk IV is also fitted with Vokes filters.

No. 88 Squadron, Attlebridge, 1941
Unusually for late 1941 this Mk IV lacks undernose guns, but has 'Stickleback' flame dampers.

No. 55 Squadron, Fuka, Egypt, 1942
In 1942 No. 55 Sqn adopted a number of camouflages for overwater operations. Z9601 wore a mix of Mediterranean blue and azure.

No. 13 Squadron, Odiham, 1942
No. 13 Sqn Mk IVs participated in the Dieppe raid of 1942 – the last major action of UK-based Blenheims in the bombing role.

Finnish Blenheim Mk IIs and Mk IVs were used to great effect, harrying invading Soviet forces during the Winter War of 1940. Here a Mk IV of Lentolaivue 46 is refuelled during operations against Soviet supply lines in March 1940. In 1944 they would see action against a different enemy as German forces were driven from the country.

No. 27 Squadron was dispatched to attack the enemy forces landing at Kota Bahru, but failed to locate the target in the heavy rain. Shortly after landing at their base they were subjected to an attack by enemy bombers. Eight Blenheims were destroyed on the ground and the scenario was repeated at other airfields, leaving the RAF strength heavily depleted in the opening moves of the battle. The remaining Blenheims were withdrawn to Kallang airport, Singapore. Six Blenheims of No. 34 Sqn mounted a desperate raid on an enemy airfield at Singara. Personnel drawn from the remnants of No. 60 Sqn manned half the aircraft and the range to the target was too great for the aircraft to return to base. Five Blenheims were shot down into the jungle and one landed back at Tengah.

Just before Singapore fell, in February 1941, a handful of surviving Blenheims withdrew to Sumatra, joining a squadron just sent out from the Middle East. The survivors of this force were forced to withdraw again within days, flying to Java, where the surviving RAF aircrew destroyed their aircraft and escaped by boat.

Several Blenheim squadrons were rushed out from the Middle East to India, where they played a vital role in the retreat from Burma, the defence of Ceylon, and in the Arakan campaign, which demonstrated that ground could be re-taken from the Japanese (smashing the myth of Japanese invincibility), and which marked the first faltering steps in the campaign which eventually drove the Japanese back to their homeland, and defeat.

The definitive Mk V

The Blenheim faded from home-based Bomber Command squadrons during 1942, with No. 2 Group re-equipping with Bostons, Venturas and Mitchells. The last Bomber Command Blenheim missions were flown in August, and Coastal Command retired the type in September. It had briefly appeared as though the last Bomber Command Blenheim Mk IVs might be replaced by the newer Blenheim Mk V, and No. 139 Squadron received the new version in June 1942. The Blenheim Mk V was underpowered, and was soon assessed as being unsuitable for operations over Europe. No. 139 Squadron withdrew its Mk Vs and replaced them with de Havilland Mosquitos, while four further Blenheim V units formed, but were immediately despatched to North Africa. Nos 13, 18, 114 and 614 Sqns flew to Algeria via Gibraltar as part of Operation Torch and supported the Allied landings in North Africa, but losses were heavy, forcing the aircraft to operate only by night, or with fighter cover, and from the spring of 1943 the type was relegated to coastal reconnaissance, ASW, Elint and calibration until April 1944, when the final unit (No. 244 Squadron at Masirah) re-equipped. The Blenheim Mk V served in the bomber role in India and Burma, though the last

Finland's Blenheim operations

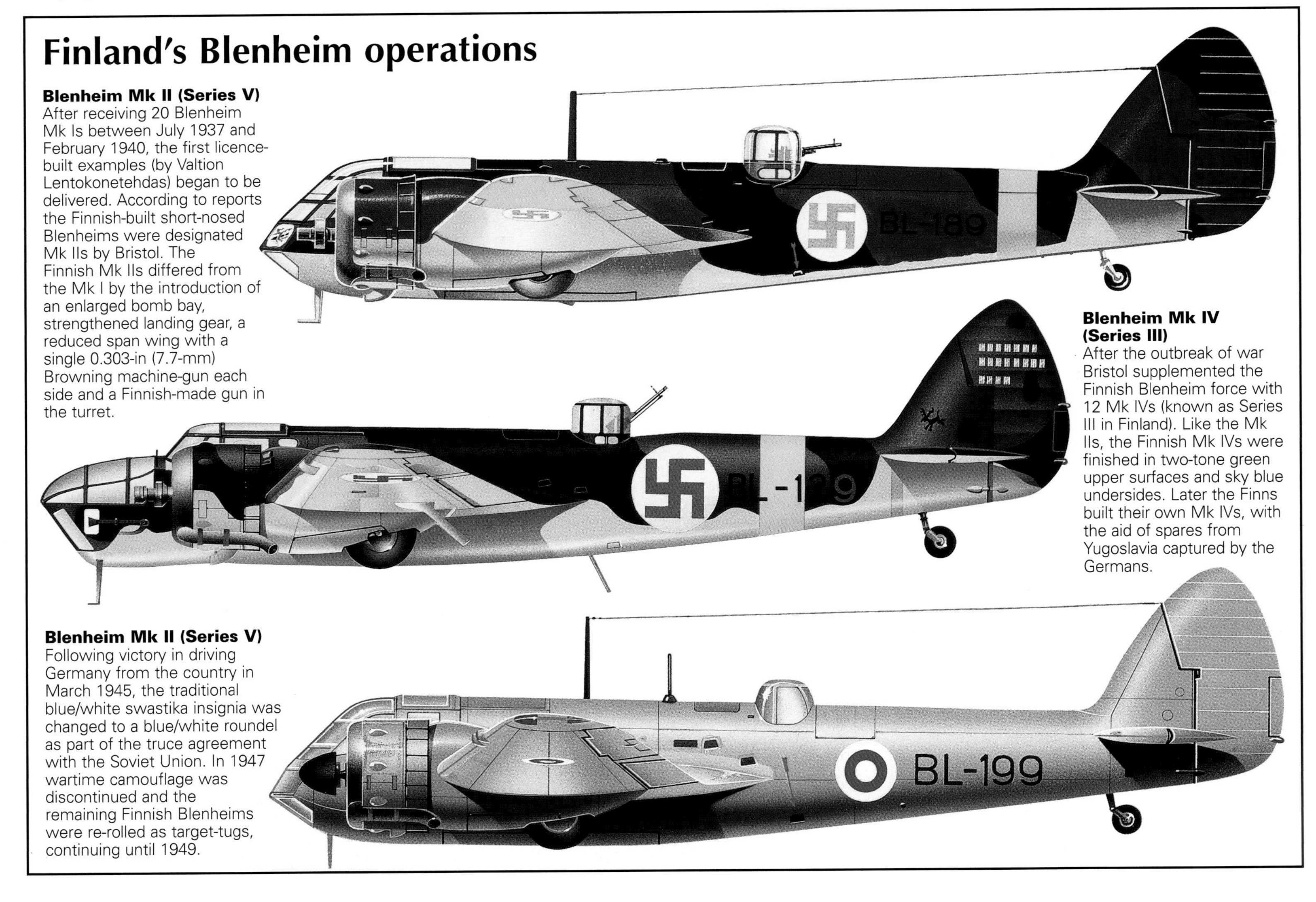

Blenheim Mk II (Series V)
After receiving 20 Blenheim Mk Is between July 1937 and February 1940, the first licence-built examples (by Valtion Lentokonetehdas) began to be delivered. According to reports the Finnish-built short-nosed Blenheims were designated Mk IIs by Bristol. The Finnish Mk IIs differed from the Mk I by the introduction of an enlarged bomb bay, strengthened landing gear, a reduced span wing with a single 0.303-in (7.7-mm) Browning machine-gun each side and a Finnish-made gun in the turret.

Blenheim Mk IV (Series III)
After the outbreak of war Bristol supplemented the Finnish Blenheim force with 12 Mk IVs (known as Series III in Finland). Like the Mk IIs, the Finnish Mk IVs were finished in two-tone green upper surfaces and sky blue undersides. Later the Finns built their own Mk IVs, with the aid of spares from Yugoslavia captured by the Germans.

Blenheim Mk II (Series V)
Following victory in driving Germany from the country in March 1945, the traditional blue/white swastika insignia was changed to a blue/white roundel as part of the truce agreement with the Soviet Union. In 1947 wartime camouflage was discontinued and the remaining Finnish Blenheims were re-rolled as target-tugs, continuing until 1949.

Long-nosed fighters – Blenheim Mk IVF

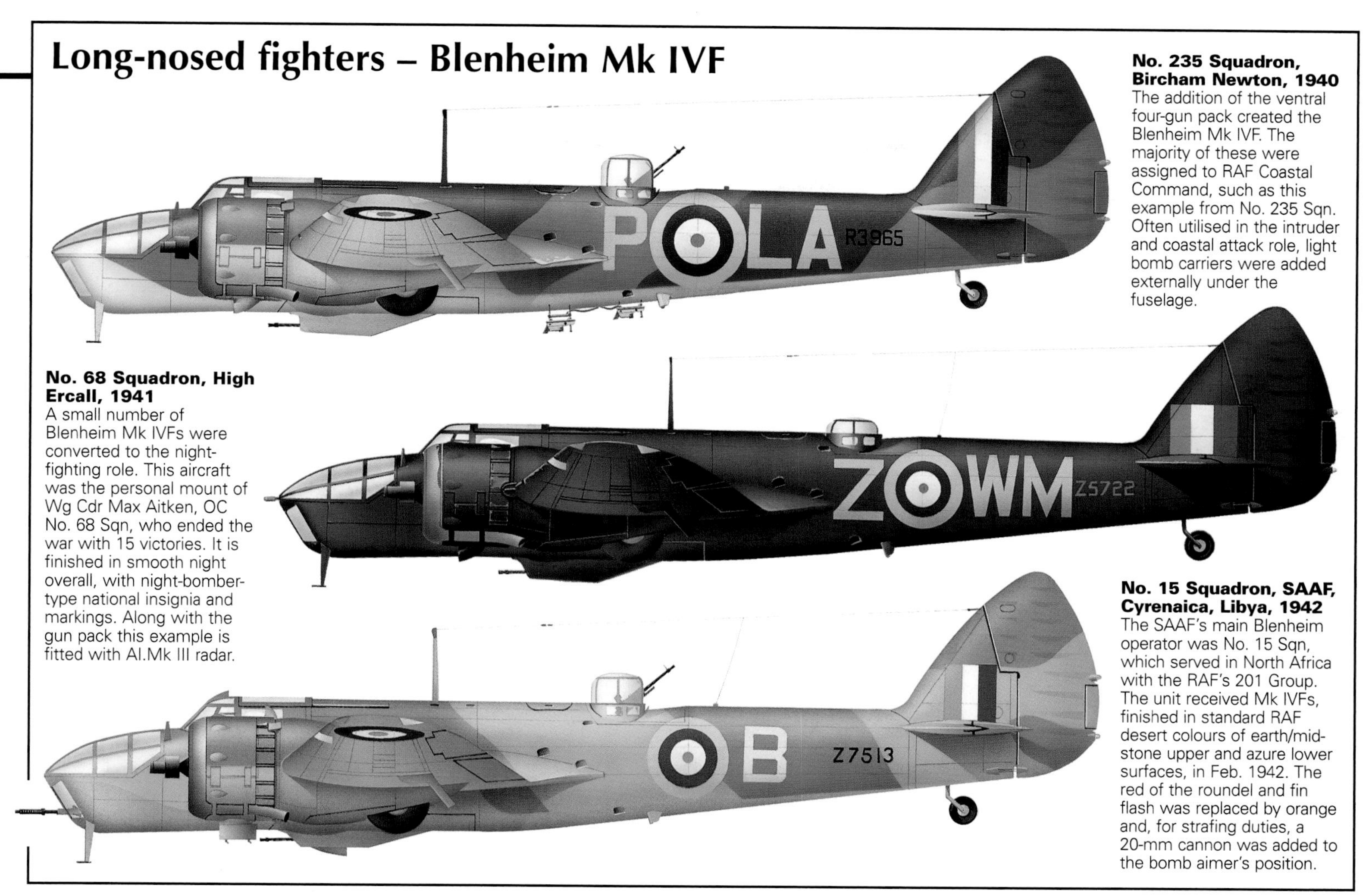

No. 235 Squadron, Bircham Newton, 1940
The addition of the ventral four-gun pack created the Blenheim Mk IVF. The majority of these were assigned to RAF Coastal Command, such as this example from No. 235 Sqn. Often utilised in the intruder and coastal attack role, light bomb carriers were added externally under the fuselage.

No. 68 Squadron, High Ercall, 1941
A small number of Blenheim Mk IVFs were converted to the night-fighting role. This aircraft was the personal mount of Wg Cdr Max Aitken, OC No. 68 Sqn, who ended the war with 15 victories. It is finished in smooth night overall, with night-bomber-type national insignia and markings. Along with the gun pack this example is fitted with AI.Mk III radar.

No. 15 Squadron, SAAF, Cyrenaica, Libya, 1942
The SAAF's main Blenheim operator was No. 15 Sqn, which served in North Africa with the RAF's 201 Group. The unit received Mk IVFs, finished in standard RAF desert colours of earth/mid-stone upper and azure lower surfaces, in Feb. 1942. The red of the roundel and fin flash was replaced by orange and, for strafing duties, a 20-mm cannon was added to the bomb aimer's position.

five units converted to Hurricane fighter-bombers between July and October 1943. One Mk V was briefly returned to active service by the Beaufighter-equipped No. 176 Squadron in May 1944, after it was found abandoned at Kangla. No. 176's CO flew two missions in the aircraft, attacking an enemy airfield, a field HQ and a vital bridge. But thereafter, the RAF used the Blenheim only in second-line roles.

Elsewhere, Finnish Blenheims fought on, and in Canada Bolingbrokes continued in use, principally in the training and target-towing roles.

Faults and failures

The Blenheim was probably the perfect aircraft for the RAF's rapid interwar expansion, offering an aircraft to which crews with experience on the Hart and Hind (or even fresh from advanced training) could convert without difficulty, but which introduced them to more modern and representative performance and handling characteristics and to modern equipment and refinements like retractable landing gear, hydraulically actuated split flaps and a power-operated gun turret. The Blenheim could operate from the RAF's existing grass airfields, yet carried double the Hart's bombload, twice as far.

But the aircraft was never suited to the demands of modern war. 'Sold' on the basis of being 'faster than the fighters of the day' the Blenheim could achieve a speed of 260 mph (418 km/h), making it marginally faster than the 235-mph (378 km/h) Gloster Gauntlet or 250-mph (402 km/h) Gladiator, but rather slower than the 316 mph (509 km/h) Hurricane Mk I, and far short of the speed which aircraft like the Spitfire and Bf 109 could achieve.

Moreover, the power settings needed to attain these maximum speeds were available only very briefly. In the Mk I, the +5 lb of boost in rich mixture was available for only five minutes – and using it sent fuel consumption spiralling to 160 gallons (727 litres) per hour or more. This made daunting inroads into the Mk I's 280-Imp gal (1273-litre) fuel capacity.

In the Mk IV, a higher +9 lb boost setting was available (but only for 3 minutes at take off or 30

Above: Fitted with Vokes tropical filters below the cowlings and undernose gun turrets, a formation of No. 14 Squadron Blenheim Mk IVs (with accompanying Kittyhawk fighter escort) attack enemy positions during the North African campaign. The Mk IV was an important weapon in the theatre, particularly in 1940-41 when some seven squadrons and two detachments were equipped with the type.

Right: No. 107 Blenheim Mk IVs fly in formation at low level during a bombing mission to Cologne in August 1941. These daylight missions against heavily defended targets were among the most hazardous flown by any RAF type. The crew relied on the element of surprise at ultra low-level and a large slice of luck to survive.

Canada's Bolingbrokes

Bolingbroke Mk IV
The largest operator of the Blenheim family after the RAF was the Royal Canadian Air Force. Fairchild licence-built the type as the Bolingbroke, of which the Mk IV was the definitive operational version. The type's main operational role was ASW and coastal patrol.

Bolingbroke Mk IVT
The majority of Bolingbrokes were built as Mk IVT general-purpose dual-control trainers and were mainly used for navigation and gunnery training. Towards the end of the war a number of Mk IVTs received this yellow/black scheme and had their dorsal turrets removed for use as target-tugs.

minutes in flight), and this imposed a 190-Imp gal (863-litre) per hour consumption rate. This required 100 octane petrol, though, leading to the practice of fuelling up with 100 octane in the outer tanks only, leaving the inners filled with 280 Imp gal (1273 litres) of 87 octane.

What made matters worse was that the Blenheim cruised at a mere 200 mph (322 km/h), only 14 mph (23 km/h) faster than the biplane Hart and Hind, or 12 mph (19 km/h) faster than the cruising speed of the Vickers Wellesley. With a full load, or with external bombracks, the Blenheim's speed dropped to a paltry 180 mph (290 km/h).

The Bolingbroke name was originally associated with the Bristol-built Type 149, which became the Blenheim Mk IV. However, Fairchild-built examples in Canada retained the Bolingbroke name, of which this aircraft (702) was the first to roll from the production line as one of 18 Bolingbroke Mk Is.

Blenheim production and serials

Type	No. produced	Serial
Bristol 142	1	K7557 (c/n 7838, ex G- ADCZ, R-12, later 2211M)
Bristol 142M	1	K7033
Bristol 143	1	R-14 (c/n7839, ex G-ADEK)
RAF Blenheim I – Filton	633	K7034-7182, L1097-1546, L4817-4822, and L4907-4934
Finnish Blenheim I (Srs IV) - Filton	12	BL-134 to BL-145
Turkish Blenheim I - Filton	30	2501-2512, 397-408, 485-490
Yugoslav Blenheim I - Filton	2	
RAF Blenheim I – Avro	250	L6594-6843
RAF Blenheim I – Rootes	250	L8362-8407, L8433-8482, L8500-8549, L8597-8632, L8652-8701, L8714-8731
Blenheim I - Ikarus (Romania)	16	
Blenheim II (Srs II) - Valmet	15	BL-146 to BL-160
Finnish Blenheim I (Srs V) - Filton	30	BL-161 to BL-190
RAF Blenheim IVL – Rootes	86	L9170-9218 and L9237-9273
RAF Blenheim IV – Filton	316	L4823-4906, N6140-6174, N6176-6220, N6223-6242, P4825-4864, P4898-4927, P6885-6934, P6950-6961
Greek Blenheim IV – Filton	12	G-AFXD to G-AFXO
RAF Blenheim IV – Avro	750	N3522-3545, N3551-3575, N3578-3604, N3608-3631, R2770-2799, Z5721-5770, Z5794-5818, Z5860-5909, Z5947-5991, Z6021-6050, Z6070-6104, Z6144-6193, Z6239-6283, Z6333-6382, Z6416-6455, Z9533-9552, Z9572-9621, Z9647-9681, Z9706-9755, Z9792-9836
RAF Blenheim IV – Rootes	2,117	L8732-8761, L8776-8800, L8827-8876, L9020-9044, L9294-9342, L9375-9422, L9446-9482, R3590-3639, R3660-3709, R3730-3779, R3800-3849, R3870-3919, T1793-1832, T1848-1897, T1921-1960, T1985-2004, T2031-2080, T2112-2141, T2161-2190, T2216-2255, T2273-2295, T2318-2357, T2381-2400, T2425-2444, V5370-5399, V5420-5469, V5490-5539, V5560-5599, V5620-5659, V5680-5699, V5720-5769, V5790-5829, V5850-5899, V5920-5969, V5990-6039, V6060-6069, V6120-6149, V6170-6199, V6220-6269, V6290-6339, V6360-9399, V6420-6469, V6490-6529, Z7271-7320, Z7340-7374, Z7406-7455, Z7483-7522, Z7577-7596, Z7610-7654, Z7678-7712, Z7754-7803, Z7841-7860, Z7879-7928, Z7958-7992
Finnish Blenheim IV (Srs VI) – Valmet	10	BL-196 to BL-205
RCAF Bolingbroke I – Fairchild	18	RCAF 702 to 719
RCAF Bolingbroke IV – Fairchild	201	RCAF 9001-9201
RCAF Bolingbroke IV-T – Fairchild	457	RCAF 9851-10256
Bristol 149CS Bisley I – Filton	1	AD657
Bristol 149HA Bisley I – Filton	1	AD661
Blenheim V – Rootes	942	AZ861-905, AZ922-971, AZ984-999, BA100-118, BA133-172, BA191-215, BA228-262, BA287-336, BA365-409, BA424-458, BA471-505, BA522-546, BA575-624, BA647-691, BA708-757, BA780-829, BA844-888, BA907-951, BA978-999, BB100-102, BB135-184, DJ702, DJ707, EH310-355, EH371-420, EH438-474, EH491-517

Its defensive armament was never adequate, and even when the turret was augmented by undernose armament, the rifle-calibre 0.303-in (7.7-mm) machine-guns served merely to add weight to the already lumbering machine.

Even before the invasion of France, Bomber Command's C-in-C, Sir Charles Portal, began to oppose continued daylight raids by Blenheims, due to the heavy losses and limited return. Once the Wehrmacht attacked, there was no alternative than to throw the Blenheims into action, and losses accelerated.

One former Hurricane pilot who watched the slaughter of the Blenheim squadrons in France later commented that: "The Blenheims were worse than nothing at all, because they took valuable aluminium, engines, variable-pitch propellers (VP props, for Christ's sake, when the Spits and Hurris still had fixed-pitch wooden props) and production capacity, and they achieved nothing at all. Except that the cream of the pre-war air force, guys who could have made a real difference had they been strapped into Spits, Hurricanes or a modern bomber, were thrown away to die in useless bloody Blenheims. Bristol should have delivered them before assembly, and we could have piled up the wings and fuselages on the beaches as anti-invasion defences!"

With the comfortable benefit of 20:20 hindsight, it can be seen that the Blenheim was conceptually flawed from the very outset, and that for a high-speed day bomber to succeed it needed to be a stripped, lightweight, unarmed type in the mould of the Mosquito, or a single-seat fighter-bomber like the Westland Whirlwind. The latter machine could carry a Blenheim-sized bombload 720 miles (1159 km) and could outrun (and in some circumstances out-turn) a Bf 109E, while packing the useful punch provided by four closely grouped 20-mm cannon. It is hard to imagine the results had squadrons of Whirlwinds been dispatched against the kind of targets which cost the RAF so many Blenheims. Losses would have been negligible, and more bombs would have hit their targets, day after day, and defending fighters might have taken a pasting too. Aircraft like the Beaufighter offered a compromise – with a two-man crew, rudimentary self defence armament and Blenheim-type range, but these too were ignored by Bomber Command. The alternatives were heavier medium bombers typified by the Ju 88 and Douglas Boston, which were too forward thinking and too radical to be considered by the inter-war Air Staff, and which were quite unsuitable for operating from peacetime RAF aerodromes, though they would prove more useful and more survivable than the Blenheim once exposed to the rigours of wartime use.

But when war broke out, the RAF had to go to war with what it had, and not with what it could or should have had. And in late 1939, that meant the Bristol Blenheim, which was available in huge numbers, equipped large numbers of squadrons, and which, by comparison with the Fairey Battle, looked very good indeed. And despite its weaknesses and inadequacies, the type was able to conduct meaningful daylight attacks against enemy targets, albeit at a heavy cost, and was able to carry the war back to the enemy and to 'hit him where it hurt', when the value of night bombing was extremely limited.

DJ702 was the first pre-production Blenheim Mk V. Like subsequent Mk Vs, the aircraft had a Direction Finder loop antenna in a clear Perspex 'teardrop' on the forward fuselage spine. Despite being the most powerful and best protected of all the Blenheim family, the Mk V was actually some 1,120 lb (508 kg) heavier than the Mk IV, and was also slower and less manoeuvrable.

Blenheim survivors

Remarkably, there is one flying Blenheim survivor. Although the last Finnish Blenheims bowed out in 1958, the last aircraft were not kept airworthy. Instead, in 1974, after they had spent almost 30 years derelict on a remote farm, Ormond Haydon Baillie brought two long-retired Canadian Bolingbrokes back to Duxford, where one was slowly restored to airworthy condition. This made its maiden flight on 20 May 1987, but was written off in a spectacular crash at Denham a month and a day later. A second Bolingbroke was subsequently restored, and this flew at Duxford on 18 May 1993, although this aircraft was also badly damaged in a landing accident in 2003 and is currently under repair.

Jon Lake

Mk V – the definitive Blenheim

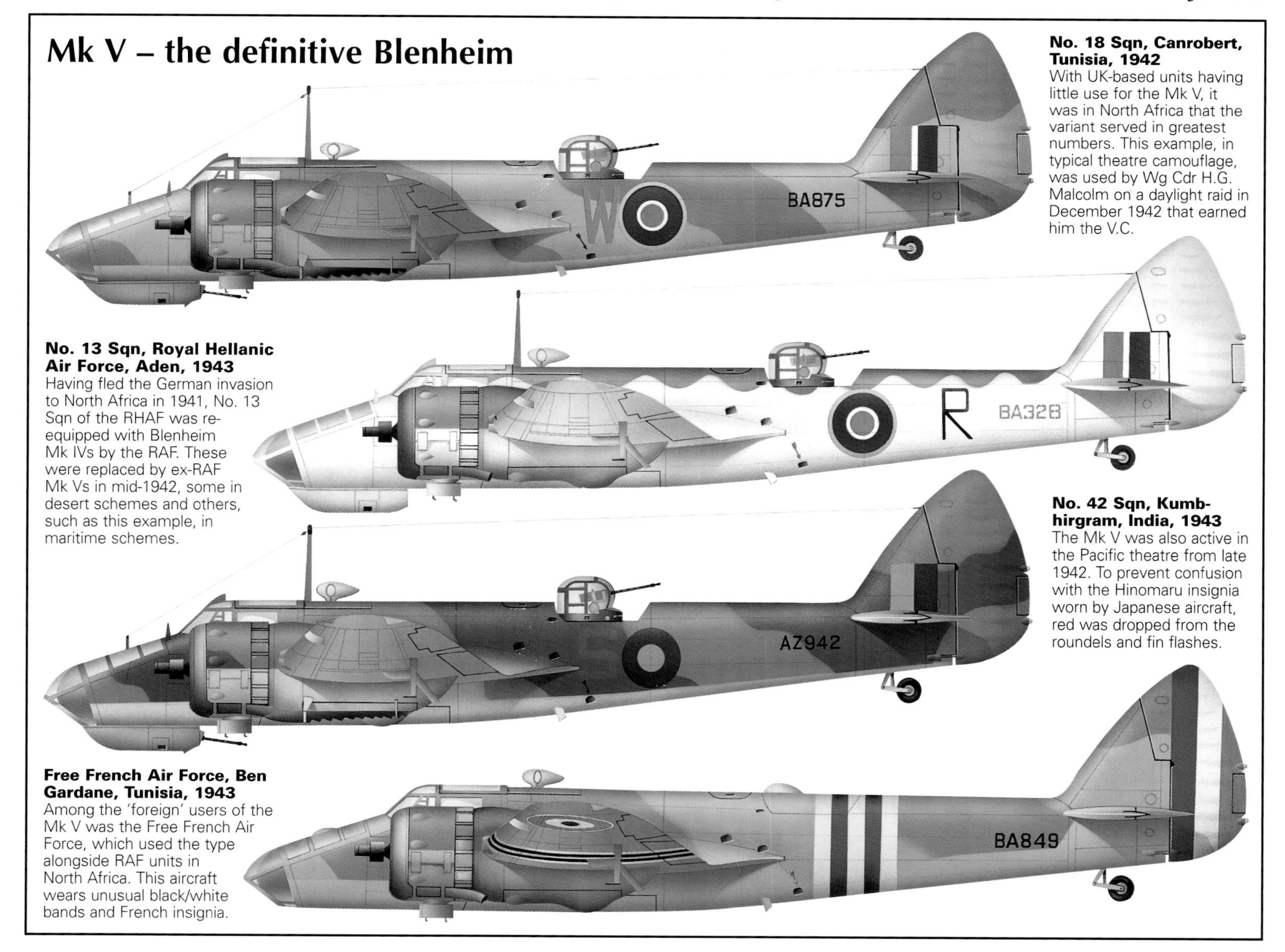

No. 18 Sqn, Canrobert, Tunisia, 1942
With UK-based units having little use for the Mk V, it was in North Africa that the variant served in greatest numbers. This example, in typical theatre camouflage, was used by Wg Cdr H.G. Malcolm on a daylight raid in December 1942 that earned him the V.C.

No. 13 Sqn, Royal Hellanic Air Force, Aden, 1943
Having fled the German invasion to North Africa in 1941, No. 13 Sqn of the RHAF was re-equipped with Blenheim Mk IVs by the RAF. These were replaced by ex-RAF Mk Vs in mid-1942, some in desert schemes and others, such as this example, in maritime schemes.

No. 42 Sqn, Kumbhirgram, India, 1943
The Mk V was also active in the Pacific theatre from late 1942. To prevent confusion with the Hinomaru insignia worn by Japanese aircraft, red was dropped from the roundels and fin flashes.

Free French Air Force, Ben Gardane, Tunisia, 1943
Among the 'foreign' users of the Mk V was the Free French Air Force, which used the type alongside RAF units in North Africa. This aircraft wears unusual black/white bands and French insignia.

Blenheim operators

United Kingdom

Squadron	Variant	Dates	Code	Bases
Royal Air Force				
6 Sqn	IV	11/41-1/42	JV	Kufra
8 Sqn	I, IV, V	4/39-1/44		Khormaksar
11 Sqn	I, IV	7/39-9/43		India, Singapore, Egypt, Greece, Crete, Iraq, Ceylon
13 Sqn	IV,V	7/41-12/43	OO	Odiham, MacMerry, N.Africa
14 Sqn	IV	9/40-9/42		Sudan, Egypt, Iraq, N.Africa
15 Sqn	V	12/39-11/40	LS	Wyton, Alconbury
18 Sqn	I, IV, V	5/39-4/43	WV	Upper Heyford, France, Malta, N.Africa
21 Sqn	I, IV	8/38-7/42	YH	Eastchurch, Watton, Lossiemouth, Luqa, Bodney
23 Sqn	IF	12/38-4/41	YP	Wittering, Collyweston, Ford, M Wallop
25 Sqn	IF, IVF	12/38-1/41	ZK	Hawkinge, Northolt, Filton, N.Weald, Martlesham, Debden
27 Sqn	IF	11/40-1/42		Risalpur, Singapore, Malaya
29 Sqn	IF	12/38-2/41	YB	Debden, Drem, Digby, Wellingore
30 Sqn	I, IF	1/38-5/41		Egypt, Iraq, Greece, Crete, N.Africa
34 Sqn	I	7/38-4/43	LB	Upper Heyford, Watton, Singapore, India
35 Sqn	IV	11/39-4/40		Cranfield, Bassingbourn, Upwood
39 Sqn	I, IV	8/39-1/41		Singapore, India, Egypt
40 Sqn	IV	12/39-11/40	BL	Wyton
42 Sqn	IV, V	2/43-10/43	AW	India
44 Sqn	I	12/37-2/39		Waddington
45 Sqn	I, IV	6/39-8/42		Egypt, Burma
52 Sqn	IV	10/42-2/43		Mosul, Kasfareet
53 Sqn	IV	1/39-7/41	TE	Odiham, France, various UK airfields
55 Sqn	I, IV	3/39-4/42		Iraq, North Africa
57 Sqn	I, IV	3/38-11/40		Upper Heyford, France, Wyton, Gatwick, Lossiemouth, Elgin
59 Sqn	IV	5/39-9/41	PJ	Andover, France, various UK airfields
60 Sqn	I, IV	6/39-8/43	MU	Ambala, Lahore, Singapore, India
61 Sqn	I	1/38-3/39		Hemswell
62 Sqn	I	2/38-1/42	PT	Singapore, Malaya
64 Sqn	IF	12/38-4/40	GR	Church Fenton, Evanton, Catterick
68 Sqn	IF, IVF	1/41-5/41	WM	Catterick, High Ercall
82 Sqn	I, IV	3/38-3/42	OZ/UX	Cranfield, Watton, Bodney,
84 Sqn	I, IV	2/39-6/42		Iraq, Egypt, Greece, Crete, N.Africa
86 Sqn	IV	12/40-7/41	BX	Gosport, Leuchars, Wattisham, N. Coates,
88 Sqn	IV	2/41-12/41	RH	Sydenham, Swanton Morley, Attlebridge
90 Sqn	I, IV	5/37-4/40, 10/41-2/42		Bicester, West Raynham, Weston on the Green, Upwood, Polebrook
92 Sqn	IF	11/39-3/40	GR	Tangmere, Croydon
101 Sqn	I, IV	6/38-5/41	SR	Bicester, West Raynham
104 Sqn	I, IV	5/38-4/40		Bassingbourn, Bicester
105 Sqn	IV	6/40-5/42	GB	Honington, Watton, Swanton Morley, Luqa
107 Sqn	I, IV	8/38-1/42	OM	Scampton, Harwell, Wattisham, Leuchars, Luqa
108 Sqn	I, IV	6/38-4/40		Bassingbourn, Bicester
110 Sqn	I, IV	1/38-3/42	VE	Waddington, Wattisham
113 Sqn	I, IV, V	6/39-10/42	VA	Egypt, Greece, Burma
114 Sqn	I, IV, V	3/37-4/43	FD	Wyton, France, UK, North Africa
139 Sqn	I, IV, V	7/37-12/41	XD	France, UK, Malta
140 Sqn	IV	9/41-8/43		Benson, Weston Zoyland, Mount Farm
141 Sqn	IF	11/39-5/40		Grangemouth, Turnhouse,
143 Sqn	IVF	12/41-9/42	HO	Aldergrove, Limavady, Thorney Island, Docking
144 Sqn	I	8/37-3/39		Hemswell, North Coates
145 Sqn	IF	11/39-4/40	SO	Croydon, Tangmere
162 Sqn	IV, V	3/42-3/44		Egypt, North Africa
173 Sqn	IV			Heliopolis
203 Sqn	I, IV	3/40-11/42		Sheikh Othman, Khormaksar, Kabrit, Heraklion, LG101
211 Sqn	I, IV, V	5/39-2/42		Ismailia, Greece, Egypt, Palestine
212 Sqn	IV	2/40-6/40		Heston
218 Sqn	IV	7/40-11/40	HA	Oakington
219 Sqn	IF	10/39-12/40	FK	Catterick, Redhill
222 Sqn	IF	11/39-3/40	ZD	Duxford
223 Sqn	I	5/41-1/42		Wadi Gazouza
226 Sqn	IV	2/41-11/41	MQ	Sydenham, Wattisham
229 Sqn	IF	11/39-3/40	RE	Digby
234 Sqn	IF	11/39-3/40		Leconfield
235 Sqn	IF, IVF	2/40-12/41	QY	N.Coates, Bircham Newton, Detling, Dyce
236 Sqn	IF, IVF	12/39-3/43	ND	Martlesham, N.Coates, Carew Cheriton, Wattisham, Oulton
242 Sqn	IF	12/39-12/39		Church Fenton
244 Sqn	IV, V	4/42-4/44		Sharjah, Masirah
245 Sqn	IF	11/39-3/40		Leconfield
248 Sqn ,	IF, IVF	12/39-7/41	WR	Hendon, N.Coates, Thorney, Dyce, Bircham Newton
252 Sqn	IF, IVF	12/40-4/41		Chivenor, Aldergrove, Malta, Crete, Egypt
254 Sqn ,	IF, IVF	11/39-7/42	QM	Stradishall, Bircham, Sumburgh, Aldergrove, Carew Cheriton
267 Sqn	I, IV			Egypt (comms)
272 Sqn	IVF	11/40-4/41		Aldergrove
285 Sqn 1	I, IV	12/41-3/42		Wrexham
287 Sqn 1	IV	11/41-2/42		Croydon
288 Sqn 1	IV	11/41-12/41		Digby
289 Sqn 1	IV	11/41-1/42		Kirknewton
404 Sqn 5	IVF	4/41-1/43		Thorney Island, Casteltown, Skitten, Dyce, Sumburgh
406 Sqn 5	IF, IVF	5/41-6/41		Acklington
407 Sqn 5	IV	5/41-7/41		Thorney Island
415 Sqn 5	IV	8/41-11/42		Thorney Island, N. Coates, Wick, Leuchars
454 Sqn 6	IV	9/42-1/43		Aqir, Palestine and Qaiyara, Iran
459 Sqn 6	IV	2/42-5/42	BP?	201 Group, Burg Al Arab
489 Sqn 7	IV	8/41-1/42		Leuchars
500 Sqn	IV	4/41-11/41	MK	Detling, Bircham Newton
516 Sqn 2	IV	5/43-12/44		Dundonald
521 Sqn 3	IV	8/42-3/43		Bircham Newton
526 Sqn 4	IV	6/43-5/45		Inverness
527 Sqn 4	IV	6/43-5/45		Castle Camps, Snailwell, Digby
528 Sqn 4	IV	6/43-9/44		Filton, Digby
600 Sqn	IF, IVF	1/39-2/41	BQ	Hendon, various UK airfields, Catterick
601 Sqn	IF	1/39-3/40	YN/UF	Hendon, Biggin, Tangmere
604 Sqn	IF	1/39-1/41	NG	Hendon, North Weald, Northolt, Manston, Middle Wallop
608 Sqn	IV	3/41-7/41	UL	Thornaby
614 Sqn	IV, V	7/41-2/44		MacMerry, Odiham, Portreath, N. Africa

In addition the following training units operated the Blenheim: Nos 1, 2, 4, 5, 13, 17, 42, 51, 52, 54, 55, 60, 63, 72 132 OTUs and No. 12 PAFU.

Fleet Air Arm

Squadron	Variant	Dates	Bases
748 Sqn (T)	IV	11/43-3/44	Chivenor, Yeovilton, Henstridge
759 Sqn (T)	IV	7/43-9/44	Yeovilton, Angle
762 Sqn (T)	IV	3/44-?	Yeovilton, Dale
770 Sqn (FR)	I, IV	3/42-6/45	Crail, Dunino, Drem
771 Sqn (FR)	I, IV	4/41-5/45	Donibristle, Twatt
772 Sqn (FR)	IV	3/44-4/45	Ayr, Ronaldsway
775 Sqn (FR)	IV	3/45-8/45	North Front, Dhekelia
776 Sqn (FR)	I, IV	1/44-4/45	Speke, Millom, Walney Island
780 Sqn (T)	I	6/43-12/43	Lee-on-Solent, Charlton Horethorne
787 Sqn (FT)	I, IV	10/42-5/45	Duxford, Wittering, Tangmere
788 Sqn (FR)	IV	1942	China Bay
798 Sqn (T)	IV	10/43-3/44	Lee-on-Solent

T: Training, FR: Fleet Requirements, FT: Fighter trials

Australia

Though the Royal Australian Air Force was a major operator of the Beaufort and Beaufighter, it is not generally recognised as a Blenheim user. In fact, at least three Blenheim Mk Is were briefly used by the RAAF's No. 456 Squadron (under RAF Command) during 1942 and 1943.

Croatia

With the invasion of Yugoslavia, newly independent Croatia became a German ally. The Croatian Air Force Legion was formed using surviving ex-Royal Yugoslav Air Force aircraft, though these were soon augmented by German aircraft types. Mainly used in support of German forces on the Eastern Front, the Croatian Air Force Legion included about eight Blenheims when it formed, and these served with the 5th Bomber Wing's 12th Bomber Squadron.

Canada

In Canadian service, the Bolingbroke served in the coastal patrol and general reconnaissance role with several Bomber Reconnaissance (BR) Squadrons, and with Operational Training (OT) and Composite (C) target facilities units. One Army Co-operation squadron briefly used Bolingbrokes in the recce role. The most important use of the Bolingbroke was as a navigation and gunnery training aircraft within the British Commonwealth Air Training Plan. Training units included No. 3 (RCAF) OTU at Patricia Bay, British Columbia, and No. 34 (RAF) OTU at Pennfield Ridge, New Brunswick.

Squadron	Base	Variant	Period	Code
No.8 BR Squadron	Sydney, NS	Bolingbroke I, IV	12/40-8/43	VO-
No.13 OT Squadron	Sea Island, BC	Bolingbroke IV	10/41-6/42	AN-
No.115 BR Squadron	Annette Is, AK	Bolingbroke I	8/41-12/41	BK-
No.119 BR Squadron	Hamilton, Ont	Bolingbroke I	7/40-8/41	DM-
		Bolingbroke IV-W	11/41-6/42	DM-, GR-
No.121 C Squadron	Dartmouth, NS	Bolingbroke IVTT	8/42-9/45	EN-
No.122 C Squadron	Patricia Bay, BC	Bolingbroke IVTT	8/42-9/45	AG-
No.147 BR Squadron	Sea Island, BC	Bolingbroke I, IV	7/42-3/44	
No.163 AC Squadron	Sea Island, BC	Bolingbroke IV (PR)	3/43-6/43	

FINLAND

Finland's Blenheims enjoyed a longer combat career than those of any other nation, and participated in three wars! They were also arguably the most successful and had with the longest overall life, lasting until 1958. The Blenheims flew 423 sorties and dropped 3,168 bombs during the Winter War, gunners claiming five kills. Seven were lost. During the Continuation War they flew 2,768 sorties scoring three further kills, but lost 30 aircraft. A final 157 sorties were flown during the Lapland War.

Some 37 Finnish Blenheims survived the war, and 31 served with PLeLv 41 and 45 until 15 September 1948, when bombers were banned under the terms of the peace treaty with Russia. Some 13 airworthy aircraft were stored, and five of these (including a Blenheim Mk I and a Mk II) were brought back into use as survey aircraft and target tugs from 1 August 1951. BL-199 made the last flight by a Finnish Blenheim on 20 May 1958.

Unit	Base	Dates of service
LAs 6		1937-39
LLv 42 (LentoR 4)	Luonetjärvi, Siikakangas	3/1940-14/2/44 to PLeLv 42
LLv 44 (LentoR 4)	Luonetjärvi, Siikakangas	1939-20/2/1943 to Ju 88
LLv 46 (LentoR 4)	Luonetjärvi	1939-6/1941 non-op to Do 17
LLv 48 (LentoR 4)	Onttola	15/11/43-14/2/44 to PLeLv 48
PLeLv 41 (LentoR 4)	Naarajarvi	4/12/44-1/1/45
PLeLv 42 (LentoR 4)	Siikakangas, Naarajarvi	14/2/44 -4/12/44 to PLeLv 41
PLeLv 45		1/1945-48
PLeLv 48 (LentoR 4)	Onttola, Vesivehmaa	14/2/44 -4/12/44 to PLeLv 45
IlmVE		1940
TLeLv 12		1944
TLeLv 17		1942-44
KoeLtue (Test Flight)	Kuorevesi	1945-1948
		2/5/1957-20/5/1958
LeR 3		1952-53
PLeLv 41 (LeR 4)	Luonetjärvi	1944-15/9/1948
	Luonetjärvi	1/8/1951-1/12/1952
PLeLv 43 (LeR 4)	Luonetjärvi	1945-46
PLeLv 45 (LeR 4)	Luonetjärvi	1/1945-15/9/1948
1. Lennosto (1st Air Command)	Luonetjärvi	1/12/1952-1/1/1957
Hämeen Lennosto	Luonetjärvi	1/1/1957-10/4/1958

FRANCE

Free French forces in Africa made extensive use of the Blenheim. Groupe Mixte de Combat 1 included one flight of Blenheims when it formed at Odiham in August 1940, before shipping out to Chad. This unit later merged with the Blenheim-equipped Escadron Topic at Maidagur to form Groupe Réservé de Bombardement at Fort Lamy in December 1940. The Groupe Réservé de Bombardement disbanded in March 1941, but formed the basis of Free French Flights Khartoum and Sudan, which were later absorbed into Groupe Lorraine in December 1941 as Escadrilles Metz and Nancy. Blenheims also served with Groupe Bretagne, the Free French Flying School at Bangui, and with Groupe de Chasse 1 Alsace. The last frontline French Blenheims were withdrawn in late 1942, but the type served on in second-line roles until 1945.

GREECE

Greece ordered 12 Blenheim Mk IVs after the war began, equipping No. 32 Sqn at Larissa, and subsequently gained six ex-RAF Mk Is. After the German invasion, in June 1941, the Royal Hellenic Air Force reformed the 13th Light Bombing Squadron at Dekheila, Egypt, as part of the RAF's 201 Group. Originally equipped with Ansons, the squadron gained Blenheim Mk IVs in November 1941, and Mk Vs in August 1942. The unit operated mainly in the convoy escort and ASW roles. It finally re-equipped with Baltimores in mid-1943.

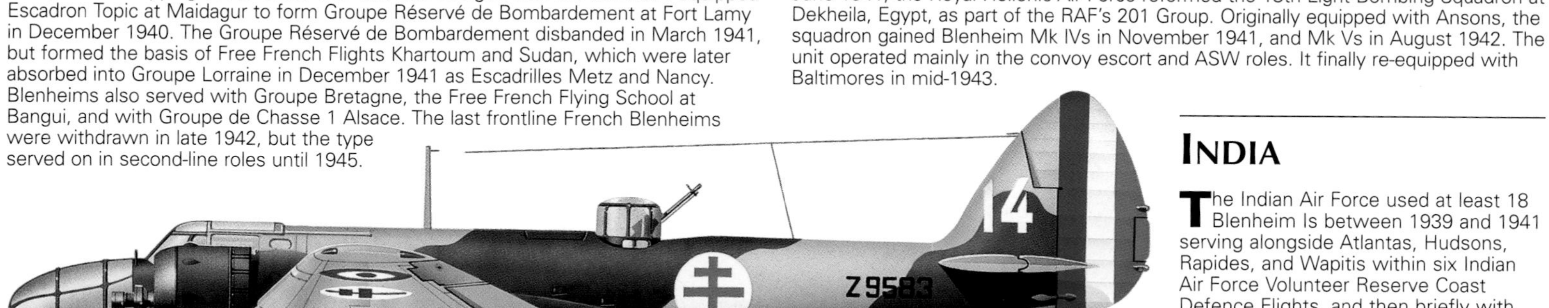

Above: A Free French Air Force Mk IV carries red Cross of Lorraine insignia during service in North Africa in late 1941.

INDIA

The Indian Air Force used at least 18 Blenheim Is between 1939 and 1941 serving alongside Atlantas, Hudsons, Rapides, and Wapitis within six Indian Air Force Volunteer Reserve Coast Defence Flights, and then briefly with IAF Squadrons.

NEW ZEALAND

Like the RAAF, RNZAF units operating under RAF command (notably No. 488 Sqn) used a handful of ex-RAF Blenheims as trainers and hacks.

PORTUGAL

In 1943 Portugal received more than 20 Blenheim Mk IVs under the terms of the Azores Treaty, together with a smaller number of Blenheim Mk Vs. Further aircraft were obtained when aircraft were impounded en route to the Middle East. The Blenheims equipped Esquadrilha ZE at Base Aérea 2 at Ota. The Portuguese Navy took 12 more Blenheim Mk IVs to equip Esquadrilha B das Forcas Aéreas da Armada at Portela de Sacavém, though these were later taken over by the air force and replaced by Beaufighters.

Below: Carrying typical 'ZE' codes, this Portuguese Mk IV served with the Esquadrilha ZE at Ota.

ROMANIA

Britain supplied 40 Blenheim Mk Is to Romania, in the vain hope that this might keep the country from joining the Axis. Three of the aircraft were lost en route to Romania, but the survivors equipped four long-range reconnaissance squadrons (The 1st-4th Long-Range Reconnaissance Squadrons).

The 2nd Long-Range Reconnaissance Squadron disbanded in 1941, and its Blenheims were distributed between the other three squadrons. The remaining units suffered heavy losses during Operation Barbarossa, losing 10 aircraft by the end of the 1941 campaign.

Despite the introduction of three ex-Yugoslav aircraft bought from Germany the number of Blenheims in service dwindled steadily, and by August 1942, only the 1st Long-Range Reconnaissance Squadron was still equipped. These were withdrawn to Romania in December 1942, and thereafter flew patrol missions over the Black Sea. Three surviving Blenheims were attached to a Ju 88D-1 squadron in August 1944, but were soon relegated to transport and training duties.

SOUTH AFRICA

The SAAF evaluated a single Blenheim Mk I from December 1938, and this aircraft was pressed into use when war broke out, joining the hunt for the *Graf Spee* and attacking and forcing an Italian ship to run aground. Despite this success, the SAAF did not place an order, and the aircraft returned to the UK. The SAAF's No. 15 Sqn, serving in North Africa with the RAF's 201 Group, acquired Blenheim Mk IVs in February 1942, and gained Mk Vs in July, before finally receiving Bostons in July 1943. In the ASW and coastal patrol role, Nos 16, 17 Sqns used Mk IVs over the Indian Ocean and Persian Gulf during 1943. The type was also used by Nos 35 and 60 Sqns.

TURKEY

Turkey purchased 40 Blenheim Mk Is from the UK from 1937, augmenting these with some three or five Blenheim Mk IVs (sources differ on the exact number) and 19 Blenheim Mk Vs in 1943. The Blenheim Mk Is and Mk IVs served until 1947, with the Mk Vs lingering on until 1948.

YUGOSLAVIA

The Yugoslav government purchased two Blenheim Mk Is in 1937, and acquired a licence for Ikarus AD at Zemun to build 50 more. Some 16 of these had been completed in 1941 by the time of the German invasion, and were augmented by 20 ex-RAF Mk Is delivered during 1940. By the time Germany attacked, Blenheims served with nine squadrons.

During the fierce 12-day war that accompanied the German invasion of Yugoslavia, the Blenheims flew bombing missions against enemy airfields and advancing tank columns, but were heavily outnumbered, and suffered correspondingly heavy losses.

About 24 of the Ikarus aircraft were destroyed before completion, to prevent them from falling into enemy hands, but 10 incomplete aircraft were captured by the Germans, along with a number of completed aircraft which survived the fighting. Eight surviving Yugoslav Blenheim Mk Is went on to serve with the 'puppet' Croat Air Force, three were sold by the Germans to Romania, and parts were sold to Finland.

'Flying Cheetahs'

No. 2 Squadron, SAAF, in Korea

Although the war in Korea was largely fought by US forces, other nations made significant contributions to the United Nations effort. Among them was South Africa, which provided an experienced squadron of volunteers to fly the fighter-bomber mission, initially with F-51Ds. No. 2 Squadron soon gained a reputation as one of the most effective units in the theatre.

Above: *Two rocket-armed 'Cheetahs' F-51Ds (and a 36th FBS F-80) await their pilots for the next mission as a C-54 takes off from Suwon (K-13). The Mustangs flew from this base in December 1950 for a short while, and later used it as a refuelling and rearming post while based in the south at Chinhae or Pusan.*

Right: *A little bit of home in a far away land – this is 'The Sugarbush Sit and Drink Club'. The South Africans were adept at making themselves comfortable while operating in the primitive conditions to be found at many bases in Korea.*

Early on Sunday morning, 25 June 1950, the 'Land of the Morning Calm' was suddenly awoken to the crash of artillery and the clanking of tank treads. The North Korean Communists were attempting to take over the entire Korean Peninsula by force of arms. The United States had a commitment to the government of South Korea, and immediately put up an aerial umbrella to both protect the evacuation of US personnel from Korea and to attempt to stop the North Korean tanks and troops through a show of force in the air. In the latter aim the air 'umbrella' failed, and by the third day of the fighting the capital of South Korea, Seoul, was in Communist hands.

After committing US airpower to Korea to help in the evacuation of non-combatants, President Harry Truman went to the United Nations and gathered support to bring about an end to the fighting, either through negotiation or by defeating the communist armies with UN forces, both on the ground and in the air. On 27 June 1950, after a lengthy debate on the floor of the Security Council, the United Nations voted to go to war against the North Korean forces. One of the first nations to commit major combat forces to the conflict was Australia, with No. 77 Squadron, Royal Australian Air Force, arriving in Korea on 30 July 1950 and being assigned to the control of the US 5th Air Force. On 4 August 1950, the South African Cabinet approved commitment of a fighter squadron, complete with ground personnel, to the Korean conflict. All members of this squadron would be volunteers. All aircraft were to be supplied by the US government. The squadron was designated as No. 2 Squadron, commonly known as the 'Flying Cheetahs'.

No. 2 Squadron had a long combat history from World War II. During that war, the unit operated in the Mediterranean Theater of Operations (MTO), flying in both North Africa and Italy. It operated both P-40 Tomahawks and P-51 Mustangs against the feared German Luftwaffe. Following the end of the war in Europe, the unit was deactivated, but on 5 September 1950, the squadron was reactivated at AF Station Waterkloof with 157 enlisted personnel and 47 officers, most of whom were pilots. But they had no aircraft: their aircraft would be waiting for them in Japan.

On 27 September 1950, No. 2 Squadron left Pretoria and boarded the troopship M.V. *Tjisdane* in Durban harbour. Following stops at many of the Commonwealth ports, including Singapore and Hong Kong, the *Tjisdane* docked in Yokohama on 5 November at 0430 hours local time. The personnel went directly to Johnson AB, Japan, to pick up equipment, supplies and a full complement of ex-USAF F-51D Mustangs. Interestingly, one of the agreements that the South African government

Above: A crowd has gathered to watch the departure of No. 2 Squadron Mustangs from Johnson AFB, Japan, to Pusan in Korea on 16 November 1950. Having settled at Pusan the squadron began combat operations on 19 November.

Below: By the end of the war No. 2 Squadron was flying the F-86F Sabre. This aircraft was the first to carry the name Bevkeneve _before it was renamed_ Black Dick. _The second_ Bevkeneve _was lost in April 1953, resulting in a third aircraft acquiring the name._

made was that they would purchase all supplies and equipment from US sources in Japan, paying for them by cash. One of the biggest problems was meals for the officers. Officers ate with USAF personnel at the Johnson AB Officers Mess, and USAF officers paid cash for their meals. But the South African personnel had very little money, and none in US currency. Of course, the American troops saw to it that their South African friends did not go hungry, although it was a little tight at times.

At Johnson AB, all No. 2 Squadron personnel were forced to go through a six-week conversion course in the F-51D Mustang, the aircraft they would be flying in combat over Korea. This rankled many of the pilots, as most already had experience in either the P-40 or the P-51. All the aircraft were ex-USAAF P-51Ds (now designated F-51D), drawn from the salvage yards in Japan and put back into flying condition. As soon as the first pilots had completed the conversion course, No. 2 Squadron was declared combat-ready, However, only 16 F-51Ds were available for operations.

Into combat

Commandant S. van Breda Theron took four F-51Ds to Pusan AB (K-9), South Korea, where 13 officers and 21 enlisted personnel were already in place and awaiting their arrival. It was 16 November and the South African flight flew its first mission three days later on the 19th, when Commandant Theron and Captain Lipawsky flew as one element in a 12th Fighter Bomber Squadron flight. The mission was a road recce, striking any moving targets along the Main Supply Routes in North Korea. The flight did not return to K-2, instead landing at K-24, commonly known as Pyongyang East, where it was attached to the 6002nd Tactical Support Wing. Mustangs staged through K-24 for several missions prior to being based there from 22 November. On 24 November 1950, the 6002nd TSW was redesignated as the 18th Fighter Bomber Group.

During the early period at both K-9 and K-24, No. 2 Squadron suffered a variety of problems, including bad communications with the Joint Operations Center in Seoul. Pyongyang East (K-24) was desperately in need of improvement just to make it habitable. The field was all grass and had no hangars, resulting in the runways being either very muddy or, when the mud finally froze solid, being very

Groundcrew worked near-miracles to keep the fighter-bombers flying. Above crew remove the covers from F-51Ds in the bone-numbing cold at Pyongyang East in November 1950, while in slightly warmer weather armourers load the 'whole nine yards' into the wing troughs of* Shy Talk *(left).

dusty during flight operations. K-24 was, however, very close to the action, so close that underwing drop tanks were not required and badly needed additional ordnance could be carried in their place. That first winter was so cold that Captain Badenhorst remarked that, "it's a pity we didn't bring a refrigerator. We could take turns sitting in the bloody thing to keep warm!" Temperatures on the ramp at K-24 often went to 30° below zero at night.

The 'Cheetahs' were not at K-24 very long, nor were they attached to the 6002nd TSW very long, either. Chinese intervention in the war was rapidly forcing all UN forces to retreat back down the peninsula, and the 'Cheetahs' were forced to abandon K-24 on 30 November and set up shop at Suwon (K-13) on 2 December. During those first few weeks of combat, from mid-November to 7 December, No. 2 Squadron consisted of a total of six F-51D aircraft, 11 pilots, and 20 ground personnel.

On 17 December the squadron moved to Chinhae (K-10), where it was permanently attached to the other F-51 unit in 5th Air Force, the 18th Fighter Bomber Group. Chinhae was located on the extreme southern coast of Korea, roughly 200 miles (320 km) from the action which had now moved to roughly the 38th Parallel. The pilots were forced to stage through Seoul City Airport (K-16) to reach the 'bomb line' and still have sufficient fuel left to loiter over the target areas. However, K-16 was overrun by the Chinese in January, which forced the Mustangs to stage through other bases, such as Suwon (to the south of Seoul), which were closer to the fighting.

Truck-hunting Mustangs

The mission of the 'Cheetahs' would be close air support (CAS) of UN forces along the main line of resistance (MLR), and road interdiction missions of the main supply routes (MSRs) throughout North Korea. Truck-hunting along the MSRs in North Korea usually involved an element of two Mustangs. The lead aircraft would drop down to an altitude of around 300 ft (90 m) and search every inch of ground. His wingman would fly at about 1,000 ft (305 m), keeping an eye out for the MiGs that were constantly roaming the skies looking for easy prey like the 'Cheetah' Mustangs. The normal combat munitions load for truck-hunting was a full load of 0.50-in (12.7-mm) calibre rounds and four 5-in (12.7-cm) HVAR rockets, plus underwing drop tanks to extend the loiter time in the target area. In January 1951, 'new' F-51Ds began arriving. They had been in storage at Kelly Field since the end of World War II. Some were literally brand new aircraft with less than 25 hours on the airframe, although they were all at least five years old and had to be completely restored to flyable condition. No. 2 Squadron began receiving these 'new' aircraft as soon as they arrived.

It was while flying out of Suwon that the squadron suffered its first loss of the war when Captain J.F.O. Davis bellied one of the F-51Ds in near Pyongyang on 15 December 1950. Although the aircraft was a total loss, Captain Davis was rescued by USAF helicopter and returned to Suwon. However, Captain Davis was later killed in action on 10 March 1951 on a mission north of the Han River. By February 1951, the squadron had flown its 1,000th mission, but success came at a cost. On 2 February 1951, No. 2 Squadron suffered its first pilot loss when Lieutenant W. Saint Elmer Wilson was hit by small arms fire in the Chosin Reservoir region of North Korea. Elmer coaxed the mortally wounded Mustang out to the Sea of Japan before crashing. Rescue forces were unable to find any trace of the aircraft or pilot. Five days later, another 'Cheetah' went down. 2nd Lieutenant Doug Leah was shot down over the Main Line of Resistance (MLR) and declared KIA (killed in action).

Initially, rotation home was set for pilots at 75 missions. The first replacement pilots began to arrive on 24 February 1951. By 4 March, the squadron had lost another five Mustangs and four pilots. The weather and base conditions at K-10 got so bad that, on 23 March 1951, the squadron was forced to evacuate Chinhae and temporarily move to Pusan East Airfield (K-9), where it was attached to the 35th Fighter Interceptor Group. The weather then cleared up and the squadron moved back to Chinhae on 29 April, where it was reunited with its comrades in the 18th FBG. On 12 April 1951, No. 2 Squadron recorded its 2,000th mission. The unit was, indeed, an integral portion of the UN battle plan for defeating the Communists in Korea. On 7 May 1951, following the rebuilding of the runways and facilities at Seoul City Airport, staging was resumed through K-16, and remained thus until 1 October, when the 18th Group and No. 2 Squadron began staging through Hoengsong (K-46). Far East Air Force moved the combat elements of the 18th Group permanently into K-46 in June 1952, which was only 60 miles (96 km) behind the MLR. The 18th Wing Headquarters remained at Chinhae.

Tangling with MiGs

On 8 July 1951, the 'Cheetah' Squadron got its first look at the vaunted MiG-15 jet fighter, when two flights of Mustangs struck the airfield at Kangdong, just north of Pyongyang. Following their bomb runs over the runways, the Mustangs formed up for the flight home to K-46. Suddenly the RT erupted with the exclamation, "MiGs!". Four Chinese MiG-15s had jumped the much slower F-51Ds of the US 12th FBS, which were part of the strike force hitting

These were the operations huts used by the constituent squadrons of the 18th FBW at Hoengsong. The blue hut at the far end was the lair of the 'Cheetahs'.

K-46 Hoengsong, in the central region of Korea, became the focal point of Mustang combat operations from October 1951, initially as a forward staging post and then as a permanent base. Here red-spinnered 'Cheetah' Mustangs prepare to depart alongside blue/white-nosed aircraft of the 39th FIS in 1952.

Kangdong. Two of the MiGs attacked the 'Cheetah' flight from the north, while a second pair came at them from the south, but the Mustang pilots simply turned into them over and over and the MiG pilots finally tired of the game and headed for home. Meanwhile, the 'Cheetah' flight leader had screamed for help and a flight of 4th Fighter Wing Sabres came to the rescue, shooting down three of the four MiGs before they could escape back across the Yalu River.

On 23 July 1951, a flight of No. 2 Squadron Mustangs was conducting a weather reconnaissance up the west coast of Korea. Just north of the Imjin River, the 'Cheetah' flight encountered heavy and accurate anti-aircraft fire, which promptly shot down two of the F-51s. Captain Freddie Bekker was killed in action when his Mustang burst into flames and exploded. Lieutenant M.I.B. Halley was also shot down but was able to bail out and, upon landing, Halley waved to his comrades that were orbiting overhead waiting for the rescue helicopter to pick him up. During the ResCAP flight, Lieutenant Rool duPlooy was also shot down and the rescue effort was abandoned. Lieutenant duPlooy and Captain Bekker were both KIA, while Lieutenant Halley was taken prisoner by the North Koreans. DuPlooy was

Right: An F-51D taxis through puddles at an airfield showing clear signs of earlier damage – probably Seoul (K-16), from where No. 2 Squadron operated during January 1951 before the Chinese counter-offensive drove them south.

Below: Most of the Sabres were named, but this was far less common on the F-51D, although* Miss Marunouchi *was an exception. The aircraft carries a typical armament of 5-in rockets and box-finned bombs.

Above: Last-minute servicing and arming is under way at Hoengsong in 1951/52.

Left: A 'Cheetah' F-51D taxis past trolleys laden with 5-in rockets, a staple weapon for the fighter-bomber role. The aircraft in the background are mostly from the 12th FBS, which used prominent shark-mouth markings.

posthumously awarded the US Air Force Silver Star for his actions during the aborted ResCAP mission.

The men of No. 2 Squadron, SAAF, celebrated their one-year anniversary of operations in Korea on 27 September 1951. The credits for that first year in combat included 4,920 sorties in 1,320 missions, 2,180 buildings destroyed, 13 bridges dropped, 458 trucks, 14 tanks, 24 artillery guns, 83 anti-aircraft positions, four locomotives and 173 rail cars destroyed. Some 188 enemy supply dumps were also destroyed, and the guns of No. 2 Squadron Mustangs were estimated to have killed some 1,634 enemy troops.

On 20 March 1952, MiGs again jumped the slower F-51Ds with the Springbok insignia, this time with disastrous results. The mission was a road recce near Sinuiju, deep in the heart of 'MiG Alley'. The MiGs were based just over the Yalu River and five of the speedy Red jets jumped the flight of 'Cheetah' Mustangs. Lieutenant David Taylor's F-51D was badly damaged by MiG gunfire and he was forced to dive for the deck and head for home. His Mustang was last seen streaming heavy smoke. As he headed back for Hoengsong, the MiGs regrouped and attacked again, but the Mustangs just kept turning into the MiGs to avoid further damage. During one of these manoeuvres, Lieutenant J.S. Enslin pulled his Mustang hard around and into the path of one of the MiGs. The MiG flew directly through his guns and Enslin pulled the trigger and watched his bullets strike the right wing of the jet. The MiG pilot promptly turned and ran for the safety of the Yalu River. Taylor's Mustang never made it back to Hoengsong and he was listed MIA. Enslin was awarded a 'damaged' credit for the hits he had scored on the MiG.

Request for the Sabre

The 'Flying Cheetahs' encountered the wily MiG again on 6 April, but neither side got any hits during the battle. However, on 12 October, Lieutenant T.R. Fryer was shot down by a MiG. The 'Cheetah' pilots had seen enough – they did not want any part of MiG-fighting in the F-51. They wanted something that was on a par with the MiG-15 in every way. In late 1952, that 'something' was made available to them in the form of brand new F-86F Sabres, but not before the pilots of No. 2 Squadron recorded a pair of exceptional 'firsts'. On 9 November, Lieutenant Brian Forsyth flew the 50,000th sortie of the 18th FBG during Korean War operations, and the same month Lieutenant Amstan Mather flew the 10,000th sortie by a No. 2 Squadron pilot.

With rockets and bombs underwing, an F-51D is waved off at the start of a mission from Hoengsong. The 13 serviceable Mustangs remaining when No. 2 Squadron converted to the Sabre were passed on to the ROKAF.

Left: ***South African personnel investigate the first F-86F Sabre to be delivered to the unit, 52-4352, which arrived at Osan in full South African insignia on 23 January 1953. The thin fin-flash subsequently gave way to a much wider marking. The large tri-colour marking was also adopted by the 18th FBW's three USAF squadrons, which had previously employed a wide blue band with four white stars superimposed, flanked by narrow bands in the individual squadron colours.***

Below: **Ruth II** ***is seen amid USAF Sabres on the line at the REMCO facility at Tsuiki. This Japanese base was the main Sabre overhaul and repair depot in the Far East theatre.***

On 18 July 1952, 5th Air Force made the decision to re-equip both the 8th and 18th Fighter Bomber Groups, including No. 2 Squadron, with F-86F Sabres. These aircraft would be brand new F-86F-30 fighter-bombers with a new strengthened wing which had additional underwing hardpoints for the carriage of bombs and much-needed external fuel tanks. On 2 November 1952, 21 No. 2 Squadron NCOs attended a familiarisation course on the North American F-86F Sabre at the Tsuiki REMCO facility in Japan. Tsuiki was the primary Sabre maintenance facility in the Far East.

On 27 December 1952, No. 2 Squadron flew its last missions in the veteran F-51Ds. All the pilots in Baker Flight completed the required 75-mission total on this day, but several of the pilots opted to stay for another tour in the new jets. The final statistics for No. 2 Squadron in F-51D Mustangs included 2,890 missions and 10,337 sorties. Pilots from the unit were awarded two USAF Silver Stars and 26 Distinguished Flying Crosses. The cost was high, with 34 pilots listed as KIA and seven confirmed as POWs. F-51D losses totalled 74, of which 45 were classed as 'M' losses (enemy action, although one of these was accidentally shot down by a USMC Skyraider).

Jet conversion

On 26 December 1952, the combat elements of the 18th Group – including No. 2 Squadron – moved from Hoengsong to the new jet air base that had been built at Osan-ni in anticipation of the arrival of the F-86s, but problems with the delivery of the Sabres held up the conversion until early 1953. Left over but flyable F-51Ds went to Osan on 11 January 1953, with all aircraft being assigned to the 67th Squadron but flown by pilots drawn from all three squadrons. The 12th Squadron and No. 2 Squadron would be the first to transition into the new F-86F. The last Mustang missions were flown on 23 January 1953.

It would be no easy task for the South African air and ground crews to transition into the Sabre. None of the pilots had ever flown a jet, nor had any of the ground crews maintained an aircraft as complicated as the F-86. Plus it would be quite a challenge to transition from a propeller-driven, 'tail-dragger' fighter aircraft like the F-51 Mustang to a jet-powered aircraft with tricycle landing gear like the F-86 Sabre. Beginning on 7 January 1953, a mobile training detachment for the F-86F came from Tsuiki to Osan to begin the conversion training. It continued eight hours a day, seven days a week, until the task was completed. Several experienced pilots from the veteran 4th and 51st Interceptor Wings were transferred into the 18th to ease the pilot transition.

On 28 January 1953, the 18th Fighter Bomber Wing received its first three F-86F Sabres. One aircraft, F-86F 52-4352, carried a distinctive orange, white and blue stripe on the vertical tail, the colours of South Africa, with Springbok insignia on the wings and fuselage. The other

Lady of Lorette ***receives some attention on the flightline at Osan in July 1951 (above), while*** **Sharonne** ***leads out a USAF Sabre past other 18th FBW aircraft (below). The USAF aircraft wear the old-style 18th fin markings, the red stripes denoting the 67th FBS.***

In the Sabre era the squadron commander's aircraft was coded 'A'. At the start of Sabre operations the aircraft was assigned to Commandant Ralph Gerneke, seen here in the cockpit (left). It was named Sherdanor II*. Subsequently, when J.S.R. Wells assumed command of No. 2 Squadron, the aircraft was renamed* Renkins *(above).*

two aircraft had 18th FBG tail bands and US national insignia. On 30 January 1953, Commandant Ralph Gerneke, commander of No. 2 Squadron, and Major J.S.R. Wells became the first two South African pilots to fly solo in the Sabre.

On 22 February, the first mission was flown, by the commanders of the three squadrons in the 18th Group. The 18th Group Sabres were assigned as part of a fighter interceptor force conducting a MiG Sweep down the Yalu River. Major Jim Hagerstrom, CO of the 67th Squadron and one of several 4th Fighter Group pilots brought into the 18th Group to help with pilot transition into the Sabre, led the flight, with Commandant Gerneke as No. 2. Colonel Maurice Martin, new CO of the 18th, was no. 3, and Major Harry Evans, CO of the 12th Squadron, flew as no. 4. Although several flights of MiGs were called out, combat with the speedy Russian jets was not joined. Hagerstrom eventually scored 6.5 MiG kills to become the first and only ace from the 18th Group. The first dive-bomb mission was flown on 14 April, and the first close support of troops along the MLR was flown on 27 April. The mission of No. 2 Squadron when equipped with the F-86F remained the same as it had been with the F-51Ds: Combat Air Support of front-line troops, and truck-busting along the North Korean MSRs. By the spring of 1953 the pilots had become quite adept at dropping bridges and sealing railroad tunnel entrances. And now, they could and did fly MiG Combat Air Patrols along the Yalu River.

Sabres from the 18th Group, with top cover from 4th and 51st Group Sabres, knocked Radio Pyongyang (Ping-Pong Radio) off the air during the May Day attack led by General Glenn Barcus, boss of 5th Air Force. Radio Pyongyang had been broadcasting disparaging remarks about the men of 5th Air Force, saying they had been dropping germ bombs, hitting hospitals, strafing schools, etc. General Glenn Barcus, commanding 5th AF, ordered an all-out attack on the radio station. Sabre fighter-bombers from the 8th and 18th Groups, including No. 2 Squadron, went north on 1 May to spoil the Communists' May Day celebrations. The large Sabre force looked like it was heading for the Yalu and another typical MiG Sweep, but about 30 miles (48 km) north of Pyongyang, the Sabre fighter-bombers suddenly turned south and struck the radio station again and again with 1,000-lb (454-kg) bombs. General Barcus then radioed the Communists on their own frequency, saying in effect – "I told you!".

At 1000 hours on 27 July 1953, the Korean Armistice agreement was signed, stopping the fighting in Korea. Throughout that final day of the war, UN aircraft roamed the skies over North Korea searching for targets of opportunity. The 'Cheetah' Sabres were among the

Above: **Imp VIII** ***and*** **Black Dick** ***get airborne out of Osan at the start of a fighter-bomber mission. Both aircraft carry a single 1,000-lb (454-kg) bomb under each wing.***

Below: A vital member of the maintenance and support team was the North American Aviation 'Tech Rep' – or Air Force Contract Technician. This is Bill Grover at Osan in 1953.

F-86F-30 Sabres flown by No. 2 Squadron, SAAF

SAAF serial/ tailcode	USAF serial	name(s)	notes
601/A	52-4352	*Sherdanor II/Renkins*	squadron CO's aircraft. To CNAF
602/J	52-4361	*Imp VIII*	to CNAF
603/B	52-4315	*Miss Cloudeyes/Ruth II*	to CNAF
604/K	52-4311	*Bevkeneve/Black Dick*	to CNAF
605/C	52-4313	*Sharonne/Naughty Nellie*	to CNAF
606/L	52-4316	*Little Phyllis/Tomtit*	to CNAF
607/D	52-4320	*Just Joan*	to CNAF
608/M	52-4348	–	w/o 28 Feb 1953
609/E	52-4321	*Malobola*	to CNAF
610/N	52-4333	*Glow Worm*	to CNAF
611/F	52-4310	–	to CNAF
612/O	52-4326	–	to CNAF
613/G	52-4327	*Bevkeneve*	w/o 21 April 1953
614/P	52-4355	–	w/o 28 Aug 1953
615/H	52-4331	*Kevric*	w/o 19 April 1953
616/Q	52-4368	*Lady of Lorette*	w/o 21 July 1953
617/I	52-4346	–	to CNAF
618/R	52-4359	*Gay Jane III*	to CNAF
619/M	52-4344	–	w/o 19 May 1953
620/H	52-4386	*Kevric II*	to CNAF
621/G	52-4312	*Bevkeneve*	to CNAF
622/Q	52-4572	–	to CNAF

many flights attempting to minimise the Communist forces that were being jammed into North Korea before the agreement took effect. 'Cheetah' Sabres flew 41 sorties on that final day. Major J.F. Nortje flew his 100th mission, and Second Lieutenant Wilmans was the last 'Cheetah' pilot to touch down back at Osan. At 2201 hours, the Armistice went into effect and all UN aircraft had to be on the ground and/or south of the bomb line. Lieutenant Wilmans's mission was the 12,067th of the war to be flown by pilots of No. 2 Squadron.

Return home

Following the end of the fighting in Korea, No. 2 Squadron began advanced training to hone its skills in the F-86. By the end of July, the squadron had flown 32 combat 'training' formation sorties. Operation Big Switch brought the South African POWs back home, beginning with Lieutenant Hector MacDonald on 4 August 1953. The training was realistic and dangerous. Lieutenant M.C. Botha was declared dead after he suffered flight control problems with his F-86 and ejected into the Yellow Sea on 28 August.

At all times during the 'training', No. 2 Squadron had to have four Sabres on 15-minute alert status, from sunrise to sunset, to be ready for any further Communist aggression, including the threat of MiGs. In the event, the threat never materialised and on 3 September, Commandant Wells informed the squadron that it was scheduled to rotate back to South Africa. It would be accomplished in three batches, with the first group of personnel leaving on 7 September, and the second group going home on the 22nd.

On 1 October 1953, No. 2 Squadron ceased all operational flying and began turning its Sabres over to 5th AF units still operational in Korea. The last aircraft were returned on 11 October, and all South African personnel had departed Korea by 29 October. The final tally for No. 2 Squadron was impressive indeed. Of the 12,067 missions flown, 10,373 were in F-51Ds and 1,694 were flown in Sabres. They had destroyed 891 vehicles, 44 tanks, 221 pieces of artillery, 147 anti-aircraft sites, 11 locomotives, 553 rail cars, 441 ammunition and supply dumps, dropped 152 bridges, and destroyed 9,837 buildings. Some 2,276 enemy troops fell to squadron weapons during the war.

Above: Seven Sabres from No. 2 Squadron taxi towards the active runway at Osan in 1953. The unit remained in Korea for a few weeks after the armistice, standing alert and conducting combat 'training' operations.

Right: The 'Cheetahs' enjoy a 'jar' in the squadron's bar at Osan, which was known as Rorke's Inn. The squadron's winged cheetah crest took pride of place above the bar and was also worn on the aircraft, with the unit's motto 'Upwards and Onwards'.

In the three years of war in Korea, No. 2 Squadron consisted of 243 officers and 545 ground personnel. Losses included 26 pilots listed as KIA, with eight more who were taken prisoner and repatriated under Operation Big Switch. Two ground personnel were killed in accidents. Of the 95 F-51D Mustangs assigned to the squadron, 74 were lost either to enemy action or operational problems. Of the 22 Sabres that were assigned to the squadron, six were lost. The squadron was awarded a US Presidential Unit Citation for actions during the period of 28 November 1951 through 30 April 1952, and a Republic of Korea Presidential Unit Citation was issued on 1 November 1951. Pilots of No. 2 Squadron were awarded 55 US Air Force Distinguished Flying Crosses, with one cluster to the DFC; two Silver Stars, three Legions of Merit, 40 Bronze Stars, and 176 Air Medals. The squadron was also awarded dozens of Republic of Korea air medals and two Commonwealth MBEs.

So impressed were the veterans of No. 2 Squadron that they campaigned to have the F-86F Sabre become the primary fighter aircraft of the South African Air Force. In 1956, the South African Air Force began receiving Canadair Sabre Mk 6s, flying them well into the 1960s before retirement. As for No. 2 Squadron, it subsequently converted to the Dassault Mirage IIICZ, then the Cheetah C/D, and is now preparing to become the SAAF's only combat unit when it receives the Saab/BAE Systems Gripen from 2008.

Larry Davis

Right:* Little Phyllis *rests at Osan between missions. The open door below the cockpit covered the ammunition containers for the nose guns.

Below: In Korea the UN units had to contend with the weather as well as the enemy. This view shows the extent of the flooding at Osan in the summer of 1953.

Convair B-36 Peacemaker

Designed initially with the aim of striking Germany from US soil, the B-36 was delayed repeatedly so that it did not fly until long after the war's end. By the time the Peacemaker finally entered service, the world was a very different place, and the giant aircraft was entrusted with the grim mission of delivering nuclear weapons to the heart of the Soviet Union. In this role it carried the biggest H-bombs that the US built, and for a few years in the 1950s was the first line of deterrence against the 'Red threat'.

The second B-36A poses on the ramp at Fort Worth, dwarfing three of its bomber forebears (B-17, B-18, B-29). The B-36 shared with the XC-99 and Bristol Brabazon the distinction of having the widest landplane wings (230 ft/70.1 m) until the advent of the Antonov An-124.

The ultimate strategic bomber of World War II made its first flight on 8 August 1946 – nearly a year after the end of the war. The Convair B-36 Peacemaker embodied all of the theory of strategic airpower as it was practised during World War II, and it did so to a greater extent than any other bomber conceived during the war. It had a bomb capacity, range and operational altitude exceeding those of other World War II bombers by several orders of magnitude. Indeed, it was featured as the ultimate weapon in Walt Disney's patriotic 1943 film adaptation of Alexander de Seversky's book, *Victory Through Airpower*. In this animated film, it is the B-36 – or an airplane configured just like it – that delivers the final blow to the Empire of Japan, but that was just a movie.

The B-36 was an amazing milestone of military aviation that was born without a war to fight, and served its entire career having never dropped a bomb in anger. The giant six-engined Peacemaker was the epitome of the wartime vision of a powerful strategic bomber. No combat aircraft ever conceived, before or since – not even the B-52 – was designed to be capable of carrying a greater conventional payload. Of course, it was conceived at a time when the only payloads were conventional. Nuclear payloads would eventually redefine strategic warfare, but in the context of the era that gave birth to the B-36, strategic airpower meant conventional high explosive bombs – and tons of them.

This idea of 'strategic airpower', the notion of taking the war to the enemy's industrial heartland, was born in World War I with airmen such as General William Lendrum 'Billy' Mitchell of the US Army Air Service. Never implemented in World War I and officially disregarded in the years after that war, the strategic doctrine languished in the background until the young forward-looking captains of World War I reached positions of authority within the US Army Air Corps. Among these was General Henry H. 'Hap' Arnold, who became chief of the US Army Air Corps in September 1938.

It was on Arnold's watch that the Air Corps (US Army Air Forces after June 1941) began to seriously nurture long-range strategic bomber programmes involving four-engined long-range aircraft such as the Boeing B-17 Flying Fortress and the Consolidated B-24 Liberator. Meanwhile, similar ideas had been developing in Britain as aircraft such as the Handley Page Halifax and the remarkable Avro Lancaster were taking shape.

As for the doctrine of strategic bombing, it would remain largely theoretical. That is, until the summer and early autumn of 1940, when the German Luftwaffe was perceived to have come within a hair's breadth of defeating the United Kingdom with strategic bombing in the Battle of Britain – and the Luftwaffe was fielding smaller twin-engined bombers.

This was the catalyst that clearly helped to justify increased funding for the manufacture and deployment of B-17s and B-24s. It also served as the catalyst for a six-engined 'intercontinental bomber'. In the conceptual planning for a strategic air campaign against Germany, the range of the B-17s and the B-24s required bases in England. If the Germans had overwhelmed the United Kingdom, such a campaign would have had to have been carried out from bases in the western hemisphere. In 1940, no bomber capable of this was in service or in the production pipeline.

Both Boeing and Consolidated, America's key builders of large warplanes, were already working on aircraft that exceeded the size and range of their B-17s and B-24s. However, even these huge aircraft, the B-29 Superfortress and the B-32 Dominator, would not have the range for round trips to Europe from western hemisphere bases.

On 11 April 1941, a secret design competition was initiated. It called for a plan for an aircraft that had a range that represented an order of magnitude beyond that of the B-29 and B-32. Specifically, the Air Corps wanted a bomber with a combat radius of 5,000 miles (8046 km) that was capable of delivering a five-ton bomb load over that distance. The capability was five times that of the B-17 and double that of the B-29 – whose first flight was still two years away.

As early as September 1940, engineers at Consolidated in San Diego had done some advance design studies for such an aircraft under the in-house designation Model 35. It had the twin tail familiar from Consolidated's B-24 and its early B-32 design. It had the pusher engines that would become familiar in the later B-36, although in the Model 35 the engines numbered four. The design work was updated and presented to the USAAF Air Matériel Division engineers at the Wright Field Aeronautical Engineering Center in October 1941 as Consolidated Model 36. Whereas the Model 35 had a wingspan of 164 ft (50 m), a length of 128 ft (39 m) and a wing area of 2,700 sq ft (250.8 m^2), the Model 36 was still larger. It had a wingspan of 230 ft (70.1 m), a length of 163 ft (49.68 m) and a wing area of 4,772 sq ft (443.3 m^2). The design initially retained the twin tail of the B-24 and the Model 35, but the engines now numbered six.

At Wright Field (now Wright Patterson AFB) near Dayton, Ohio, the Experimental Division was then headed by Brigadier General George Kenney, who went on to serve as commander of the Fifth Air Force during World War II and as the first commander of the post-war Strategic Air Command. The man who would eventually command the operational B-36 fleet was the same man who formally advocated the Consolidated Model 36 to General Arnold.

Based on Kenney's recommendations, the USAAF awarded contract W535-AC-2232 to Consolidated for the purchase of two Model 36 prototypes under the coincidental military designation XB-36. The date on the contract was 15 November, three weeks before the attack on Pearl Harbor.

Getting under way

In San Diego early in 1942, Consolidated's general manager, Isaac 'Mac' Laddon, assembled an XB-36 development team that included men who had already distinguished themselves on what would be the two biggest programmes in company history – the B-24 bomber and the PBY Catalina flying boat. The design problems that were encountered by the team were literally without precedent. This was because of the sheer size of the aircraft. At the time, almost no engineer had ever dealt with structure to gross weight ratios dictated by such a massive airframe.

Preliminary design was under the direction of Ted Hall, who would achieve notoriety after the war for creating a flying automobile, the Convair Model 118. Heading the engineering group was Harry Sutton, while Ralph Bayless headed the aerodynamics group, Bud Woerschel directed the powerplant group, Ken Ward finalised the external shape of the aircraft and Robert Widmer began preparations for the long wind tunnel programme.

Initially, these wind tunnels tests would be conducted at the Massachusetts Institute of Technology (MIT) and at the California Institute of Technology (Caltech) in Pasadena. Later they moved to the large National Advisory Committee for Aeronautics (NACA) tunnels at Langley Field in Virginia, and at the Ames Laboratory at Moffett Field, just south of San Francisco.

B-36s are worked on round the clock at Convair's Fort Worth plant (RB-36E in the foreground). The B-36 represented a massive effort by the facility, which not only encompassed the production of 383 aircraft (not including prototypes) between 1947 and 1954, but also numerous modification and depot-level overhaul programmes.

As well as North American 'pre-strike' bases in Alaska (Ladd, Eielson), Maine (Limestone), Puerto Rico (Ramey) and Labrador (Goose Bay), B-36s were routinely deployed overseas to destinations such as Thule (Greenland), Andersen (Guam), Nouasseur (French Morocco), Yokota (Japan) and Dhahran (Arabia). B-36 TDY locations in England included Burtonwood, Fairford, Lakenheath, Sculthorpe and Upper Heyford. There were also wartime recovery bases, including Adana in Turkey. Only one B-36 ever visited continental Europe, for display at the 1955 Geneva airshow. Around 20 B-36s can be seen here during a 1957 TDY deployment to Yokota in Japan. Parked on a taxiway in the middle of the main group is a Martin RB-57D high-altitude reconnaissance aircraft, serving with the 'Black Knights' of the 4025th SRS.

The XB-36 cruises serenely over the Texan countryside. The giant bomber's first flight was accomplished on 8 August 1946, with Beryl Erickson and Gus Green leading a crew of eight on a 37-minute flight. Although the maiden flight encountered few problems, the early flight test programme was beset with numerous technical difficulties, such were the innovations incorporated in the design.

The XB-36 is seen during taxi tests, which were conducted in late July/early August 1946. These culminated in high-speed runs at up to 97 mph (156 km/h), just below lift-off speed. The massive tyres – the largest ever fitted to an aircraft – were necessary to provide sufficient braking power yet still fit into the wheel well in the wing. Advances in brake technology allowed the development of the production four-wheel bogie, albeit one which required an overwing bulge to accommodate it.

Meanwhile, a full-scale plywood mock-up was taking shape at Consolidated's plant in San Diego. Largely complete by late spring, it was not approved by the USAAF Air Matériel Command until September 1942. Nevertheless, this mock-up was merely a rough draft as the XB-36 design would go though extensive changes over the course of the coming months.

As the airframe work was proceeding, Woerschel's powerplant group was working with Pratt & Whitney, who were developing a 3,000-hp (2238-kW) engine to be used for the big bombers. The result would be the R-4360 Wasp Major, which was essentially a pair of 14-cylinder R-1830 Twin Wasps combined into a single unit. The Wasp Major would ultimately have a long post-war career, being used to power numerous aircraft types in addition to the B-36.

New plant

The plan was for the two XB-36s to be built in San Diego, and to be delivered in May and November 1944. Thereafter, the production series would also be built at San Diego. However, the demand in 1942 and 1943 for the B-24 and the PBY was so great that Consolidated made the decision to shift B-36 production entirely to the new government-owned Plant 4 at Fort Worth, Texas, that was then being built to handle part of the B-24 work load.

By the end of 1942, not long after the arrival of the XB-36 programme in Fort Worth, this vital, top-secret programme gradually started to be downgraded in importance. Delays began to be experienced in the development programme as wind tunnel tests were downgraded in priority. Additional months were lost in the wing trailing edge redesign, and even the move contributed a loss of three to four months.

In September 1942, General Arnold had ordered the "highest priority" to both the Northrop XB-35 and the Consolidated XB-36. From Bataan to Guadalcanal, the United States had been taking heavy losses in the Pacific at the hands of the Japanese, and an Allied counter-strike against the Germans in Europe or North Africa was still a pipe dream. It could not be clearly seen that the tide would be reversed any time soon. However, by the end of the year, things had begun to change. The Allies landed throughout North Africa in Operation Torch, and they had secured a foothold against the Axis. In the Pacific, the tide turned at Guadalcanal, and the pressure was lifted.

The winning of the war was not on the horizon in the last days of 1942, but on the other hand, American and other Allied planners could look at the unprecedented and almost unbelievable industrial mobilisation that had taken place, and they could see that the war would not be lost.

During 1943, as the fall of China seemed imminent, the priority of the XB-36 programme was moved to the fore again, because if China fell to the advancing Japanese, the USAAF could not base long-range B-29s there, as they were planning to do in 1944. They would need either an aircraft with the intended range of the XB-36, or they would need other long-range bases for the B-29s. Eventually, the island-hopping campaign in the Pacific brought bases within B-29 strike distance, but the reconquest of Guam and the Marianas could not be foretold with certainty in mid-1943.

The most difficult obstacle to the progress of the project became the shortage of manpower as the war grew in intensity. The lack of experienced and qualified engineering personnel was felt especially hard on a project as large and complex as the XB-36. In 1943, the average aeronauti-

Thanks to its airliner-style flight deck and lack of operational equipment, the XB-36 was a very 'clean' aircraft by comparison with production machines. Representing Convair's most famous wartime product, the company-owned C-87 Liberator (a dedicated transport version of the B-24D) often flew 'chase' during XB-36 tests, providing engineers a close-up look at the new bomber's behaviour in the air.

cal engineer had only 15 months of experience. By mid-1943, the accelerated B-24 programme was siphoning off personnel and selective service attrition was taking its toll.

Meanwhile, Consolidated was experiencing changes of its own. During 1943, the founder, Reuben Fleet, stepped down (or, as some stories go, was forced out) and the company merged with the Vultee Aircraft Company. The new company was to be known as Consolidated Vultee, and the abbreviation 'Convair' came into use.

Another factor that delayed the XB-36 also impacted other programmes. This was the difficulty that all manufacturers were having in getting urgently needed parts to complete existing orders. One of the big problems with the XB-36, however, was that the USAAF had so far committed itself on paper to only two XB-36s. Consolidated Vultee's new president, Thomas Girdler, complained to Robert Patterson in the War Department that it was difficult to get sub-contractors to work on an order for only two airplanes. He suggested that suppliers would be more interested in the XB-36 if the programme held some promise for large-scale production.

First orders for the B-36

In July 1943, after discussing this with General Arnold, Patterson directed General Oliver Echols, the USAAF's head of procurement, to issue the letter of intent to build 100 B-36 bombers at Fort Worth. This commitment, plus the increased priority prompted by reversals being suffered in the mainland China campaign, gave new motivation to the lagging XB-36 programme, but again, as the situation in the Pacific theatre improved, the programme slipped into competition for scarce resources. It was over a year before the letter of intent was replaced with a firm contract.

In August 1944 the USAAF finally ordered 100 B-36 production-model airplanes. By this time, two months after the Normandy invasion, the war against Germany was imagined to be almost over, but planners envisioned the war against Japan dragging out until 1947, so no priority was assigned to production of the 100 B-36s. In the latter days of the war, in the interest of achieving increased progress, it was decided that as much as two years could be cut from the developmental time of the B-36 by beginning production even before the two experimental models were rolled out.

With a lightplane nestling under its protective wing, the XB-36 sits on the Fort Worth ramp. During testing the main problems encountered were engine cooling deficiencies and propeller vibration. On one flight a propeller fell off – harmlessly into farmland. On the 16th flight, just after take-off, the retraction cylinder for the starboard main undercarriage burst, leaving the gear in a critical state and damaging the inner engine nacelle. The XB-36 circled for five hours to burn off fuel, and all but the two pilots bailed out before Erickson and Green nursed the aircraft back to a safe landing.

Following the capture of Guam and the Marianas Islands in October 1944, however, B-29s – as well as the small number of Consolidated Vultee B-32s – could bomb the Japanese homeland without the necessity of having bases in China. With this in mind, Consolidated Vultee was ordered to step up the B-32 programme, and once again, the XB-36 stood down for a lesser airplane.

As early as July 1944, a letter to Consolidated Vultee President Harry Woodhead from General B.E. Meyers, of the USAAF Air Matériel Command, directed "positive and vigorous action to place the B-32 on a Number One priority from an engineering, tooling and production standpoint, without reference to the effect this action may have on the XB-36 and B-36 programmes."

The 'Convair' name

Confusion over correct terminology often arises in discussions of the name of the manufacturer of the B-36. Technically, the company name was Consolidated Aircraft Corporation until 1943, and Consolidated Vultee Aircraft Corporation thereafter. The term 'Convair' came into use during World War II, and it became universally used in the late 1940s, but then only as an acronym for Consolidated Vultee Aircraft.

Consolidated Aircraft Corporation was organised by Major Reuben Fleet in 1923 in East Greenwich, Rhode Island. The company moved to Buffalo, New York, in 1925, and to San Diego in 1933 with a staff of only 900. By 1939 it had grown to 6,000 employees. As expansion of the American aircraft industry took place, it stepped up to around 40,000 employees by the time that the United States entered World War II.

During World War II, Consolidated operated 13 divisions throughout the country. There were manufacturing plants at San Diego in California, as well as at other sites, including Louisville, Allentown, New Orleans, Miami and Wayne, Michigan. Modification plants were also located at Tucson, Arizona, and Elizabeth City, New Jersey. A research division was also centred at Dearborn, Michigan. Aside from the home base at San Diego, Consolidated's most important facility was the massive facility known as Plant 4. Built and owned by the United States government, the factory was operated by Consolidated and located across the runway from the Fort Worth Army Air Field. Plant 4 had a mile-long assembly line that was specifically constructed for production of Consolidated B-24 Liberators.

The Fort Worth factory delivered its first B-24 in April 1942 and would be the centre of production for both the B-36 Peacemaker and the B-58 Hustler. When it opened, this facility was the world's largest integrated aircraft factory. Completely windowless, it was so long that people on bicycles became as commonplace as the many jigs, benches, desks, cabinets and assembly fixtures necessary for the giant task of building bombers. At the turn of the century it was still turning out Lockheed Martin (formerly General Dynamics) F-16s and is preparing to assemble the F-35 Joint Strike Fighter.

During the war, Consolidated also operated a now forgotten transpacific airline called Consairway for the USAAF Air Transport Command.

In March 1943, Consolidated merged with the Vultee Aircraft Corporation of Downey, California, and the company became the Consolidated Vultee Aircraft Corporation. Through this merger, manufacturing plants at Downey and Nashville, Tennessee were added.

Further mergers would result in Convair becoming the Convair Division of General Dynamics Corporation in 1954. It was now officially Convair, but it was the Convair Division of the larger corporate entity. Convair would survive in this form until 1995, when General Dynamics closed the Convair Division permanently.

In early 1950 the XB-36 was used to test an experimental track undercarriage. The complex arrangement dramatically increased the aircraft's footprint and would, theoretically, have allowed operations from weak runway surfaces. The aircraft only flew in this configuration once, on 26 March 1950, the take-off being rough and noisy. After a circuit, the aircraft landed safely, although the undercarriage began to break up during the roll-out. The gear was never intended for production, the trial aiming to prove only its feasibility for use on a large aircraft.

Below right: A Convair engineer lends scale to the single mainwheel which was initially fitted to the XB-36 and YB-36. The tyre measured 9 ft 2 in (2.79 m) in diameter and 3 ft 10 in (1.17 m) in width. Only three runways in the US were strong enough to handle it at the time of the B-36's debut.

Below: In June 1948, on its 32nd flight, the XB-36 took the four-wheel bogie undercarriage into the air for the first time. The prototype was retired in January 1952, and was subsequently incrementally destroyed on the Carswell AFB fire dump.

By the end of 1944, the XB-36 experimental factory was 18 months behind schedule, and this blow further complicated the matériel delays, armament revisions and minor redesigns of system equipment.

In the two years that elapsed from the time the XB-36 programme moved to Fort Worth, and the summer of 1944, when it was pushed behind the B-32 in importance, the intercontinental bomber had evolved in a series of fits and starts, but it had evolved. One of the first major refinements in the original Model 36 design came in the fall of 1943. The USAAF Air Matériel Command was doing the static tests of the twin-tail configuration, and after a considerable amount of research, and conferences with Consolidated, it was decided to abandon the twin tail in favour of a single fin and rudder. Static tests of the twin-tail, with its large vertical fins and rudders and associated control brackets, showed that these could shear off in a hard landing or severe flight conditions. The design had been successful on the B-24 and in Consolidated Vultee's early flying boats, and there was some influence felt by what had been seen as Reuben Fleet's preference for it as a sort of 'trademark' look. However, by 1943, Fleet was out of the picture and the extreme size of the XB-36 mitigated against twin tails.

In another part of the airplane, a single-wheel landing gear was incorporated because the state of the art at that

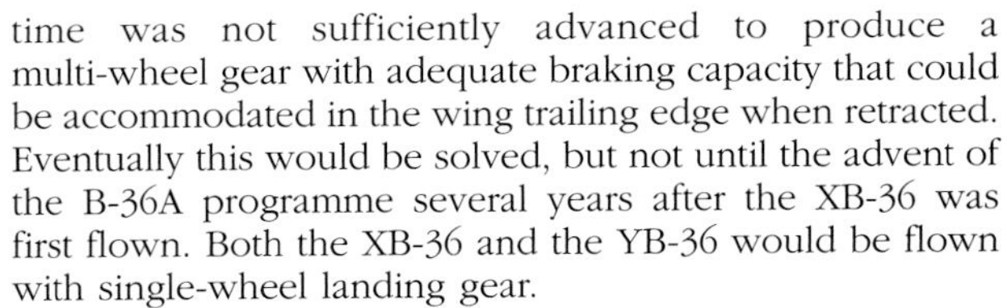

time was not sufficiently advanced to produce a multi-wheel gear with adequate braking capacity that could be accommodated in the wing trailing edge when retracted. Eventually this would be solved, but not until the advent of the B-36A programme several years after the XB-36 was first flown. Both the XB-36 and the YB-36 would be flown with single-wheel landing gear.

The first Pratt & Whitney R-4360-5P Wasp Major test engine was to have been delivered to Fort Worth in May 1943, but design improvements delayed it until October. Pratt & Whitney's Wasp Major, like the B-36 itself, represented new and untested technology that had evolved as a result of the advances that had occurred during World War II.

By December, the full-scale wind tunnel tests were begun in Wright Field's 20-ft (6.1-m) wind tunnel. This series of tests was directed by Consolidated's Dalton Suggs, in close association with the Air Matériel Command Powerplant Laboratory and Pratt & Whitney engineers. These tests continued through 1944 and into 1945. In August 1945, under the direction of Suggs, the engine tests were moved to the full-scale 40 x 80-ft (12.2 x 24.4-m) wind tunnel at the Ames Laboratory at the Moffett Field Naval Air Station south of San Francisco.

While there had been a good deal of technical progress on the XB-36 in the years from 1942 and 1945, the USAAF's official ambiguity toward the once-vital project caused it to languish on a back burner with no clear sense of when – if ever – a real aircraft would ever be built and flown. In the same building at Fort Worth where B-24s had been completed, rolled out and flown away by the hundreds – a new aircraft every few hours for years – the single XB-36 prototype still languished, unfinished, on the factory floor.

Procurement cut-backs

In the months following the surrender of Germany in May 1945, warplane procurements were cut back and many contracts were re-examined. Among heavy and very heavy bombers, B-17, B-24 and B-32 production ended, with only the B-29 still being manufactured. However, in reviewing the progress of the war, USAAF planners were sobered by the tremendous price that had been paid in lives and matériel during the campaign for bases in Guam, the Marianas and, especially, for Iwo Jima. In August 1945, the advent of nuclear weapons provided yet another argument in favour of the development of the intercontinental bomber. In a future nuclear war, retaliation would have to be immediate, and could not wait for the conquest of over-

seas bases. In early August 1945, an Air Staff Group conference recommended that four groups of B-36s be included in the mobile task force of the proposed 70-group post-war Air Force.

Although the XB-36 was now back in favour with USAAF brass, the programme did not resume with ease, nor did the product of USAAF indifference please the USAAF now that it was no longer indifferent. Four years of fits and starts, as well as the lack of official interest, had taken a toll. In August 1945, General Lawrence Craigie, the commander of the Air Matériel Command Engineering Division, inspected the XB-36 at Fort Worth and complained to Harry Woodhead about what he saw as inadequate quality control, but in Consolidated Vultee's defence, this was a programme for which the USAAF had expressed little enthusiasm over the preceding year, and from which the company had been compelled to divert resources in order to address the needs of higher priority wartime programmes.

There were also a number of changes that the USAAF had imposed on the programme during the winter of 1944-1945. These were received by Consolidated Vultee against a backdrop of uncertainty over whether the USAAF was going to cancel the programme. The crew was increased to 14, and in early 1945 the USAAF had requested armament changes which added nearly five tons to the gross weight. The original requirement for eight 0.50-in (12.7-mm) calibre machine-guns and a half dozen 37-mm cannon had been stepped up to 10 0.50-in and five 37-mm guns, but the USAAF had changed its mind again. The Air Matériel Command decided to use eight General Electric powered, remotely controlled turrets housing two 20-mm cannon each, with six of these turrets now specified to be retractable.

During the summer of 1945, it still seemed that the war against Japan would drag on for another year, so there was still seen to be a future for the B-36 in World War II. Then, in early August, a pair of B-29s delivered a pair of heretofore secret nuclear weapons against the Chrysanthemum Empire, and the conflict abruptly ended. Japan officially signed the surrender documents on 2 September.

World War II was over, and for the first time since the preliminary design days of 1941, the XB-36 programme could now have the experienced engineering and production personnel that it needed. Reorganisation and closer supervision by Ray Ryan brought new efficiency and a new mood. This and other changes helped to re-establish confidence in Consolidated's development of the airplane. In August, Ryan was moved up to assistant division manager, and Wesley 'Wes' Magnuson came from the Downey Division as the third experimental shop manager. Hinkley became assistant chief engineer for production design, and Henry Growald, a very competent engineer, took his place as project engineer.

With the XB-36 in the background, the engineless YB-36 is wheeled out at Fort Worth in May 1947. It did not fly until 4 December, initially with the single-wheel undercarriage but introducing the domed flight deck. This was adopted to provide a better view for the pilots, and also to provide a position for nose guns. After its test work was completed in May 1949, the aircraft was outfitted for active duty as an RB-36E. At the end of its service career it was placed on display at the USAF Museum, but the arrival of another B-36 meant that the RB-36E was redundant. It was broken up, but its remains were bought by a local scrap dealer, and they remain at his property today.

During World War II, the B-36 programme had been embraced and rejected time and again. The United States government, which had been so enamoured by the notion of the great intercontinental bomber in 1941, would be a very fickle lover.

XB-36 prototype

In September 1945, just three weeks after VJ Day, the XB-36 was rolled out of the experimental shop on its own landing gear. After several design changes and many refinements, the XB-36 was a reality at last. The biggest obvious design change, the huge single tail, now rose nearly five storeys above the ground. The tailplane had nearly the same span as the wingspan of a C-47 transport, and as such, it was larger than any aircraft yet conceived.

The XB-36 had the same dimensions as the Model 36 proposal, with a tail height of 46 ft 10 in (14.27 m). It weighed 131,740 lb (59757 kg) empty, and had a gross weight of 276,506 lb (125423 kg). It would be powered by six 3,000-hp (2238-kW) Pratt & Whitney R-4360-25 Wasp Major air cooled radial engines, and it would have a top speed of 346 mph (557 km/h) at 35,000 ft (10668 m) with a service ceiling of 36,000 ft (10973 m).

The sun had finally shone on the Consolidated Model 36 – which had been begun before the United States entered World War II – just a month after the end of the war. The airplane which might have made the bloody campaigns on Iwo Jima and Okinawa unnecessary, had spent the entire war in a windowless room. The programme had also spent the entire war in a metaphorical windowless room in the priority file of USAAF planners. Now it was out in the light, both literally and figuratively.

When the war ended, no one knew better than the USAAF hierarchy how huge a role in the victory had been played by strategic air power. As the war wound to a close, the B-36 programme was once again at the top of the list. However, the sheer size of the bomber and the infrastructure required to build it made the planners nervous.

Building large numbers of a massive aircraft requires an equally impressive production facility. B-36 construction was undertaken in the huge Forth Worth plant: here RB-36Ds progress down the Final Assembly line. The aircraft had to be angled towards the end of the line once the wingtips had been fitted (below), while to move them they had to be tipped up so that the tail would clear the roof (below left).

Proudly wearing its 'buzz' number BM-004, the first production B-36A is seen during its first flight (right) and during its second and last flight (above), a ferry trip to Wright Field. The aircraft had flown before the YB-36, and after its initial test flight from Fort Worth was sent to Ohio for various ground trials, including cabin pressurisation and static load tests. In the course of these it was systematically destroyed – it had only flown for 7 hours and 36 minutes.

The emergence of the prototype had been accompanied by apprehension and conjecture. Critics had conceded that progress was being made in size, but what about military effectiveness and necessity? Would the B-36 be a milestone or a millstone?

Through the first post-war winter, the refinements to the still-unflown XB-36 continued. The XB-36 programme, if not the XB-36 prototype itself, moved forward. The USAAF was eager to get the XB-36 into the air at the quickest possible date and had met with Consolidated to determine ways of expediting the initial flight. Just as the design process had progressed sporadically during the war, so did the final work on the prototype after the war. Quality control problems were smoothed out, but two union strikes – in October 1945 and February 1946 – affected and interrupted both engineering and production.

The engines were ground-tested and eventually hung, but when Consolidated began engine tests on 12 June 1946, the wing flaps literally disintegrated. The magnesium alloy, fabric-covered flaps could not withstand the tremendous punishment caused by propeller turbulence. The structure experienced rib cracking and fabric ruptures up to the torque box, and it was necessary to delay further testing for six weeks while stronger aluminum alloy flaps were fabricated.

The crew chosen to take the big airplane on its initial flight were two veteran test pilots from the B-24 and B-32 programmes. Beryl Erickson would be the chief test pilot for the B-36 programme, while G.S. 'Gus' Green, Consolidated Vultee's chief of flight test and Erickson's boss, would fly as co-pilot in the initial stages for testing. Having done developmental flying for both the B-24 and B-32, Erickson was one of the few pilots checked out in the Douglas XB-19, which was the largest landplane in the world – until the XB-36. Flight engineers for the early stages of the programme would include William Easley and James McEachern. There would also be two flight test analysts, W.H. Vobbe and A.W. Gedeman, and flight test engineer Robert Hewes, aboard for the early flights.

To check out the flight systems, Erickson, Green and the engineering team conducted a series of taxi tests that lasted from 21 July through 7 August. During the 37-minute first flight, which began at 10:00 am on the following day, Erickson and Green took the XB-36 up to 3,500 ft (1067 m) and tested it at speeds up to 155 mph (249 km/h). A 43-minute second flight followed on 14 August, with Major Stephen Dillon, the USAAF XB-36 flight acceptance officer aboard. Dillon was a wartime B-17 pilot who coincidentally had been Erickson's high school classmate in Venice, California, before the war.

As the flight test programme continued into the autumn of 1946, several problems began to appear. One such difficulty was with inadequate engine cooling, which prevented extended operations above 30,000 ft (9144 m). Another major issue was propeller vibration. A third involved the use of aluminum, rather than copper, wiring. This was used in order to save weight, but there were numerous failures and short circuits.

The case for the B-36

"The responsibility for a great portion of our ounce of prevention [against war] lies squarely on your shoulders. The tense situation between the world powers points out very clearly the importance of our having in hand that prevention now. During the early stages of World War II, with our furthermost bases in Australia and India, and a little later in the Solomon Islands and New Guinea, the need for a 10,000-mile bomber was easily apparent if we were to strike the enemy in home territory. In order to meet these requirements, the B-36 was brought into being and immediate action was taken to produce the XB-36. However, the Japanese situation deteriorated more rapidly than we had expected. With bases closer to Japan, we could then use B-29s. The priority for production of the B-36 was dropped in favour of the other planes.

"We continued our work on the B-36 after the war, despite the natural reluctance of the American public to continue to spend money for such programs as B-36 production. In view of the unrest in the world of today, this line of action appears to have been well justified. Recently, it has been possible to prove the potentialities of the B-36 under actual flight conditions.

"We believe in the adage to the effect that an ounce of prevention is worth more than a pound of cure. We are vitally concerned now with the production of B-36s because the B-36 is our ounce of prevention."

Lieutenant Colonel Beverly Warren, Chief of the Air Force Procurement Field Office, Fort Worth (1948)

The first serious incident came during the 16th test flight on 26 March 1947 when the starboard main landing gear hydraulic retraction cylinder ruptured, releasing the landing gear. Erickson and Green cut power to the nearby engines, regained control and began circling to use up fuel while the rest of the crew parachuted to safety. Several of the men, including Major Dillon, suffered serious injuries in hard parachute landings. Despite his injury, Dillon insisted on going to the Fort Worth control tower to help talk the plane down. The heroic effort on the part of Erickson, Green and Dillon resulted in a successful landing and surprisingly little damage to the XB-36. Loss of the aircraft would have been a serious blow to Consolidated Vultee and to the programme.

The company continued to test the XB-36 through June 1948, when it was officially turned it over to the US Air Force – which had been formed in September 1947 when the USAAF was formally divorced from the US Army. The flight test analysis equipment was removed, and 3,500-hp (2611-kW) Pratt & Whitney R-4360-41 engines and new four-wheel landing gear were installed. The aircraft would now be used by the Strategic Air Command for crew training. By this time, the first production series B-36A aircraft had been flowing from the Fort Worth factory for nearly a year.

YB-36

The initial XB-36 contract, dated 15 November 1941, had called for two airplanes, both to be built in San Diego, with the first to be delivered by May 1944, and the second to be delivered in November 1945. In fact, no XB-36s were delivered in 1945, and only one XB-36 would be built. The second aircraft appeared much later than the first, and it was delivered under the designation YB-36. While the 'X' prefix meant 'experimental', the 'Y' prefix identified the second B-36 as a 'service test' aircraft, which officially earmarked it as the flight test airplane for the joint use of Consolidated Vultee and military personnel.

When Japan was defeated in the summer of 1945, the USAAF began to rapidly decelerate most development programmes, but the contrary was true with the B-36. As noted above, the USAAF showed renewed interest in the first XB-36. Fabrication of basic components for the second prototype, then also still an XB-36, was actually pushed ahead faster. By the spring of 1946, design and engineering changes that had come about during the development of the first prototype were incorporated into the second, and it was redesignated as YB-36 before it was actually completed.

As it took shape during the winter of 1946-1947, the YB-36 began to differ from the XB-36 in a number of ways, principally by its redesigned flight deck. The XB-36's limited visibility had been noted by Beryl Erickson and Gus Green during early flight testing, but engineering studies had already begun on an improved layout. In fact, a full-scale mock-up of a revised flight deck had existed as early as June 1945.

B-36A

Proudly wearing the shield of the 7th Bomb Wing on its nose, the last B-36A to be built is seen on display at an air show. Notable are the bomb bay doors, which slid round to lie flush with the fuselage sides. From the B-36D four-section folding doors were adopted, which could snap open in around 2 seconds.

The first production variant was built in four production blocks. It featured the domed canopy first seen on the YB-36 and a redesigned four-wheel main landing gear system, which replaced the single wheel main landing gear of the XB-36 and YB-36. The first example of this variant made its initial flight on 28 August 1947. All but one of the B-36As were later modified to RB-36E standard.
Quantity produced: 22 **Tail numbers:** 44-92004 to 44-92025

The ninth B-36A takes off from Fort Worth. Convair pilots usually undertook around three or four 'shake-down' flights with a new aircraft before it underwent its Air Force acceptance test flight. On 26 June 1948 the 12th aircraft (BM-015, named* City of Fort Worth*) was the first to be handed over to the 7th Bombardment Group, which was located across the runway from the Convair plant at Carswell AFB.

Another consideration in this redesign was the desire by the USAAF to put a nose turret in the B-36, because experience during the war had shown bombers to be vulnerable to head-on attacks by fighters. Nose and chin turrets of various designs were worked out in the early days of the war and applied to B-17s and B-24s, while the B-29 was built with a four-gun turret atop the forward fuselage.

In the new and improved forward cabin design, visibility was greatly increased by the greenhouse-dome-type canopy covering the pilot, co-pilot and flight engineer. There was improved efficiency in this layout, because the flight engineer was now facing aft, and was able to scan the six pusher engines in flight.

The YB-36 made its first flight on 4 December 1947, 16 months after the first flight of its sister ship. Again it was Beryl Erickson who was at the controls. With its more efficient turbo-superchargers and other engine equipment changes, the YB-36 easily outperformed the XB-36, and only 10 days after its first flight, the YB-36 topped the XB-36's highest altitude mark by 3,000 ft (914 m), going to 40,000 ft (12192 m).

B-36Bs from the 7th BG (Wing from 1951) thunder across Texas. This unit provided a five-ship flypast at President Truman's inauguration in January 1949.

B-36B - the first operational model

The second production variant was the first that was fitted out for combat operations with all defensive armament installed. It also had 3,500-hp (2611-kW) R-4360-41s fitted in place of the 3,000-hp (2238-kW) R-4360-25s of the B-36A. The first B-36B flew on 8 July 1948 and deliveries began to the USAF on 30 November. The 7th BG was the first recipient.
Quantity produced: 62 **Tail numbers:** 44-92026 to 44-92087
59 B-36Bs were later modified to B-36D standard. The last 34 aircraft of the initial B-36B order were switched to the subsequently cancelled B-36C (44-92065 to 44-92098). They were completd as B-36Bs, B-36Ds or RB-36Ds.

Above: BM-026 was the first B-36B, and is seen before installation of cannon armament – one of the main features which distinguished the operational bomber variant from the B-36As, which were mainly used for service test and crew training duties.

Force deployment was a cornerstone of SAC's deterrence policy, and to facilitate rapid movement to forward-based airfields Convair developed a two-pod system to transport complete R-4360 engines. The two pods were connected by a simple plank cross-piece, which was attached to the aircraft's no. 1 bomb bay. The bay doors were opened before the pods were attached. The pods are seen here on B-36Bs, including the first aircraft (right).

To procure or not to procure

That was the question. It was asked with varying answers through World War II, and it continued to be asked after the war ended. Even as the XB-36 was making its first flights, there was a constituency within the USAAF (US Air Force after September 1947) that favoured acquisition of the big bomber, and a constituency who felt that its World War II-era technology was obsolescent.

In March 1946, the USAAF had reorganised themselves into 'major' commands. Excluding the occupation forces in Europe and the Far East, three USAAF major commands

would be operational commands that operated and maintained combat aircraft. Air Defense Command (ADC) would have interceptors, while Tactical Air Command (TAC) controlled tactical aircraft, such as fighter-bombers, attack bombers and medium bombers. The strategic bombers, as well as a number of escort fighters, were assigned to the Strategic Air Command (SAC). This organisation inherited most of the strategic bombers in the USAAF, as well as existing production contracts for future strategic bombers, such as the order for 100 production B-36s already placed.

To head the new Strategic Air Command, the USAAF chose General George C. Kenney, who had commanded the Far East Air Forces (FEAF) during World War II. As such, Kenney had commanded virtually all USAAF assets in the Pacific theatre except the B-29 strategic force that had brought Japan to its knees. The B-29s had been flown exclusively by the Twentieth Air Force, which was commanded by USAAF Chief General 'Hap' Arnold, and managed in the field by General Curtis LeMay.

Strategic Air Command's new commander had been an early supporter of the intercontinental bomber concept, but his experience had been with tactical air power. When he assumed leadership of Strategic Air Command at Bolling Field on 21 March 1946, he inherited an organisation with a mission unlike that which he had overseen in World War II.

General Kenney recommended the Boeing B-50, an improved variant of the B-29, with bigger engines and a range of over 4,600 miles (7400 km) with a full bomb load. Although this was less than the B-36, Kenney argued that this was adequate because the B-50 was based, in large part, on proven technology – the B-29. He said that the B-50 was superior to the B-36 in all respects except those of bomb capacity and range. He told Spaatz that he would like to consider the B-50 to be the standard heavy bomber, with the as-yet-untested six-jet Boeing B-47 as its eventual replacement. He also believed that strategic bombing operations would always require the use of advance bases within striking range of enemy targets.

Meanwhile, the US Air Force had expressed a concern – ironic in light of the vast arsenals that exist today – that the supply of atomic bombs might be limited. With this in mind there was a perceived need for an all-purpose 'workhorse' bomber to be used against targets which might have to be destroyed by conventional bombs from bases close to a potential enemy. One suggestion had been to refurbish the large number of B-29s that were still in service from the wartime deliveries that had topped 3,000 aircraft.

Kenney was one of the sceptics within the USAAF who imagined little future for the B-36, but his boss, the Chief of Staff, General Carl 'Tooey' Spaatz, disagreed. Spaatz had commanded the USAAF Strategic Air Forces in Europe during World War II and he could see a need for a massive bomber with an unprecedented bomb load.

This page: Eighteen B-36Bs from the 7th Bomb Wing were assigned to Project Gem in 1948/49, in which B-36s were outfitted to carry Fat Man, Little Boy and Mk 4 atomic bombs, and were tested for operations against the Soviet Union from Limestone AFB in Maine, Goose Bay in Labrador, and Ladd AFB and Eielson AFB in Alaska. Their mission profiles would have taken them 'over the Pole', and they were given high-conspicuity red markings (fluorescent paint had not then been developed) for Arctic operations. Gem represented the first real operational capability for the B-36 force.

In spite of the predisposition of Spaatz in favour of the B-36, Kenney recommended to him that the 100-plane B-36 production order inked in 1944 be cancelled. He added that maybe a small number of service test YB-36s would be acceptable. The former tactical air commander allowed that he was concerned about resting the weight of the Strategic Air Command's unproven reputation and the hard-won struggle for strategic air power on an airplane which was experiencing difficulties during its flight test programme. He mentioned that available data indicated that the B-36 would have a useful range of only 6,500 miles (10460 km).

General Nathan Twining, who had commanded the Fifteenth Air Force heavy bombers during World War II, and who was then head of the Air Matériel Command, also disagreed with Kenney. Twining said that the XB-36 could not be judged totally by experimental tests. He cited the fact that both the B-17 and the B-29 had encountered considerable difficulty before they became satisfactory performers. In addition, a new and more powerful R-4360-41 engine had been developed for installation in the 23rd production B-36A airplane, and the new four-wheel landing gear would be installed on those airplanes currently in production, enabling use of any field suitable for B-29s. Twining agreed with Kenney that the all-jet Boeing B-52, then in the earliest design stage, would probably be a better airplane, but everyone knew that it would not be available until 1953 or later, and there were still nagging doubts over the long-range capabilities of jet engines.

Twining argued that, with the Cold War brewing in Europe, the USAAF needed an airplane of the B-36's range, and they needed it in the 1940s and the early 1950s. He pointed out that if, in the past, the USAAF had waited for the 'best' rather than the 'best available', such a policy would have lost the B-17, B-29, P-47 and P-51, all of the aircraft that are remembered as the USAAF's pivotal combat aircraft in World War II.

B-36D and RB-36D - jet augmentation

The B-36D was essentially a B-36B modified by the addition of four General Electric J47-GE-19 turbojet engines mounted in two outboard wing pods. The D also introduced new folding bomb bay doors in place of the earlier sliding doors. The first B-36Ds were completed as such at Fort Worth, including the last four aircraft from the original 1944 order. Of the total of 81 aircraft, 36 B-36Ds were converted to Featherweight III standard. The conversions were from B-36Bs, the majority of the work being performed at San Diego
Quantity produced: 26 **Tail numbers:** 44-92095 to 44-92098, 49-2647 to 49-2668
Quantity converted: 59 (from B-36B)

The RB-36D was the reconnaissance variant of the B-36D. The first example made its initial flight on 18 December 1949. Of the total of 24 aircraft, 11 were converted to Featherweight III standard and 10 were converted as FICON Fighter Conveyors and redesignated as GRB-36Ds.
Quantity produced: 24 **Tail numbers:** 44-92088 to 44-92094, 49-2686 to 49-2702

Above: 44-92057, built as a B-36B, served as the B-36D prototype. Rather than wait for the intended General Electric J47-GE-19 turbojets to be ready, the prototype was completed with Allison J35-A-19s instead, housed in slimmer nacelles on an unbraced pylon. It flew on 26 March 1949 with these engines, and on 11 July with the J47s.

Above right: This view of a B-36D highlights the enormous wing area, and also the new quick-action bomb bay doors.

Spaatz agreed with Twining and overruled Kenney's suggestion. The B-36 programme continued as scheduled, but Kenney still grumbled, and his comments would supply ammunition for other critics.

Based on preliminary performance estimates deduced from the performance of the XB-36, the B-36A was believed to be vulnerable to enemy fighter attack because of its relatively low speed. However, since its great range gave the crew some chance of getting back, it was considered a better means of delivering nuclear weapons than medium-range bombers on one-way missions.

By the time that the US Air Force became independent of (and equal to) the Army under the Department of Defense on 18 September 1947, the long-range strategy implications of wartime events began to come into better focus. For instance, there was a new examination of the role of strategic bombing and the means of carrying it out. Since August 1945, strategic air power, employing atomic weapons, was capable of dealing such powerful blows that it had altered the traditional relationship between the military services.

Meanwhile, there was a great deal of talk about a 'Nuclear Pearl Harbor', a sneak attack by 'an enemy' (read Soviet Union) using nuclear weapons. It was now perceived by many that the very survival of the nation might depend on the readiness of the forces needed to deliver a counter-attack, and in mid-1947 the B-36 appeared to be the only weapon which would enable the new US Air Force to launch such an attack without first acquiring bases overseas. It was like a *déja vu* flashback to the autumn of 1941, when the B-36 programme had been initiated.

Based on the global strategic situation as it existed in 1947, the Air Force would decide to acquire the B-50 and to continue with the procurement of 100 B-36s as originally planned. This number was considered adequate for a special-purpose bomber, in view of the limited availability of atomic bombs. The long-range plan was that eventually the B-36 would be replaced by an all-jet heavy bomber, such as the Boeing B-52.

Going into production

Designing the XB-36 and the YB-36 had been a long and complex process, clouded by indecision on the part of the USAAF procurement staff over whether and when they actually wanted the big bombers in service. Considering how often the USAAF changed its mind, it is amazing that the initial production contract for the aircraft occurred roughly two years before the XB-36 made its first flight.

In August 1944, when it seemed like the fast Allied advances in western Europe would bring about Germany's defeat before Christmas, USAAF planners began to draft a blueprint for rearranging American resources for the big assault on Japan. This included the order for 100 B-36A production aircraft based on the still-untested XB-36. They imagined that it could take until 1947 to beat Japan and they wanted to pull out all the stops. The USAAF-owned Fort Worth plant, which had been built to build B-24s, was now challenged to manufacture an airplane roughly twice the size and complexity of the Liberator.

However, in an abrupt about-face, the USAAF had reversed itself and ordered Consolidated Vultee to begin producing large quantities of B-32 Dominators. They were

Above: Servicing the B-36 remained a challenge throughout its career. Wheeled 'workshops' were devised which allowed engine maintenance to be undertaken with some protection from the elements.

Right: "Six turning, four burning". This B-36D wears the SAC badge and 'Milky Way' sash on the nose. The yellow markings indicated a squadron assignment within the particular wing.

A classic Peacemaker view shows RB-36D 49-2688 on a pre-delivery flight. The reconnaissance B-36 carried a camera suite in its no. 1 (forward) bomb bay and a row of three ECM antennas on the no. 4 bomb bay. These were later moved to the aft fuselage to allow the RB-36 to carry a larger bombload. There was an additional radome between the nose glazing and nav/bomb radar, and a row of stick antennas on the lower forward fuselage sides. The shiny sections of the airframe were aluminium, while the dull portions were magnesium. The first aircraft to be built as an RB-36D first flew in December 1949, but initially lacked the jet engines.

large bombers similar to the Boeing B-29 Superfortress that would ultimately fly a few missions against Japan, but would not be present in time or in sufficient numbers to make a difference in the war. But that, as they say, is another story.

B-32 production had ceased before the programme built up any momentum. When Japan was finally defeated, the partially-assembled B-32s on the production line were cut up for scrap, and the mile-long main assembly building at Fort Worth became a cavernous, vacant 'bowling alley'. The big room would not remain vacant for long. Soon it would be producing an aircraft worthy of its own scale, the largest bomber that would ever be built anywhere.

The mile-long B-36 assembly line was divided into three segments: major components, major mating, and final assembly. Pratt & Whitney, the engine builder, had its own crews on hand in the latter section through 1948 to help install the six new-technology R-4360 engines and to develop future maintenance and handling procedures for this marriage of engine and airplane.

Reliability of the R-4360 engine was still unproven, and Pratt & Whitney kept a crew of personnel at Fort Worth giving certain flight test airplanes the sort of attention that came to be called "the white glove treatment." The engine contractor had a big investment in the future of the B-36 programme, so, because of the complexity of the engine, Pratt & Whitney wanted to be on hand to assist Convair and Air Force personnel in learning every facet of its operation, and to see that it would obtain the proper high-altitude performance. The crew from Pratt & Whitney monitored the Flight Testing Programme throughout 1947 and 1948, developing procedural systems for proper maintenance and operation.

Putting the team in place

Consolidated Vultee Executive Vice President 'Mac' Laddon assembled the team that would make the production B-36 a reality. Many had been in touch with the programme ever since the Fort Worth plant opened. Fort Worth Chief Engineer Sparky Sebold and his assistant, Jack Larson, headed the B-36 engineering department. Robert Widmer led Advanced Development, and Roland Mayer would serve as the Fort Worth Division Manager until he was succeeded by his former assistant, Ray Ryan, in September 1948. Supervising the factory operation was Factory Manager C.H. 'Cliff' White. Meanwhile, 'Mac' Laddon continued his close supervision of the programme, as he had done since the early days, when the company first submitted the B-36 design in the 1941 USAAF heavy bomber design competition.

On the USAAF side, Air Matériel Command set up a B-36 Project Office at Wright Field (now Wright Patterson AFB), under Colonel (later General) Donald Putt and his assistant, J.A. 'Art' Boykin. Putt had been project officer on the XB-17 Flying Fortress, and he had been severely burned attempting to extricate the crew when the XB-17 prototype crashed

Proud defenders of America – a B-36D flies alongside a Lockheed F-94B from the Oregon Air National Guard in the mid-1950s (the F-94 was only operated by the 123rd FIS between October 1955 and June 1957). B-36s often engaged in mock attacks on the US to exercise Air Defense Command's interceptors. This also gave the B-36 crews experience in penetrating the kind of well-defended airspace that they would have faced if they had gone to war against the Soviet Union. Note the anti-thermal reflective white paint which was added to the undersides in an attempt to shield the aircraft from the effects of nuclear flash.

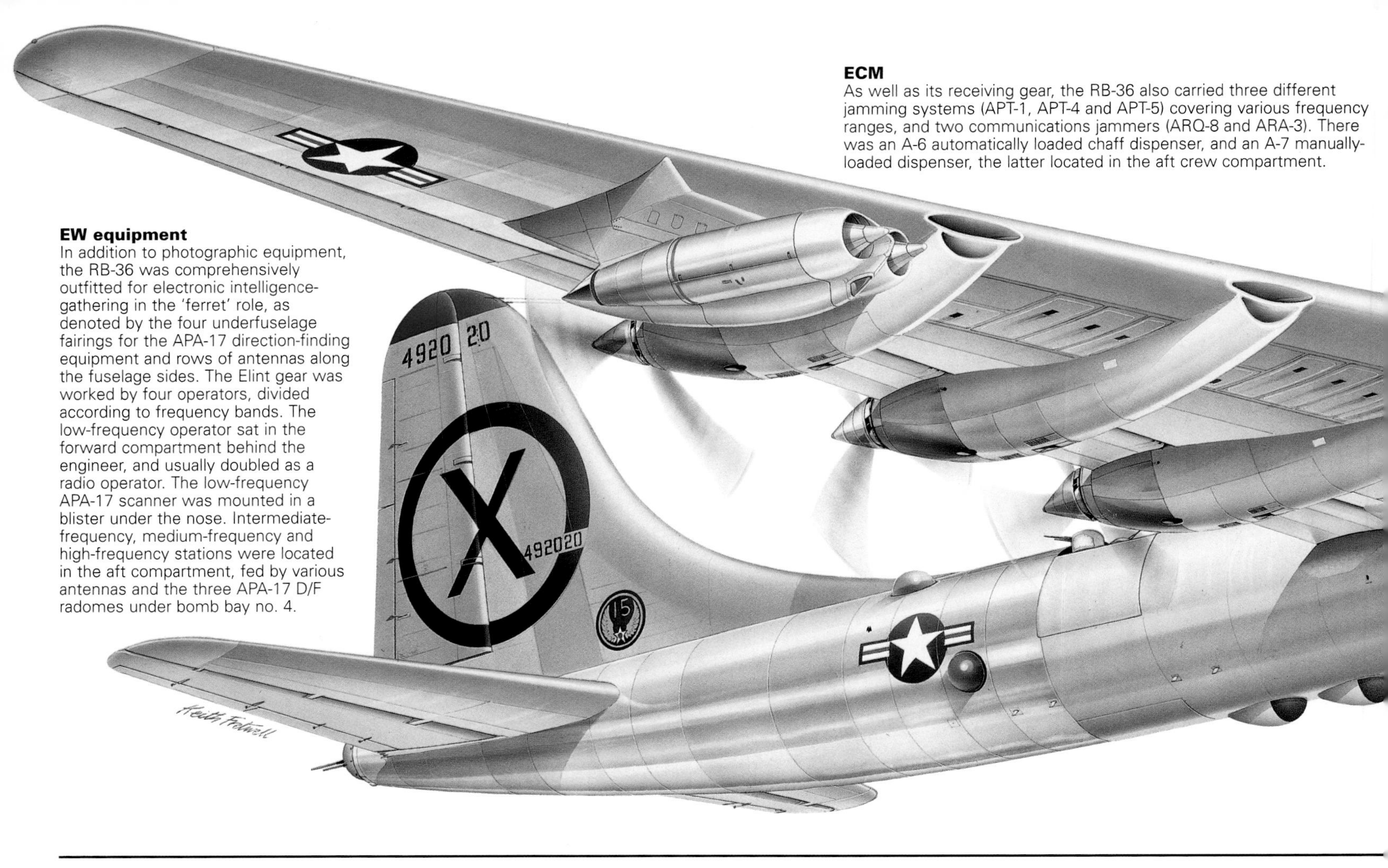

ECM
As well as its receiving gear, the RB-36 also carried three different jamming systems (APT-1, APT-4 and APT-5) covering various frequency ranges, and two communications jammers (ARQ-8 and ARA-3). There was an A-6 automatically loaded chaff dispenser, and an A-7 manually-loaded dispenser, the latter located in the aft crew compartment.

EW equipment
In addition to photographic equipment, the RB-36 was comprehensively outfitted for electronic intelligence-gathering in the 'ferret' role, as denoted by the four underfuselage fairings for the APA-17 direction-finding equipment and rows of antennas along the fuselage sides. The Elint gear was worked by four operators, divided according to frequency bands. The low-frequency operator sat in the forward compartment behind the engineer, and usually doubled as a radio operator. The low-frequency APA-17 scanner was mounted in a blister under the nose. Intermediate-frequency, medium-frequency and high-frequency stations were located in the aft compartment, fed by various antennas and the three APA-17 D/F radomes under bomb bay no. 4.

44-92088 was the aircraft converted as an ERB-36D to carry the Boston camera. The camera was mounted in the forward fuselage and could be rotated to peer downwards, or obliquely sideways through a window in the port side.

at Wright Field in October 1935. He had recovered and went on to serve as project officer on the XB-29 programme as well.

In July 1947, the first B-36A aircraft was the third Peacemaker off the line, but it would actually make its first flight on 28 August, three months ahead of the YB-36. Ironically, it was designed not to be flown. It was the static stress test aircraft, but was completed as a flyable aircraft because it was easier to fly it to Wright Field than to ship it. On its second flight, Beryl Erickson took the B-36A to the Ohio base, where it would be literally ripped apart to study its ability to withstand structural pressures. Under this scenario, this aircraft was designated as YB-36A.

Initial deliveries and deployment

The first production series B-36A, appropriately christened *City of Fort Worth*, was delivered on 26 June 1948, nearly a year after the debut of the static test airframe. Like all of the B-36A aircraft, it was delivered to Strategic Air Command's 7th Bombardment Wing. This unit was based at Carswell AFB in Fort Worth, which was conveniently located just across the runway from the factory where B-36s were built.

Also located at the base was the headquarters of the parent organisation of the 7th Bombardment Wing, the Eighth Air Force. During World War II, the Eighth Air Force had been the keystone of the United States Strategic Air Forces in Europe and the principal element in USAAF strategic operations in that theatre.

Formerly known as Fort Worth Army Air Field, the base had become Carswell AFB on 30 January 1948, named in honour of Major Horace Carswell Jr. of Fort Worth. A 1939 graduate of Texas Christian University, Major Carswell was a B-24 pilot during World War II, who sank a Japanese cruiser and destroyer in October 1944. During a low-altitude attack on the Japanese convoy, his plane was hit by anti-aircraft fire and severely damaged, causing the loss of two engines. The remaining two engines later failed, and the plane crashed into a mountain and burst into flames. For his heroic action, he received the Congressional Medal of Honor, Distinguished Flying Cross, Air Medal and the Purple Heart. Now, the Air Force field in his home town would be his memorial.

During the early months of their assignment to Carswell, the B-36 fleet was cursed by the lack of hangars large enough for the big aircraft. Mechanics were forced to work on the ramp in the intense Texas heat during the summer of 1948. The few available portable air-conditioner units were kept busy moving along the ramp, pumping cool air into the planes so men could keep working. At the other extreme, the cold during the ensuing winter was just as bad

ERB-36D - carrying the world's largest aerial camera

The Boston Camera was designed by Dr James Baker of Harvard University and manufactured for the USAF by Boston University in 1951. To this day it remains the largest aerial camera ever built. The only B-36 to ever carry it was the first RB-36D (44-92088). Modifications to carry it began on 7 December 1953 and it was delivered to the USAF on 2 March 1954. Total weight of the camera and its aircraft mount was 6,500 lb (2948 kg). In 1955 it was removed from the RB-36D and installed in a C-97. The Boston Camera was never used operationally and with rapid improvements in both optics and film was rendered obsolete by smaller cameras of even greater resolution. It was finally retired to the USAF Museum in 1964 (left).

The camera had an f/8 lens with a focal length (lens to film) of 240 in (6096 mm). Practical requirements dictated the use of two mirrors to reduce its size for aerial applications. It incorporated a focal-plane, fixed-slit, pneumatic drive, electrically-tripped shutter having a speed of 1/400th second. Using a negative frame size of 18 x 36 in (45.7 x 91.4 cm), the camera was so powerful that it was capable of producing large photos with great magnification. The lens was focused manually to compensate for variations in temperature and atmospheric pressure, and had a high contrast resolution of 28 lines per mm. It was thus reportedly capable of photographing a golf ball from an altitude of 45,000 ft (13716 m). Because of cold temperatures at high altitude the camera was wrapped in an electric blanket.

Convair RB-36E

5th Strategic Reconnaissance Wing Travis AFB, California

RB-36s were built with broadly similar configurations and were initially tasked almost exclusively with reconnaissance. In 1954 they were re-roled as bombers, and underwent modification accordingly. Many received the Featherweight weight-reducing modifications to increase their over-target altitude.

Bomb carriage
From 1952 all RB-36s were modified so that the no. 2 bay could carry nuclear weapons. Bomb carriage was later expanded to the aft bays, resulting in the movement of the D/F antennas and other 'ferret' gear to the fuselage behind bay no. 4.

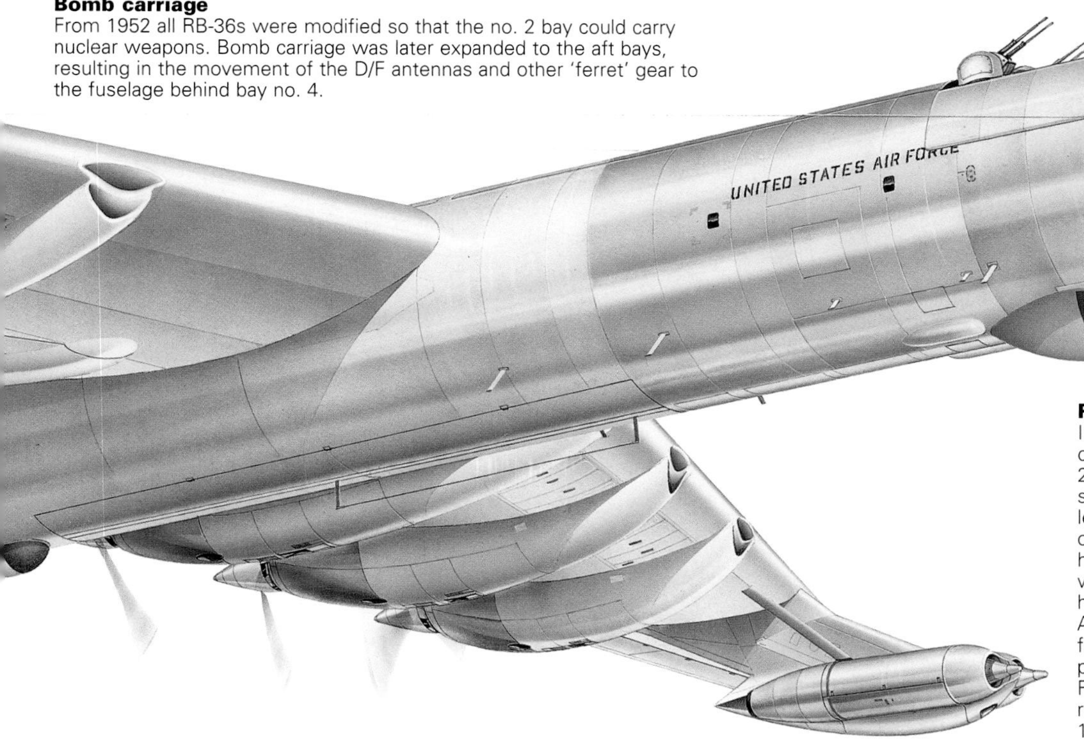

Reconnaissance equipment
In its initial form, the RB-36 had a pressurised camera compartment in former bomb bay no.1. As well as up to 23 cameras, which were K-17C, K-22A, K-38 and K-40 sensors with lenses of up to 48 in (1219 mm) focal length, the compartment had a darkroom to allow onboard processing by the photo-technician. Bay no. 2 housed 80 T86 photo-flash bombs for night photography, while bay no. 3 usually had an extra fuel tank. Bay no. 4 housed the electronic intelligence equipment. The APQ-24 nav/bomb radar and its operator were retained for target location, but the bombardier was replaced by a photo-navigator and a weather observer was also carried. RB-36s routinely carried drop-sondes for meteorological reconnaissance. The initial RB-36 crew complement was 18, but subsequently rose to 22 or 23 as extra technicians and EW operators were added.

because the Air Force still lacked adequate work-stand shelters to put over the B-36 engines where mechanics worked with numb fingers.

The B-36A was 162 ft 0.25 in (49.38 m) long, and, like the XB-36, it had a wingspan of exactly 230 ft, a wing area of 4,772 sq ft and a tail height of 46 ft 10 in. Like the XB-36, it was equipped with six Pratt & Whitney R-4360-25 Wasp Major air cooled radial engines. It had a higher empty weight of 135,020 lb (61245 kg), and a maximum weight of 310,380 lb (140788 kg). Its service ceiling was increased by 4,100 ft (1250 m) to 39,100 ft (11918 m). A combat radius of 3,880 miles (6244 km) had been calculated and the ferry range was 9,136 miles (14702 km). Several long-endurance flights demonstrated the aircraft's range. In June 1948, one B-36A made a 33-hour flight covering 8,062 miles (12974 km). Another flight of the same duration covered 6,922 miles (11139 km), part of it with a five-ton bomb load. The B-36A fleet at Carswell was used for training purposes for two years: if not so much training as crew conversion, because most crews had previous experience with other large bomber types.

Produced through February 1949, a total of 22 aircraft received the B-36A designation. These, plus a further 78 B-36B aircraft (of which 11 were eventually completed as B/RB-36Ds), would comprise the 100 aircraft on the original 1944 order. The total would be reduced by five in order to defray the cost of the improved R-4360-51 variable-discharge turbine version of the Pratt & Whitney engine that was discussed for possible use in a future B-36C.

During 1950 and 1951, all but one of the B-36As were reconfigured as reconnaissance aircraft and redesignated as RB-36E. These aircraft were retrofitted with camera equipment and re-engined with R-4360-41 engines, such as those that were installed in the B-36B.

Uprated engines

The B-36B had the same dimensions as the B-36A and differed primarily in its being equipped with six Pratt & Whitney R-4360-41 Wasp Major air-cooled radial engines each delivering 3,500 hp (2611 kW). This gave it a top speed of 381 mph (613 km/h) at 34,500 ft (10516 m) and a cruising speed of 202 mph (325 km/h). The service ceiling

RB-36E

Virtually identical to the RB-36Ds, the RB-36Es were all conversions: one was the YB-36 and 21 were former B-36As. Of these, six RB-36Es were converted to Featherweight III standard.
Quantity converted: 22
Tail numbers: 42-13571 (YB-36); 44-92005 to 44-92025 (B-36A)

This RB-36E served with the 5th Strategic Reconnaissance Wing at Travis AFB. Until around 1953 B-36s usually wore geometric tail markings. The circle was for 15th AF ('X' for the 5th SRW) while 8th AF aircraft used a triangle.

B-36F and RB-36F - increased power

The B-36F was a variant of the B-36D equipped with Pratt & Whitney R-4360-53 radial engines delivering 3,800 hp (2835 kW) each and replacing the R-4360-41 engines of the B-36D. It also carried four General Electric J47-GE-19 turbojet engines. The first example of this variant made its initial flight on 11 November 1950, and four B-36Fs were later converted to Featherweight III standard.
Quantity produced: 34 **Tail numbers:** 49-2669 to 49-2675; 49-2677 to 49-2683; 49-2685; 50-1064 to 50-1082

The RB-36F was the reconnaissance version of the B-36F. One was later converted to Featherweight III standard. One was later converted as a FICON Fighter Conveyor and redesignated as GRB-36F.
Quantity produced: 24 **Tail numbers:** 49-2703 to 49-2721; 50-1098 to 50-1102

Top: This B-36F was from the 6th Bomb Wing at Walker AFB, near Roswell in New Mexico. The wing badge on the nose depicted a pirate.

Above: A B-36F receives attention on the ramp. In the background are some of the Project Gem B-36Bs.

Right: This is an RB-36F seen towards the end of its career. It has received the Featherweight III mods which removed all but tail armament, and the Elint radomes have been removed from under the rear fuselage. From June 1954 onwards the RB-36 force was increasingly tasked with regular bombing duties, with reconnaissance becoming a secondary role.

was bumped up to 42,500 ft (12954 m). The combat radius was 3,740 miles (6019 km) with a ferry range of 8,820 miles (14194 km). It had a gross weight of 328,000 lb (148781 kg), up nearly five tons from the B-36A.

Designed to carry more than 40 tons of bombs, it was also the first Peacemaker configured for operational offensive missions. It was also the first variant equipped with the defensive package of eight remotely-controlled gun turrets, six of them retractable, and all equipped with a pair of M24A1 20-mm cannon. Parenthetically, this gun package could not be fully tested for about a year after initial deliveries because of lack of ammunition. Most of the B-36 fleet would later be stripped of all but their tail turrets in an effort to save weight under the Featherweight modification programme. The B-36B also carried the Sperry K-1 bombing system and the related APQ-24 bombing and navigation radar, supplanting the APG-23A system used on the B-36A training aircraft.

The first flight of a B-36B occurred at Fort Worth on 8 July 1948, less than two weeks after the delivery of the first B-36A. By the end of the year, 35 of the big aircraft had been delivered to the 7th Bombardment Wing at Carswell AFB. The last of the 62 in the series had been delivered by September 1950.

As with the B-36A during this era, the B-36Bs were used for long-endurance training and demonstration flights. One such mission in May 1948, with Beryl Erickson and A.S. 'Doc' Witchell at the controls, lasted more than 35 hours and involved an 8,062-mile (12974-km) non-stop round trip to Hawaii in which a 10,000-lb (4536-kg) bomb load was carried on the outbound leg.

The B-36C and the R-4360-51

Despite the support that the B-36 programme had received from General Spaatz and General Twining, the issue of its slow speed was a constantly recurring theme for its critics. Against jet interceptors, the big bombers would be like sitting ducks as they neared their targets. The first major technical effort to resolve this problem unfolded during 1947 and 1948, even before the B-36A and B-36B were delivered. The proposal was to power a B-36 variant with the new-technology Pratt & Whitney R-4360-51 variable-discharge turbine (VDT), an untested version of a very new engine type.

One of the more unusual tasks entrusted to the B-36 was the ferrying of the fifth B-58 Hustler airframe from Fort Worth to Wright-Patterson AFB for structural loads testing in February 1952. Minus engines, fin, nosecone and other elements, the B-58 weighed around 40,000 lb (18144 kg) – easily within the lifting capability of the B-36F (49-2677) chosen to carry it.

The B-36B had flown with the improved 3,500-hp R-4360-41 engine, and the plan that evolved was a B-36C that would be fitted with the still more powerful version of the R-4360. Introduced in March 1947, this R-4360-51 turbo-compound engine would have a variable-discharge turbine and be capable of delivering 4,300 hp (3208 kW). The variable-discharge turbine can be described as using the exhaust from the turbo-supercharger as a form of 'jet assist'. Such an engine had been proposed by the Air Force for the Boeing B-50C (later redesignated as B-54), and it was placed into development. Convair knew this, and, with Pratt & Whitney's encouragement, proposed the R-4360-51 for the B-36C. Convair and Pratt & Whitney estimated that, with the new engine, the B-36C could have a top speed of 410 mph (660 km/h), a service ceiling of 45,000 ft (13716 m), and a 10,000-mile (16093-km) range with a five-ton bomb load.

Tractor propellers

However, in order to accommodate the R-4360-51 in the B-36, it would be necessary to change the propeller and engine interface from the pusher type to the tractor type. Although the engine would remain aft of the wing spar, its position would have to be reversed to face forward. The propeller shaft would have to extend through the entire wing, and then another 10 ft (3 m) forward of the wing's leading edge. Such an arrangement would present a complex engineering challenge.

Under such a scenario, exhaust gases from the engine would pass through a General Electric CHM-2 turbo-supercharger with a clamshell nozzle controlling the jet thrust by varying the size of the turbine exit. The variable-discharge nozzle was to be operated by automatic control, activated by a manifold pressure sensing device. Cooling air would be ducted through wing leading-edge inlets flanking the nacelle extension. It was a cumbersome and complex design change, to say the least.

The proposal to develop the new R-4360-51 for the B-36 was first presented by the Air Matériel Command in Washington in March 1947. Various conferees opposed the suggestion because research and development funds were then too limited to be spent on the modernisation of existing weapons systems, which they felt should be reserved for developing future weapons, such as the Boeing XB-52 and the new jet-powered medium bombers, such as the North American Aviation XB-45, the Boeing XB-47 and Convair's own XB-46.

Two months later, a follow-on meeting was chaired by General Curtis LeMay, the future head of Strategic Air Command, who was then the director of research and development. Again the agenda included the proposed variable-discharge turbine engine retrofit and installation. It was suggested that the cost of adapting the R-4360-51 to the B-36 could be met largely out of procurement funds

With the B-58 airframe in place, the B-36's undercarriage could not be retracted, and the inboard propellers had to be removed. The laborious five-hour flight to Wright-Patterson was accomplished with the gear down and the jet engines running all the way. It was standard practice to only use the jets at critical times in the mission, such as take-off and climb, and over the target.

A view from the flight deck shows Featherweight B-36Hs on a low-level training mission. A general consensus among B-36 pilots was that it was a safe aircraft with fine handling, a healthy thrust margin and a good stability. At the upper reaches of its ceiling envelope it could also out-turn the jet fighters of the day. Apart from the concerns when taxiing and the slow control response caused by the sheer size of the aircraft, crosswind landings were perhaps the most challenging aspect of flying the B-36 due to the large fin and wide-span wings, the latter preventing any wing-down landing techniques for fear of hitting the jet pods on the ground.

B-36H and RB-36H – refined versions

The B-36H was produced in larger numbers than any other variant. It was essentially an improved version of the B-36F with new bombing and tail gun systems. Its six R-4360-53 piston engines were augmented by four General Electric J47-GE-19 turbojet engines. The first example made its initial flight on 5 April 1952. Of the total, 64 aircraft were modified to Featherweight III standard. **Quantity produced:** 83 **Tail numbers:** 50-1083 to 50-1097; 51-5699 to 51-5742; 52-1343 to 52-1366

The strategic reconnaissance variant of the B-36H, the RB-36H had the second highest production total of any version of the Peacemaker. Of this total, 22 aircraft were modified to Featherweight III standard. **Quantity produced:** 73 **Tail numbers:** 50-1103 to 50-1110; 51-5743 to 51-5756; 51-13717 to 51-13741; 52-1367 to 52-1392

***Above:** Three RB-36Hs of the 28th SRW are seen on the Ellsworth AFB ramp.*

***Above right:** An RB-36H cleans up after take-off. The sliding door for the forward upper gun turrets is open, a standard RB-36 take-off procedure to provide a second potential egress route for the crew in the photo compartment.*

This B-36H carries the badge of the 11th BW. The H introduced a range of new features, including a second flight engineer facing aft on the upper deck, a revised bombing system (Blue Square) and the APG-41 tail radar set, which had two antenna dishes and automatic target tracking.

instead of research and development funds. With this scheme removing the main objection to the proposal, General Spaatz approved the decision to install the R-4360-51 engine in one prototype each of the B-36 and the B-54.

Convair suggested building an additional quantity of B-36s with the R-4360-51 under a new contract and the designation B-36C. However, when this proposal went before the Air Force's new Aircraft & Weapons Board in August, it was rejected because of the ongoing assumption that no more than 100 B-36A and B-36B aircraft would be built. Created by Deputy Chief of Staff General Hoyt Vandenberg, the board served as a forum at which senior officers would exchange opinions, present factual data, and consider recommendations on matters relating to weapons selection.

Convair countered with a suggestion to complete the last 34 B-36Bs (44-92065 to 44-92098) with the new engine. Convair even went so far as to offer to deliver the last of the 34 new B-36Cs in May 1950, a delay in completion of the contract of only six months. The differing viewpoints expressed at the Aircraft & Weapons Board meeting were reminiscent of the exchange of views between General Kenney and General Twining a year earlier, but in 1947, the discussion was given a sense of urgency by the deterioration in the international political situation and the onset of the Cold War, and the fact that most of the countries in eastern Europe were firmly under total Soviet domination.

Since the main concern in the B-36 programme now was to increase the speed of the aircraft over the target, the offer to install a more powerful engine at very little additional cost seemed very attractive. Convair went a step further by agreeing to a reduction in the contract quantity from 100 to 95 airplanes in lieu of additional payment. Convair also offered to study the possibility of retrofitting the B-36A and B-36B models to the B-36C standard.

After an uphill battle lasting half a year, Convair finally got the go-ahead for the B-36C in December 1947, but ultimately the engine would turn out to be a costly failure and both the B-36C and the entire Boeing B-54 programme were terminated.

The US Air Force, meanwhile, considered terminating the entire B-36 programme after delivery of the 22 B-36As and a varying number of B-36Bs. The varying numbers discussed ranged from zero to the originally planned 78 aircraft. As noted above, the Air Force eventually settled on 73 B-36Bs, with the funds originally earmarked for the other five being used to pay for the ill-starred R-4360-51 engine project.

The catalyst for the decision not to eliminate or substantially reduce the B-36B programme was a result of increased global tensions in the wake of the Soviet blockade of the Western sectors of Berlin, which began in June 1948. Of course, it did not hurt the Peacemaker's prospects that the long-endurance flights conducted by Beryl Erickson and 'Doc' Witchell during the spring and summer of 1948 clearly demonstrated the intercontinental range of the big bomber.

Convair B-36H-25-CF Peacemaker

11th Bombardment Wing, Heavy Carswell AFB, Texas

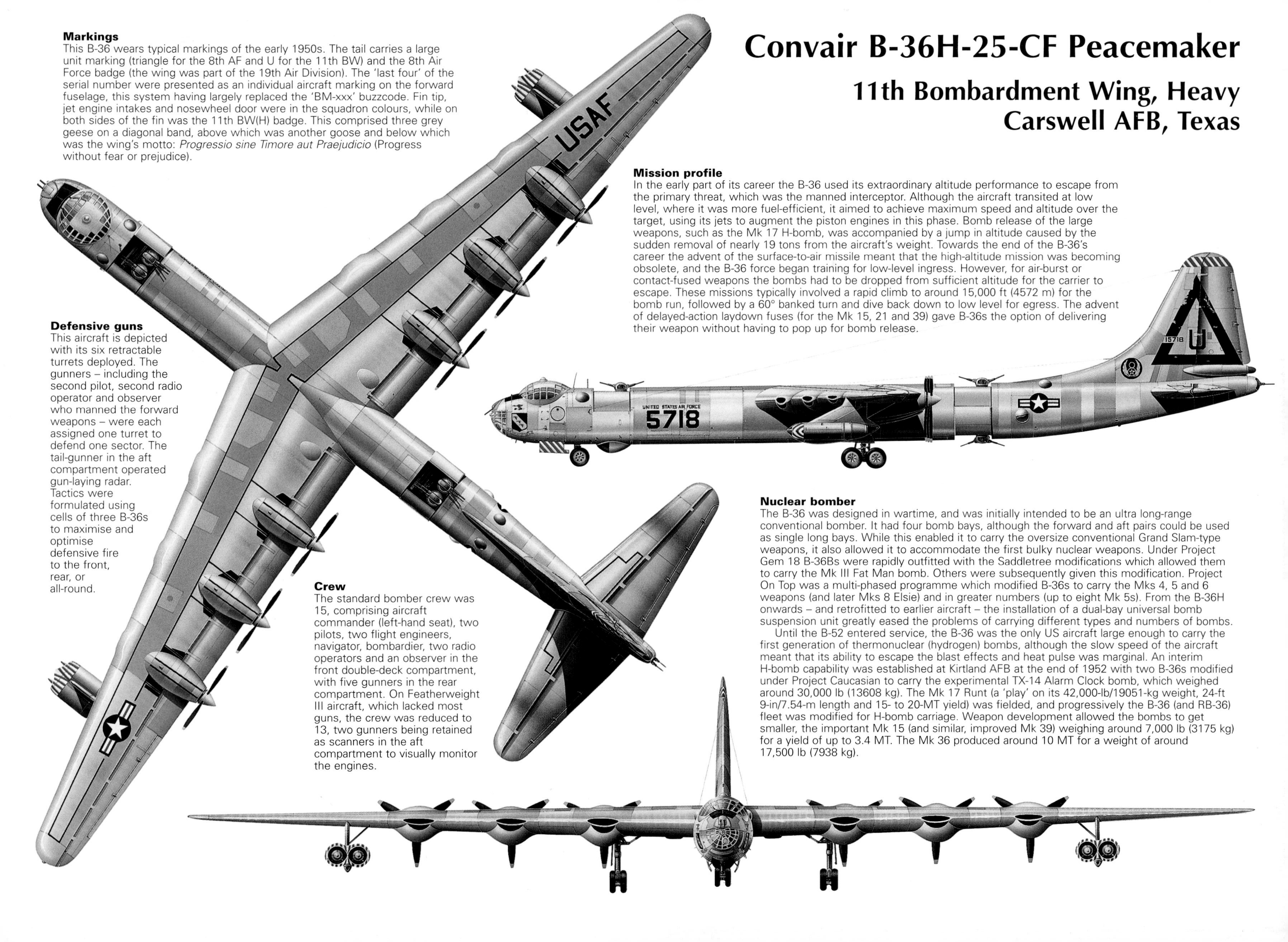

Markings
This B-36 wears typical markings of the early 1950s. The tail carries a large unit marking (triangle for the 8th AF and U for the 11th BW) and the 8th Air Force badge (the wing was part of the 19th Air Division). The 'last four' of the serial number were presented as an individual aircraft marking on the forward fuselage, this system having largely replaced the 'BM-xxx' buzzcode. Fin tip, jet engine intakes and nosewheel door were in the squadron colours, while on both sides of the fin was the 11th BW(H) badge. This comprised three grey geese on a diagonal band, above which was another goose and below which was the wing's motto: *Progressio sine Timore aut Praejudicio* (Progress without fear or prejudice).

Mission profile
In the early part of its career the B-36 used its extraordinary altitude performance to escape from the primary threat, which was the manned interceptor. Although the aircraft transited at low level, where it was more fuel-efficient, it aimed to achieve maximum speed and altitude over the target, using its jets to augment the piston engines in this phase. Bomb release of the large weapons, such as the Mk 17 H-bomb, was accompanied by a jump in altitude caused by the sudden removal of nearly 19 tons from the aircraft's weight. Towards the end of the B-36's career the advent of the surface-to-air missile meant that the high-altitude mission was becoming obsolete, and the B-36 force began training for low-level ingress. However, for air-burst or contact-fused weapons the bombs had to be dropped from sufficient altitude for the carrier to escape. These missions typically involved a rapid climb to around 15,000 ft (4572 m) for the bomb run, followed by a 60° banked turn and dive back down to low level for egress. The advent of delayed-action laydown fuses (for the Mk 15, 21 and 39) gave B-36s the option of delivering their weapon without having to pop up for bomb release.

Defensive guns
This aircraft is depicted with its six retractable turrets deployed. The gunners – including the second pilot, second radio operator and observer who manned the forward weapons – were each assigned one turret to defend one sector. The tail-gunner in the aft compartment operated gun-laying radar. Tactics were formulated using cells of three B-36s to maximise and optimise defensive fire to the front, rear, or all-round.

Crew
The standard bomber crew was 15, comprising aircraft commander (left-hand seat), two pilots, two flight engineers, navigator, bombardier, two radio operators and an observer in the front double-deck compartment, with five gunners in the rear compartment. On Featherweight III aircraft, which lacked most guns, the crew was reduced to 13, two gunners being retained as scanners in the aft compartment to visually monitor the engines.

Nuclear bomber
The B-36 was designed in wartime, and was initially intended to be an ultra long-range conventional bomber. It had four bomb bays, although the forward and aft pairs could be used as single long bays. While this enabled it to carry the oversize conventional Grand Slam-type weapons, it also allowed it to accommodate the first bulky nuclear weapons. Under Project Gem 18 B-36Bs were rapidly outfitted with the Saddletree modifications which allowed them to carry the Mk III Fat Man bomb. Others were subsequently given this modification. Project On Top was a multi-phased programme which modified B-36s to carry the Mks 4, 5 and 6 weapons (and later Mks 8 Elsie) and in greater numbers (up to eight Mk 5s). From the B-36H onwards – and retrofitted to earlier aircraft – the installation of a dual-bay universal bomb suspension unit greatly eased the problems of carrying different types and numbers of bombs.

Until the B-52 entered service, the B-36 was the only US aircraft large enough to carry the first generation of thermonuclear (hydrogen) bombs, although the slow speed of the aircraft meant that its ability to escape the blast effects and heat pulse was marginal. An interim H-bomb capability was established at Kirtland AFB at the end of 1952 with two B-36s modified under Project Caucasian to carry the experimental TX-14 Alarm Clock bomb, which weighed around 30,000 lb (13608 kg). The Mk 17 Runt (a 'play' on its 42,000-lb/19051-kg weight, 24-ft 9-in/7.54-m length and 15- to 20-MT yield) was fielded, and progressively the B-36 (and RB-36) fleet was modified for H-bomb carriage. Weapon development allowed the bombs to get smaller, the important Mk 15 (and similar, improved Mk 39) weighing around 7,000 lb (3175 kg) for a yield of up to 3.4 MT. The Mk 36 produced around 10 MT for a weight of around 17,500 lb (7938 kg).

Above: An RB-36H rests at a TDY base. The RB had a different bomb bay door arrangement to the standard bombers, and was built with a single set of doors covering bays 2 and 3. They were later reworked with one small set forwards for bay 2 and a long set for bays 3 and 4. Both B-36s and RB-36s regularly flew and deployed with nuclear weapons loaded, and towards the end of the Peacemaker's career were flying airborne alerts.

Below: This RB-36H of the 28th BW appears to have had most of its guns removed under the Featherweight III programme, but the underfuselage Elint radomes have not been moved back to allow the use of the no. 4 bomb bay.

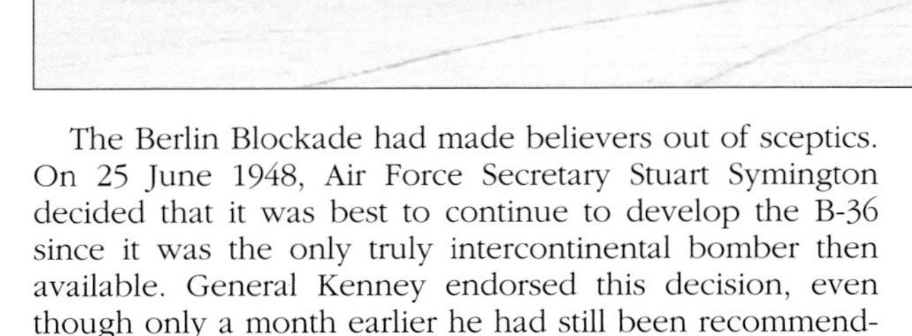

The Berlin Blockade had made believers out of sceptics. On 25 June 1948, Air Force Secretary Stuart Symington decided that it was best to continue to develop the B-36 since it was the only truly intercontinental bomber then available. General Kenney endorsed this decision, even though only a month earlier he had still been recommending that the entire B-36 programme be halted.

LeMay in command

By the autumn of 1948, General George Kenney's term as commander of Strategic Air Command was winding down, and General Hoyt Vandenberg, the new Chief of Staff, had his eyes on a new man, the tough, iron-jawed Lieutenant General Curtis Emerson LeMay. The field commander of the all-B-29 Twentieth Air Force during World War II, LeMay had served as post-war commander of the US Air Forces in Europe (USAFE), where he had just organised the spectacular Berlin Airlift.

On 16 October 1948, LeMay assumed control of Strategic Air Command, beginning his unprecedented nine-year span of leadership. With the wartime experience of the Twentieth Air Force, he was a true believer in the doctrine of strategic air power. During those next nine years, he would build Strategic Air Command into the most powerful air weapon in history. In the early years, the B-36 would be a key part of LeMay's master plan.

LeMay took over the Strategic Air Command as it shifted from Andrews AFB, near Washington, to Offutt AFB near Omaha, Nebraska, closer to the heart of the nation. One of his first tasks was to take stock of the B-36 Operational Training Programme, just beginning. Being a man of action and few words, LeMay laid down a schedule of training missions to put aircraft and crews through their paces, so he could evaluate what they could do.

LeMay was keen for demonstrations of power. The 300-aircraft – and later 600-aircraft – raids on Japan in 1945 were one example. Another would come in 1957, when the B-52 programme was in danger in Congress. LeMay responded to this by flying three B-52s around the world non-stop and making sure that the event, code named Power Flight, was well-publicised and that it ended up as a LIFE magazine cover story. A half century later, the B-52 would still be in the Air Force.

LeMay's first B-36 demonstration came – appropriately – on 7 December 1948, the seventh anniversary of Pearl Harbor. This mission to Hawaii gained a great deal of publicity in all sectors because of obvious implications to the public, as well as to knowledgeable military analysts. LeMay sent a B-36B of the 7th Bombardment Wing, piloted by Major John Bartlett, from Carswell AFB to Honolulu. Flying at an average speed of 319 mph (513 km/h), over a distance of 4,406 miles (7090 km), the bomber penetrated Hawaiian air space with a 10,000-lb dummy payload without detection and returned to Carswell without refuelling. When asked by *The Honolulu Star-Bulletin* about the apparent failure to detect the B-36, the commander of the forces directly responsible for the defence of Hawaii against air or surface attack replied "No answer." The forces in Hawaii sniffed that they had not been informed that the B-36 flight was to be made.

Another show of the B-36B's power came on 29 January 1949 at Muroc AFB (later Edwards AFB), when Major Stephen Dillon dropped two dummy Grand Slams – each weighing 43,000 lb (19505 kg) – from a B-36B from 30,000 and 40,000 ft (9144 and 12192 m). It was the largest conventional payload yet carried by a bomber. The B-36 conventional arsenal would also include the Bell VB-13 Tarzon (later ASM-A-1) a 12,000-lb (5443-kg) bomb that

Below: Along with 51-5726, 52-1357 and 52-1358, this B-36H (51-5731) was designated EB-36H (E = exempt) and assigned to the 4925th Test Group (Atomic) at Kirtland AFB, New Mexico, as a bomb drop test and recording platform. The quartet was redesignated as JB-36Hs in 1956. Two of the aircraft were painted in high visibility markings so that they could be optically tracked during shape drops. A number of other B-36s were involved in nuclear tests, including two JRB-36Hs assigned to the 4925th as camera platforms. The first B-36 live weapons release was on 16 November 1952 (Project Ivy, shot King), which tested the Mk 18 fission weapon. B-36s later conducted the high-altitude Mk 5 airburst test (Teapot), a Mk 7 drop (Climax) and two low-level Mk 12 airbursts (Wasp and Wasp Prime).

Right: Three B-36Hs – 50-1085, 51-5706 and 51-5710 – were converted to EDB-36H standard to test the Peacemaker as a launch aircraft for the GAM-63 Rascal missile. This aircraft, 51-5710, had earlier been used to test a probe/drogue refuelling system under Project Tanbo.

used Range and Azimuth guidance (RAZON) technology for precision targeting.

On 20 January 1949, five B-36Bs from Carswell AFB participated in a flyover of the United States Capitol in Washington, DC, to celebrate the inauguration of Harry Truman as President of the United States. Inside the Capitol, a storm was already brewing that would result in yet another of those patches of turbulence through which the B-36 had been flying since the day it was first proposed.

The Admirals revolt

The generals had failed to kill the B-36, and in 1949, it would be the turn of the admirals to try to do the deed. Setting the stage for this was the biggest inter-service squabble that had been seen since before World War II.

Prior to enactment of the National Security Act of 1947, the Department of the Army and the Department of the Navy had been separate cabinet-level departments. In 1947, they were downgraded to subsidiary departments within the newly-created Department of Defense, which also now included the separate subsidiary Department of the Air Force. Meanwhile, the dramatically shrinking budgets imposed upon the armed services after World War II led to intensifying competition between them for dollars. One of the most bitter rivalries was between the Navy and the Air Force over power projection. Traditionally, the US Navy's fleets had been America's first line of defence. Now, with the advent of long-range, nuclear-armed bombers, the primacy had shifted away from the Navy.

To counter this, the Navy proposed a new class of super aircraft-carriers – not unlike those which would be joining the fleet two decades in the future – from which nuclear strikes could be launched. They had already laid the keel for such a new vessel, the 65,000-ton USS *United States*. Meanwhile, the US Air Force had requested funds to build additional Peacemakers under the designation B-36D. The two programmes went head-to-head in the battle for funding on Capitol Hill. It was widely understood that the Department of Defense could not afford both of these weapons.

B-36J - the ultimate Peacemaker

The last Peacemaker production variant, the B-36J was an improved version of the B-36H with extra fuel in the wings and a strengthened undercarriage for heavier take-offs (410,000 lb/ 185976 kg). The first J flew on 3 September 1953. As with the B-36H, its six R-4360-53 piston engines were augmented by four General Electric J47-GE-19 turbojet engines. All of the B-36Js were either built or retrofitted to Featherweight III standard.

Quantity produced: 33 **Tail numbers:** 52-2210 to 52-2226; 52-2812 to 52-2827

Above: 52-2827 was a Featherweight B-36J and the last B-36 to be built, first flown from the factory on 16 August 1954. Here it is taking off from Fort Worth on its delivery flight to the USAF. The final B-36 wing was the 95th BW at Biggs AFB, Texas, which flew its last Peacemaker mission on 12 February 1959 when 52-2827 was flown back to Fort Worth for display.

Until March 1949, the United States Department of Defense had been headed by former Navy Secretary James Forrestal. However, Forrestal was replaced as Secretary of Defense by Louis Johnson who, coincidentally, had earlier served on the board of directors at Convair. When Johnson proceeded to both endorse additional B-36Ds and cancel the USS *United States*, a wave of outrage surged forth from the US Navy. They called foul, accusing Johnson of improper favouritism and reiterating all of the negative things that had been said previously about the B-36, centring on its slow speed and its potential vulnerability to Soviet jet fighters.

The Armed Services Committee of the United States House of Representatives investigated and their hearings in October 1949 were marked by some of the most serious inter-service mudslinging in American history. The Navy's top brass, including Chief of Naval Operations Admiral Louis Denfield, denounced the Air Force and its bomber,

The 95th Bombardment Wing at Biggs operated all of the B-36Js. The badge consisted of a red feather over a white cross on a blue background. Featherweight III aircraft, like this one, could reach around 60,000 ft (18288 m), requiring the crew to wear partial pressure suits for operations above 50,000 ft (15240 m). These were initially the T-1 suit and K-1 helmet, but by 1959 improved versions – the MC-4A partial pressure suit and MA-2 helmet – had become standard high-altitude issue to SAC crews.

Forty-four production B-36s (and a single F-86) can be seen in this view of the massive Fort Worth production facility. On the taxiway in the foreground are the two YB-60 prototypes, while parked on waste ground in the lower right corner is the XB-36 prototype. This scene dates from May 1954, shortly before the two jet bombers were scrapped in July. Production of the B-36J was in full swing, and other B-36s have been returned to the factory for inspection and depot-level maintenance under Project SAM-SAC. On the opposite side of the runways to the Convair plant was Carswell AFB, home to the 19th Air Division and its two wings of B-36s (7th and 11th Bomb Wings).

and fought for the Navy's role in the American nuclear force. In the end, Johnson was cleared of impropriety in his controversial decision, and the B-36D programme went forward. The Navy lost its super-carrier, and they would not have an important strategic nuclear role until the advent of the Polaris nuclear submarine force a decade later.

As for the budgets, less than a year later, in June 1950, the Korean War began and funding for all of the armed services increased four-fold or more.

The B-36 becomes a jet bomber

As noted above, the signature criticism of the B-36 during its early years – whether that reproachment came from General Kenney or any number of admirals – was that it was too slow. Throughout its development, those who maligned the B-36 had urged that the Air Force wait for the advent of jet bombers with intercontinental range. On 5 October 1948, even as the first B-36As were starting to come off the assembly line, Convair took the dramatic step of proposing a variant in which the six piston engines would be augmented by jet engines. Convair proposed using two pairs of General Electric J47 turbojets, similar to those which were then being incorporated into the prototype North American XB-45 and Boeing XB-47 all-jet medium bombers.

In January 1949, on the eve of the showdown with the admirals, the US Air Force authorised Convair to retrofit a B-36B with four jet engines as a prototype for a new variant to be provisionally designated as B-36D. For this exercise, Convair would use Allison J35 turbojet engines such as it was then using for its own prototype all-jet medium bomber, the XB-46.

The jet-augmented B-36B made its first flight on 26 March 1949. The results of the flight tests were so successful that the Air Force requested funding to not only convert more B-36Bs to the B-36D standard, but to build a new series of B-36Ds. It was at this point that the programme flew into the firestorm of the admirals' denunciation.

Having defeated the admirals on Capitol Hill, the Air Force took delivery of its first operational B-36Ds in August 1950. The Air Force proceeded with its new jet-piston hybrid, and the first B-36B converted to B-36D configuration made its debut in November 1950.

The B-36D had the same dimensions as its predecessors, and like the B-36B, it was equipped with six Pratt & Whitney R-4360-41 Wasp Major air-cooled radial engines delivering 3,500 hp. In addition, the four General Electric J47-GE-19 turbojets each delivered 5,200 lb (23.14 kN) of thrust. This gave it a top speed of 439 mph (706 km/h) at 32,120 ft (9790 m), compared to 381 mph at 34,500 ft for the B-36B. With 10 engines, the B-36D was found to have more than adequate cruising power. Indeed, the turbojets came to be used sparingly, for a heavily loaded take-off and for the increased speed that would have been required over the target.

The service ceiling of the B-36D increased from the 42,500 ft of the B-36B to 43,800 ft (13350 m), but the combat radius was reduced to 3,528 miles (5678 km) from 3,740 miles, with a ferry range of 8,820 miles (14194 km). The gross weight increased from 328,000 pounds to 370,000 lb (167832 kg). Eventually, the maximum bomb load would be increased to 43 tons.

Featherweight B-36s

During the early to mid-1950s, the B-36D was one of several Peacemaker models that was stripped of defensive armament under the Featherweight programme. This was aimed at increasing the bomber's service ceiling by stripping out varying amounts of defensive armament and other features in order to save weight. The result was an aircraft that could fly as high as 60,000 ft (18288 m). By this time, surface-to-air missiles, rather than fighters, were seen as the greatest potential threat, so flying high was a more valuable defensive measure than carrying gun turrets.

The final stage in the programme, designated as Featherweight III, involved removal of all defensive armament except for the tail turret. The weight savings also included the weight of two of the gunners, who were no longer needed.

In the electronics systems, some B-36Ds were equipped with the new K-3A bombing system which incorporated improved Western Electric APS-23 bombing and navigation radar, and the Sperry A-1A bombing computer. It should be noted that in 1951, Sperry would also introduce the world's first commercial electronic computer, the Universal Automatic Computer (UNIVAC).

Hollywood and the B-36

The B-36 had made its screen debut in 1943, three years before its actual first flight, in *Victory Through Airpower*, the Walt Disney feature based on the book of the same name by Russian emigré and airpower advocate Alexander de Seversky. Directed by James Algar and Clyde Geronimi, the animated film uses stylised B-36s to deliver the final blow against Japan.

The Peacemaker was first featured live in the 1949 film *Target: Peace* (right). A documentary verging on docu-drama, this movie featured scenes of 7th Bombardment Wing crews in action with their aircraft, as well as scenes from the Fort Worth factory and dramatised views of wholesome American families watching the big birds thundering overhead.

Certainly the best view of the B-36 on Hollywood film is the 1955 motion picture *Strategic Air Command*, directed by Anthony Mann and starring film legend Jimmy Stewart and June Allyson as his wife. Stewart, who had been a World War II bomber pilot, and who was also an Air Force reserve officer at the time of the filming, portrays Lieutenant Colonel Robert 'Dutch' Holland, a baseball star and former World War II bomber pilot who re-enlists to fly Boeing B-47s as well as B-36s (right). In the climactic scene, his B-36 is forced down in the Arctic. The picture also stars veteran television actor Harry Morgan, who is well known for his supporting roles in *Dragnet* and *MASH*. During the production of *Strategic Air Command*, Hollywood camera crews shot a great deal of extraordinary colour air-to-air footage of both the B-47 and B-36, making the film a must for fans who enjoy seeing these aircraft in action.

During World War II, the USAAF had started converting a certain number of existing long-range bombers for the strategic reconnaissance role. During the post-war period Strategic Air Command continued the practice, but also ordered variations on its strategic bombers to be factory-built as reconnaissance platforms. In addition to the 22 B-36Ds that were manufactured as bombers under a new contract, Convair also produced 17 RB-36D reconnaissance variants, in which as many as 23 cameras were carried – mainly in the forward bomb bay – and the centre bomb bay was filled with nearly 80 100-lb (45-kg) T86 photoflash bombs for aerial photography at night. The two aft bomb bays could be fitted with a fuel tank for increased range or electronic countermeasures equipment.

The first flight of an RB-36D occurred on 18 December 1949, and the first operational aircraft entered service with SAC the following June. An additional seven aircraft that were ordered as B-36Bs were completed at the factory as RB-36Ds, and 11 of the resulting 24 RB-36Ds underwent Featherweight modifications to increase their range and ceiling.

Following the success of the turbojet-augmented B-36D hybrid, the Air Force requested more. In 1950 and 1951, Convair undertook the RB-36E programme, which essentially involved turning B-36As into clones of the RB-36D. The single YB-36 and 21 B-36As were converted to reconnaissance configuration and retrofitted with the B-36D powerplant package, involving six R-4360-41 piston engines, as well as the four General Electric J47-GE-19 jet engines.

In addition to the conversions, Convair would supersede the B-36D hybrid with the more powerful, all-new B-36F family. First flown on 18 November 1950, the B-36F retained the two pairs of General Electric J47-GE-19 turbojets, but boasted the Pratt & Whitney R-4360-53 piston engines, each of which delivered 3,800 hp (2835 kW). The aircraft had a top speed of 417 mph (671 km/h) at 37,100 ft (11308 m) and cruised at 235 mph (378 km/h). The service ceiling was 44,000 ft (13411 m), before the Featherweight modification that would occur in 1954. As delivered, the B-36F had a combat radius of 3,230 miles (5198 km) with five tons of bombs and a ferry range of 7,743 miles (12461 km). A total of 34 B-36Fs and 24 RB-36Fs was delivered to Strategic Air Command between March 1951 and the end of 1952.

The final production-series Peacemakers

Ordered during the massive defence build-up that coincided with the Korean War, the H-Model was the definitive Peacemaker. More B-36H and RB-36H aircraft would be built than any other Peacemaker variant. A total of 83 B-36Hs and 73 RB-36Hs was rolled out at Fort Worth. First flown on 5 April 1952, these aircraft were delivered to Strategic Air Command in 1952 and 1953.

The B-36H family had the same dimensions as its predecessors since the B-36A, and it had the same 10-engine powerplant package as the B-36F family. One of the major differences was the addition of the APG-41A defensive radar system for the tail turret. The B-36H and RB-36H had a combat radius of 3,113 miles (5010 km) with five tons of bombs and a ferry range of just under 7,700 miles (12390 km).

The last Peacemaker model was the B-36J, of which 33 were built. Unlike the case of the B-36D, B-36F and B-36H, this number included no reconnaissance version. Beginning in 1956, the Strategic Air Command no longer classified RB-36s as distinct from B-36s. Theoretically, each type could be adapted for the opposite role.

As with its immediate predecessor, the B-36J retained the same dimensions as the B-36F, as well as the same powerplant system, consisting of the two pairs of General Electric J47-GE-19 turbojets and the six Pratt & Whitney R-4360-53 piston engines. A major difference was the addition of two additional fuel cells in the wings, which helped increase

Most of the YB-60's fuselage was identical to that of the B-36F – the domed flight deck glazing appearing out of place on an otherwise sleek bomber. A new, more streamlined nose section was fitted, as was a massive swept tail. Although the wing span was less, the area was increased to 5,239 sq ft (486.7 m²). Take-off weight and ordnance load were the same as for a B-36J. Compared to the B-52, the YB-60 had a much thicker wing, one of the factors that contributed to it being considerably slower.

YB-60 – all-jet bomber

The YB-36G began life as a swept-wing version of the B-36F, initially with six turboprop engines. The advent of Pratt & Whitney's excellent J57 turbojet, however, saw a change to an eight-jet configuration, although delays with the jets required Convair to continue the examination of turboprop-powered aircraft until the last of the eight J57-P-3s finally arrived in April 1952. On the 18th Beryl Erickson took the aircraft aloft for a 66-minute maiden flight. In January 1953 the B-60 was cancelled in favour of the much faster and more promising B-52. The first YB-60 only flew 24 times, while the almost complete second aircraft did not even gain engines. Convair proposed using the aircraft as testbeds for advanced turboprops, and even as the basis for a jet-powered airliner, but in July 1954 the two airframes were scrapped.

Then the world's largest landplane, the XC-99 takes shape at Convair's San Diego plant. The XC-99 (Model 37) married the wings, tail, undercarriage and engines of the XB-36 with a cavernous double-deck fuselage that could accommodate up to 400 troops or large cargo loads, the latter being uploaded by winches through lower hatches. This capacity was put to good use during the Korean War when the XC-99 ferried supplies to West Coast ports. It was retired in March 1957, by which time it had gained weather radar in the nose.

The XC-99 cruises over Point Loma, with its birthplace – San Diego's Lindbergh Field – visible above its spine and NAS North Island below the nose. A C-99 production version would have had a pressurised upper cabin, front and rear loading ramps and a domed flight deck. After retirement the XC-99 remained at Kelly AFB until 2004, when it was moved to the USAF Museum for restoration.

the ferry range to 9,440 miles (15192 km) and the combat radius to 3,990 miles (6421 km).

The B-36J made its first flight in July 1953 and the aircraft were delivered to Strategic Air Command between October 1953 and August 1954. The last 14 were delivered from the Fort Worth factory as Class III Featherweights, meaning that they never carried any defensive armament other than a tail turret.

The B-36G becomes the B-60

The USAAF took delivery of its first jet fighters and its first light jet bombers during World War II, and it was natural to begin projecting forward to the advent of an all-jet heavy strategic bomber with intercontinental range. As we now know, that bomber would be the Boeing B-52 Stratofortress, which made its first flight in April 1952.

In August 1950, however, Convair had formally proposed its own all-jet intercontinental bomber. The Korean War-era military build-up had yet to gain steam, so budgetary constraints were still an issue. To help sell the Air Force, Convair presented their new aircraft as a B-36 variant that would have as much as possible in common with the existing Peacemaker fleet. Indeed, the two aircraft types had 72 percent of their parts in common. The Air Force was intrigued, and ordered a pair of the new all-jet aircraft under the designation B-36G in March 1951.

Both B-36Gs began life as heavily modified B-36F airframes. The fuselage was essentially the same, although the nose was streamlined and retrofitted for the test programme with a nose probe. The wings, however, were a different matter, being swept back 37°. This was almost identical to the 36.5° sweep that Boeing was using for the B-52. The engines were the same that Boeing would specify for the B-52 prototype, the Pratt & Whitney J57-P-3 turbojets. As with the B-52, there were eight of these engines grouped in two pods under the leading edge of each wing.

As the B-36G evolved, it became obvious that it would be radically different from the B-36 family, so the aircraft was redesignated as YB-60. The two YB-60s were 175 ft 2 in (53.39 m) long, 13 ft (4 m) longer than a B-36, and had a wingspan of 206 ft 5 in (62.92 m). This was 24 ft (7.3 m) less than the B-36 series, but the wing area of 5,239 sq ft (486.7 m^2) was 10 percent great than that of the B-36. By comparison, the first Boeing B-52 was 152 ft 8 in (46.53 m) long, with a wing span of 185 ft (56.39 m) and a wing area of just 4,000 sq ft (371.6 m^2). The tip of the YB-60's tall, swept tail stood 60 ft 6 in (18.44 m) above the ground, dwarfing the B-52 tail's 48 ft 3 in (14.71 m). The new Convair bird was the largest jet aircraft in the world when it rolled out at Fort Worth in April 1952.

Beryl Erickson took the first YB-60 up for its debut flight on 18 April 1952, just three days after the first flight of the XB-52 from Boeing Field in Seattle. Unfortunately for Convair, the YB-60 would continue to figuratively eat the B-52's exhaust. In terms of performance, the YB-60 demonstrated a top speed of just over 500 mph (805 km/h), while the XB-52 was more than 100 mph (160 km/h) faster. The YB-60 would have a ferry range of 6,192 miles (9965 km), while that figure would be 7,015 miles (11289 km) for the XB-52 and 7,856 miles (12643 km) for the production model B-52C. The gross weight of the YB-60 matched the 410,000 lb (185976 kg) of the B-36J, but this would be less than that of early production model B-52s.

The US Air Force decided between the two types rather quickly, terminating the YB-60 flight test programme in January 1953 after just 66 hours. By this time, the second YB-60 was completed and waiting for its engines. They were never installed and the aircraft was scrapped before it flew.

Convair tried a number of gambits to revive the programme, from offering to finish all the remaining B-36s as B-60s at no extra charge, to converting the prototypes to jetliners for any interested airline. The proposals fell on deaf ears and within 18 months of cancellation, both YB-60s had been reduced to scrap.

The B-36 in service

The largest bomber of World War II never saw combat. Had the B-36 project evolved with the same dispatch as that of its older sister, the Consolidated XB-24, then the XB-36 would have been in the air at the time of the Battle of Midway, and a B-36A could have been available to strike any target in Hitler's Reich at the time of the Operation

Husky invasion of Italy in 1943. USAAF B-29s were finally able to begin launching attacks against Japan in the summer of 1944, flying extraordinary missions from bases in China that were at the end of the most cruel and most difficult aerial supply lines in the world. Had B-36s been pushed into squadron service with the same urgency as the B-29, they could have been attacking Japan from Hickam Field in Hawaii by 1945. What sweet irony that would have been: to be routinely bombing Japan from a base adjacent to Pearl Harbor.

But the B-36 programme was not nudged along with the same sense of importance as the Consolidated B-24 Liberator or the Boeing B-29 Superfortress. The B-36 was a vastly more complex airplane, and the B-36 programme suffered from constantly shifting notions of its priority in the halls of the USAAF.

The USAAF viewed the B-36 programme, through their myopic lens, as a bomber to hit the Reich from North America, and so long as there were bomber bases available in Britain, there was no need to bomb Germany from bases in North America. In the portfolio of priorities in the top levels of the American military establishment, other uses for the big bomber could be imagined but not considered.

The largest bomber ever to see service with SAC was never used in combat. Though they were in the Strategic Air Command inventory during the Korean War, a conscious decision was made not to use them in combat, but to retain them for use in a possible major nuclear or conventional strike against the Soviet Union. In the vernacular of the times, they were SAC's 'Sunday Punch'. They served as a demonstration of SAC's strength. To paraphrase President Theodore Roosevelt's admonition to "Speak softly and carry a big stick," they were SAC's 'big stick'.

General Curtis LeMay put the B-36 to use as a power projection demonstrator for Strategic Air Command. Though the B-36 never flew over Korea, such missions included several overseas deployments during the Korean War. The first Peacemaker mission flown outside the United States involved a half-dozen B-36Ds deploying briefly to RAF Lakenheath in England in January 1951. Later in the year, B-36s returned to the United Kingdom to represent the 7th Bombardment Wing and 11th Bombardment Wing in a Royal Air Force bombing competition at RAF Sculthorpe. Other 1951 missions involved non-stop flights from Carswell AFB to the base at Sidi Slimane in French Morocco, from where SAC operated during the 1950s.

In 1952, RB-36s of the 28th Strategic Reconnaissance Wing won the Strategic Air Command Reconnaissance Competition. The following year, B-36s of the same unit won the Strategic Air Command Bombing Competition and thereby earning the fifth Fairchild Trophy to be awarded to the year's most outstanding bomber unit. The following year, B-36 units took the top three slots, with the 11th Bombardment Wing taking home the Fairchild Trophy. In 1955, the Fairchild Trophy was won by a B-47 unit, but in 1956, it went home aboard a B-36 of the 11th Bombardment Wing. In their last two appearances in the Strategic Air Command Bombing Competition in 1957 and 1958, the B-36 outfits lost to B-47 units.

FICON parasite fighter trials began with converted GRB-36F 49-2707, and a straight-wing F-84E Thunderjet with a hook mounted on a nose gantry. After 170 successful hook-ups, mostly conducted from Eglin AFB, Florida, the YF-84F prototype for the swept-wing Thunderstreak (49-2430) was modified with a much neater retractable hook arrangement and angled-down tailplanes to continue trials from May 1953. Here the YF-84F parasite takes off as the GRB-36F swoops overhead with its trapeze lowered. Subsequently, an RF-84F Thunderflash (51-1828) was converted to GRF-84F standard for trials of the photo-recon version.

In August 1953, shortly after the armistice in Korea, B-36s of the 92nd Bombardment Wing conducted Operation Big Stick, a 30-day deployment to the Far East, specifically to Guam, Okinawa and Japan. In October 1954, the 92nd Bombardment Wing became the first B-36 unit to conduct an extended overseas deployment. The unit deployed from Fairchild AFB near Spokane, Washington, to Andersen AFB on Guam for a 90-day rotation.

XC-99

During World War II, when the XB-36 was still on the drawing boards, Convair's 'Mac' Laddon had suggested that the USAAF consider a transport version of the bomber. With the war raging, there was still an urgent need to move large numbers of troops and their supplies. The USAAF agreed and in 1942 ordered such an aircraft under the designation XC-99.

Below and bottom: These views show part of the parasite recovery procedure, as practised by the GRB-36F and YF-84F. The fighter manoeuvred gingerly toward the cradle of the trapeze to engage it first with its nose hook, before the trapeze assembly was levelled so that the two arms of the 'H'-shaped cradle straddled the fighter, and pins locked into the parasite's fuselage sides. The trapeze was then retracted so that the fighter was partially recessed in the former nos 2 , 3 and 4 bomb bays.

FICON

The FICON project was originally intended to allow B-36s to carry their own fighter escorts, initially in the form of three F-85 Goblins carried in the bomb bays with their wings folded for stowage. Problems with the XF-85 test programme, which used a B-29 as a carrier, led to the adoption of the F-84. In the event, FICON became operational as a means of providing low-level reconnaissance, in effect greatly extending the range of the RF-84 Thunderflash.

B-36 weapons

Defensive armament

The B-36 had six retractable turrets, each aimed from associated glazed blister sighting stations. The fuselage turrets offered around 100° traverse in azimuth either side of the broadside position and +89°/-24° (or vice versa for lower guns) in elevation. Fire interruptors were installed to prevent the guns being fired at airframe parts, and the turrets had contour followers to prevent the barrels from hitting the surrounding airframe. With gun barrels traversed fully forward, the turrets retracted into the fuselage by rotating inwards, before being covered by sliding panels. The turrets are seen on an RB-36H (above and above right). The nose turret fired through +30°/-28.5° in elevation, and 30° either side of the centreline in azimuth. The tail turret (being loaded, right) had a field of fire covering 40° either side of the centreline in elevation and 45° either side in azimuth. Each turret was armed with two 20-mm cannon (usually M24A-1s) providing a muzzle velocity of 2,400 ft (732 m) per second. The nose guns had 400 rounds per gun, whereas the other 14 weapons had 600 rounds each.

Free-fall bombs

Nuclear bombs were the B-36's main armament. Fission weapons were the Mks III, 4, 5, 6 (left), 8 and 18. From 1953 the B-36 carried the first thermonuclear hydrogen bombs, beginning with the Mk 14, and subsequently Mks 15, 17, 21, 36 and 39. The Mk 17 Runt (right) weighed in at 42,000 lb (19051 kg) and was the largest US bomb. The yield was around 15-20 MT. They were only carried by the B-36.

The capacious bomb bays of the B-36 could handle a wide variety of conventional free-fall weapons. The largest of these were the 12,000-lb (5443-kg) T-10 and 22,000-lb (9979-kg) T-14 'earthquake' bombs based on the Grand Slam/Tallboy weapons built by the UK during World War II, and the similarly shaped 43,000-lb (19505-kg) T-12 bomb. The B-36 could carry four T-10s, three T-14s or two T-12s, an example of which is seen here being loaded on to a B-36B for a test at Edwards. At the other end of the scale, the B-36 could carry 132 500-lb (227-kg) weapons.

Specification

Dimensions: wing span 230 ft (70.1 m); length 162 ft 1 in (49.40 m); height 46 ft 8 in (14.22 m); wing area 4,772 sq ft (443.32 m^2)
Weights: empty (B-36J) 171,035 lb (77581 kg); gross take-off 357,500 lb (162162 kg), or 410,000 lb (185976 kg) for B-36J
Fuel: total 31,648 US gal (119800 litres) of which 30,630 US gal (115947 litres) useable, plus 3,090 US gal (11697 litres) in optional bomb bay tank; B-36J had additional outer wing tanks for another 2,770 US gal (10486 litres)
Weapon load: varying conventional and nuclear bombloads in four bays up to a normal maximum of 72,000 lb (32659 kg), or 86,000 lb (39010 kg) when carrying two T-12 outsize conventional weapons
Performance (B-36J): maximum speed 411 mph (661 km/h) at 36,400 ft (11095 m); service ceiling 39,900 ft (12161 m); absolute ceiling (Featherweight aircraft) approx. 60,000 ft (18288 m); typical range 6,800 miles (10943 km) with 10,000-lb (4536-kg) bombload
Powerplant: six Pratt & Whitney R-4360 Wasp Major four-row 28-cylinder engines, rated at: 3,000 hp/2238 kW (R-4360-25 – B-36A), 3,500 hp/2611 kW (R-4360-41 – B-36B/D/E) or 3,800 hp/2835 kW (R-4360-53 – B-36F/H/J); four General Electric J47-GE-19 turbojets rated at 5,200 lb (23.14 kN) thrust (B-36D/E/F/H/J)

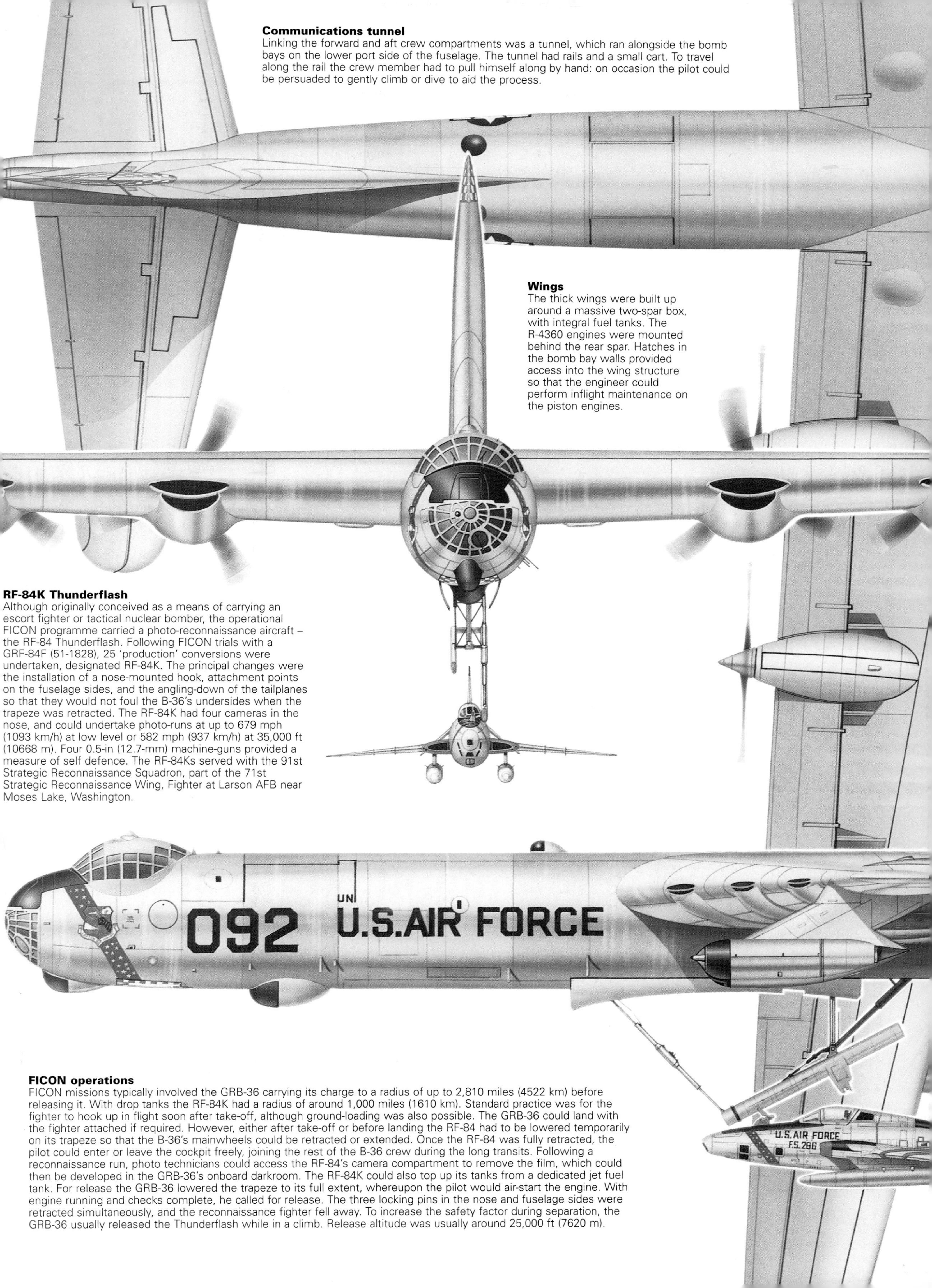

Communications tunnel
Linking the forward and aft crew compartments was a tunnel, which ran alongside the bomb bays on the lower port side of the fuselage. The tunnel had rails and a small cart. To travel along the rail the crew member had to pull himself along by hand: on occasion the pilot could be persuaded to gently climb or dive to aid the process.

Wings
The thick wings were built up around a massive two-spar box, with integral fuel tanks. The R-4360 engines were mounted behind the rear spar. Hatches in the bomb bay walls provided access into the wing structure so that the engineer could perform inflight maintenance on the piston engines.

RF-84K Thunderflash
Although originally conceived as a means of carrying an escort fighter or tactical nuclear bomber, the operational FICON programme carried a photo-reconnaissance aircraft – the RF-84 Thunderflash. Following FICON trials with a GRF-84F (51-1828), 25 'production' conversions were undertaken, designated RF-84K. The principal changes were the installation of a nose-mounted hook, attachment points on the fuselage sides, and the angling-down of the tailplanes so that they would not foul the B-36's undersides when the trapeze was retracted. The RF-84K had four cameras in the nose, and could undertake photo-runs at up to 679 mph (1093 km/h) at low level or 582 mph (937 km/h) at 35,000 ft (10668 m). Four 0.5-in (12.7-mm) machine-guns provided a measure of self defence. The RF-84Ks served with the 91st Strategic Reconnaissance Squadron, part of the 71st Strategic Reconnaissance Wing, Fighter at Larson AFB near Moses Lake, Washington.

FICON operations
FICON missions typically involved the GRB-36 carrying its charge to a radius of up to 2,810 miles (4522 km) before releasing it. With drop tanks the RF-84K had a radius of around 1,000 miles (1610 km). Standard practice was for the fighter to hook up in flight soon after take-off, although ground-loading was also possible. The GRB-36 could land with the fighter attached if required. However, either after take-off or before landing the RF-84 had to be lowered temporarily on its trapeze so that the B-36's mainwheels could be retracted or extended. Once the RF-84 was fully retracted, the pilot could enter or leave the cockpit freely, joining the rest of the B-36 crew during the long transits. Following a reconnaissance run, photo technicians could access the RF-84's camera compartment to remove the film, which could then be developed in the GRB-36's onboard darkroom. The RF-84K could also top up its tanks from a dedicated jet fuel tank. For release the GRB-36 lowered the trapeze to its full extent, whereupon the pilot would air-start the engine. With engine running and checks complete, he called for release. The three locking pins in the nose and fuselage sides were retracted simultaneously, and the reconnaissance fighter fell away. To increase the safety factor during separation, the GRB-36 usually released the Thunderflash while in a climb. Release altitude was usually around 25,000 ft (7620 m).

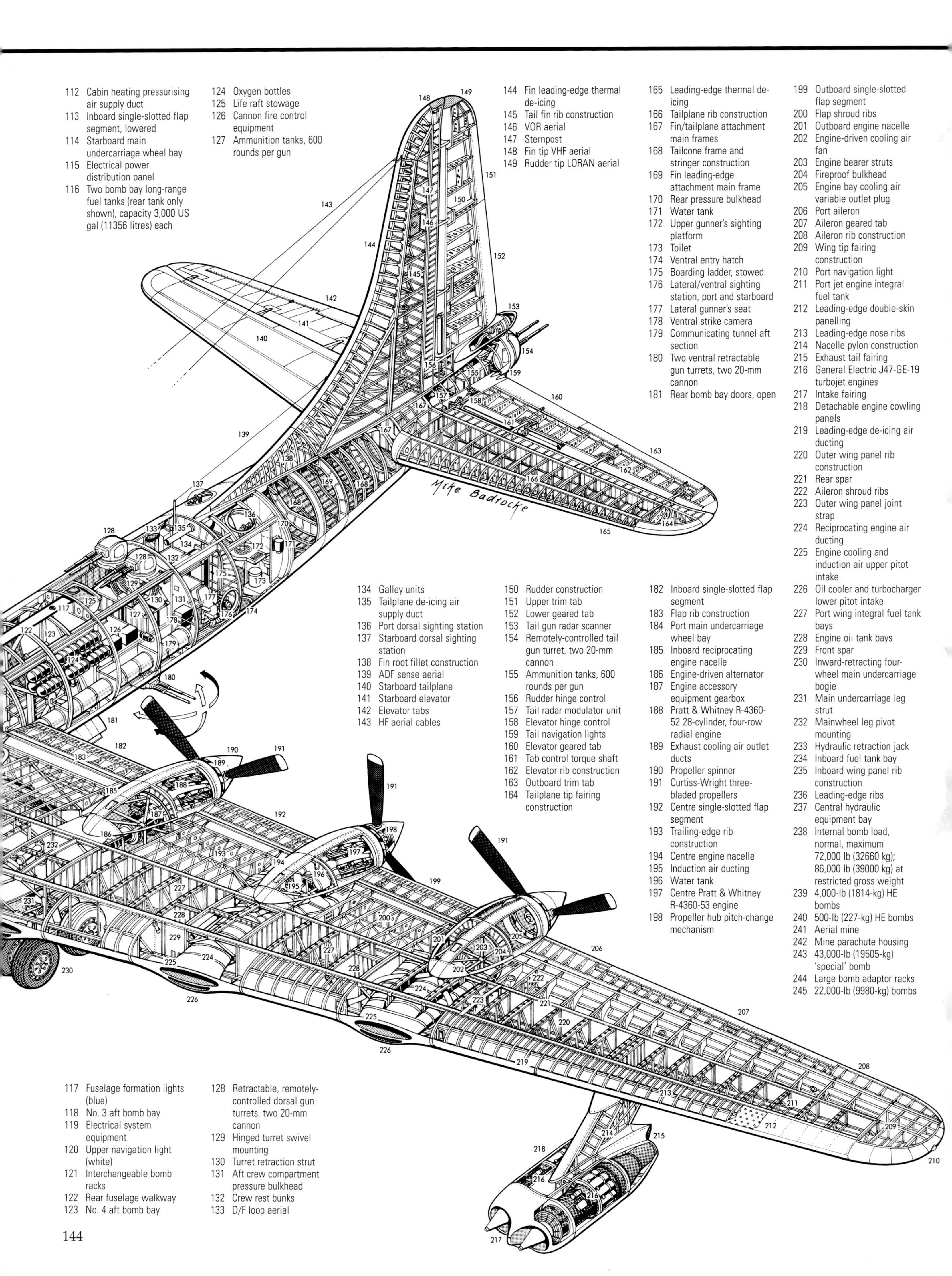

112 Cabin heating pressurising air supply duct
113 Inboard single-slotted flap segment, lowered
114 Starboard main undercarriage wheel bay
115 Electrical power distribution panel
116 Two bomb bay long-range fuel tanks (rear tank only shown), capacity 3,000 US gal (11356 litres) each

124 Oxygen bottles
125 Life raft stowage
126 Cannon fire control equipment
127 Ammunition tanks, 600 rounds per gun

144 Fin leading-edge thermal de-icing
145 Tail fin rib construction
146 VOR aerial
147 Sternpost
148 Fin tip VHF aerial
149 Rudder tip LORAN aerial

165 Leading-edge thermal de-icing
166 Tailplane rib construction
167 Fin/tailplane attachment main frames
168 Tailcone frame and stringer construction
169 Fin leading-edge attachment main frame
170 Rear pressure bulkhead
171 Water tank
172 Upper gunner's sighting platform
173 Toilet
174 Ventral entry hatch
175 Boarding ladder, stowed
176 Lateral/ventral sighting station, port and starboard
177 Lateral gunner's seat
178 Ventral strike camera
179 Communicating tunnel aft section
180 Two ventral retractable gun turrets, two 20-mm cannon
181 Rear bomb bay doors, open

199 Outboard single-slotted flap segment
200 Flap shroud ribs
201 Outboard engine nacelle
202 Engine-driven cooling air fan
203 Engine bearer struts
204 Fireproof bulkhead
205 Engine bay cooling air variable outlet plug
206 Port aileron
207 Aileron geared tab
208 Aileron rib construction
209 Wing tip fairing construction
210 Port navigation light
211 Port jet engine integral fuel tank
212 Leading-edge double-skin panelling
213 Leading-edge nose ribs
214 Nacelle pylon construction
215 Exhaust tail fairing
216 General Electric J47-GE-19 turbojet engines
217 Intake fairing
218 Detachable engine cowling panels
219 Leading-edge de-icing air ducting
220 Outer wing panel rib construction
221 Rear spar
222 Aileron shroud ribs
223 Outer wing panel joint strap
224 Reciprocating engine air ducting
225 Engine cooling and induction air upper pitot intake

134 Galley units
135 Tailplane de-icing air supply duct
136 Port dorsal sighting station
137 Starboard dorsal sighting station
138 Fin root fillet construction
139 ADF sense aerial
140 Starboard tailplane
141 Starboard elevator
142 Elevator tabs
143 HF aerial cables

150 Rudder construction
151 Upper trim tab
152 Lower geared tab
153 Tail gun radar scanner
154 Remotely-controlled tail gun turret, two 20-mm cannon
155 Ammunition tanks, 600 rounds per gun
156 Rudder hinge control
157 Tail radar modulator unit
158 Elevator hinge control
159 Tail navigation lights
160 Elevator geared tab
161 Tab control torque shaft
162 Elevator rib construction
163 Outboard trim tab
164 Tailplane tip fairing construction

182 Inboard single-slotted flap segment
183 Flap rib construction
184 Port main undercarriage wheel bay
185 Inboard reciprocating engine nacelle
186 Engine-driven alternator
187 Engine accessory equipment gearbox
188 Pratt & Whitney R-4360-52 28-cylinder, four-row radial engine
189 Exhaust cooling air outlet ducts
190 Propeller spinner
191 Curtiss-Wright three-bladed propellers
192 Centre single-slotted flap segment
193 Trailing-edge rib construction
194 Centre engine nacelle
195 Induction air ducting
196 Water tank
197 Centre Pratt & Whitney R-4360-53 engine
198 Propeller hub pitch-change mechanism

226 Oil cooler and turbocharger lower pitot intake
227 Port wing integral fuel tank bays
228 Engine oil tank bays
229 Front spar
230 Inward-retracting four-wheel main undercarriage bogie
231 Main undercarriage leg strut
232 Mainwheel leg pivot mounting
233 Hydraulic retraction jack
234 Inboard fuel tank bay
235 Inboard wing panel rib construction
236 Leading-edge ribs
237 Central hydraulic equipment bay
238 Internal bomb load, normal, maximum 72,000 lb (32660 kg); 86,000 lb (39000 kg) at restricted gross weight
239 4,000-lb (1814-kg) HE bombs
240 500-lb (227-kg) HE bombs
241 Aerial mine
242 Mine parachute housing
243 43,000-lb (19505-kg) 'special' bomb
244 Large bomb adaptor racks
245 22,000-lb (9980-kg) bombs

117 Fuselage formation lights (blue)
118 No. 3 aft bomb bay
119 Electrical system equipment
120 Upper navigation light (white)
121 Interchangeable bomb racks
122 Rear fuselage walkway
123 No. 4 aft bomb bay

128 Retractable, remotely-controlled dorsal gun turrets, two 20-mm cannon
129 Hinged turret swivel mounting
130 Turret retraction strut
131 Aft crew compartment pressure bulkhead
132 Crew rest bunks
133 D/F loop aerial

Crew comfort
Regular B-36 bombers were considered relatively roomy and comfortable, although the space could be rapidly taken up with flight bags and rations. The aircraft could fly for over 40 hours on a maximum endurance mission, requiring a lot of food boxes. The RB-36s carried seven or eight more personnel, plus additional Elint equipment, and were rather more cramped in some areas, although the camera compartment provided a roomy space big enough to accommodate a bunk or two. Those aircraft which had been 'Featherweighted' had many crew comfort items (including armrests and the oven) deemed 'superfluous' and removed accordingly as part of the weight-reduction measures.

Convair GRB-36D

348th Strategic Reconnaissance Squadron
99th Strategic Reconnaissance Wing
Fairchild AFB, Washington

Originally scheduled for completion as a B-36B, then as a B-36C, 44-92092 was the 89th production Peacemaker, and was one of the seven RB-36D-1s built as such from the last of the initial 100 (95)-aircraft wartime order. Along with nine other RB-36Ds (44-92090, 44-92094, 49-2687, 49-2692, 49-2694/6, 49-2701/2) it was modified under AF Contract 41-(608)-6464 to become a fighter conveyor 'mothership'. The first aircraft (49-2694) arrived at Fort Worth for conversion in December 1953 and the last was delivered back to the USAF in March 1955. The GRB-36Ds went back to Convair for further modifications between August 1955 and May 1956. They were used for the first operational fighter recoveries from early 1956, but by the end of the year the programme had been terminated.

Weapon bays
The B-36 had four enormous and versatile bomb bays arranged in two pairs. Each pair was covered (in the pure bomber versions) by a pair of doors that were 32 ft 4.5 in (9.87 m) long. 'Small' conventional weapons up to 4,000 lb (1814 kg) were usually carried on racks and were winched up into the bay by hoists, which were temporarily mounted on top of the aircraft's spine with the cables deployed through ports in the upper surface. Large conventional and nuclear bombs were raised into the bay from below on lifts, and were carried on hooks or in cable slings with additional sway bracing. A 3,000-US gal (11356-litre) auxiliary fuel tank could be carried in bay no. 3.

FICON modifications
The most obvious modifications for the FICON role were the new 'fillet' bomb bay doors and concave panels covering the aft three bomb bays – which also entailed moving the three APA-17 direction-finding antennas further aft – and the addition of a 30-ft retractable trapeze/cradle assembly which held the parasite fighter roughly 20 ft below the bomber when deployed. A dedicated trapeze operator was carried, with a station in the forward camera compartment, and the airborne hook-up process was further aided by observers in the lower aft scanning blisters. In the rear bomb bay (no. 4) a 1,000-US gal (3785-litre) tank was installed to carry jet fuel for the RF-84K parasite. A blister fairing on the upper forward fuselage housed the antenna for an APX-29 rendezvous beacon. Operational GRB-36Ds retained cameras in the no. 1 bay camera compartment, and also had a full photo/EW crew complement to operate the RB's own reconnaissance systems. The GRB-36Ds operated in Featherweight III configuration, with all guns deleted apart from those in the tail turret.

Inside the B-36

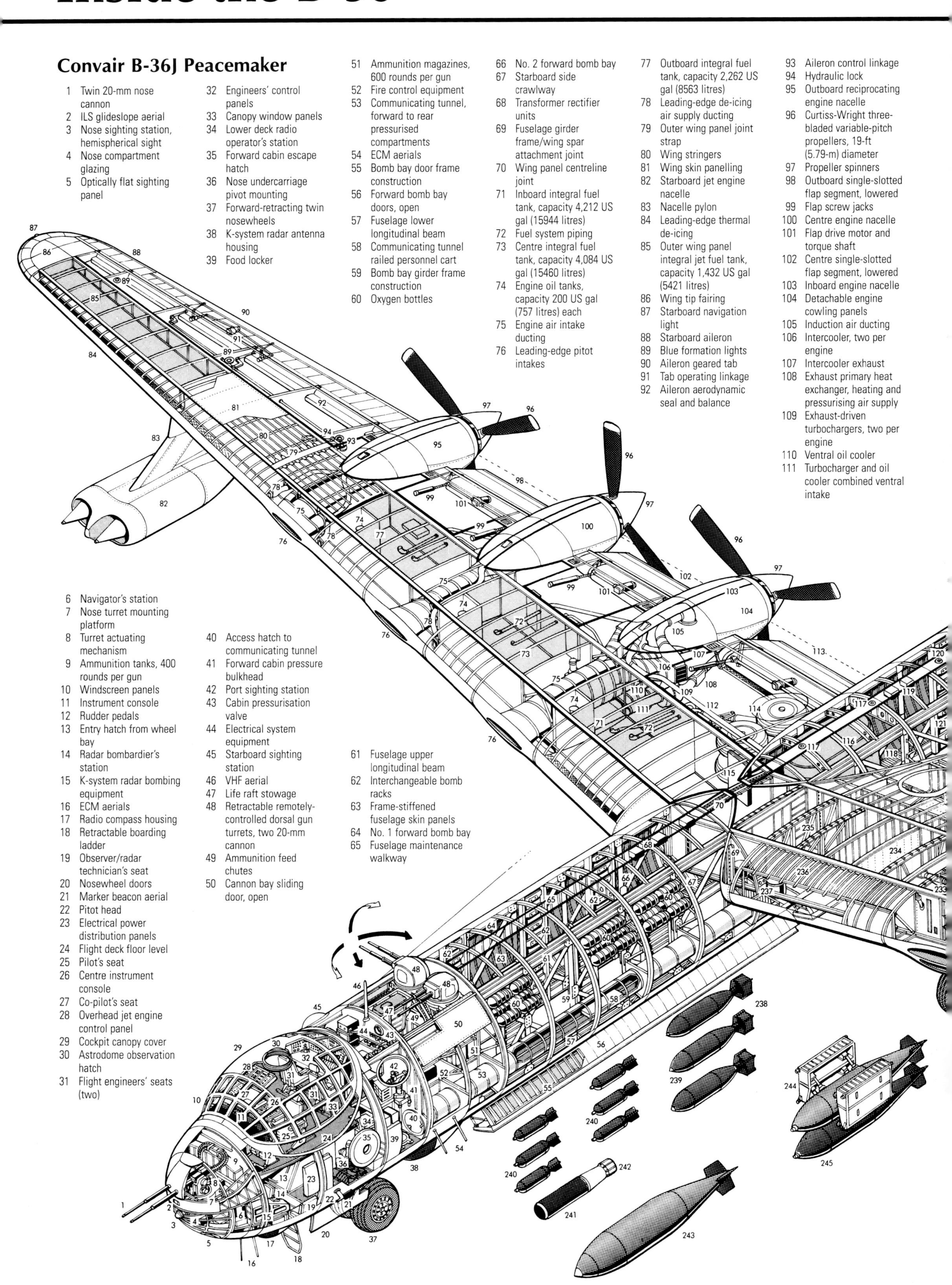

Accommodation

Right: Access to the forward compartment was made through this hatch. The aft compartment also had a circular hatch with a drop-down ladder.

Below: One or two flight engineers (depending on version) faced aft on the upper flight deck. From here they could see the engines.

Below: This view shows the lower deck, looking forward. The navigator, radar-bombardier, observer and one of the radio operators usually occupied this space. The navigator sat in the front right position, and operated the nose guns.

Above: The B-36's flight deck offered an excellent view of the world with near 360° coverage, particularly useful when taxiing. B-36s usually carried three pilots (aircraft commander, pilot, co-pilot), the latter also acting as right forward gunner.

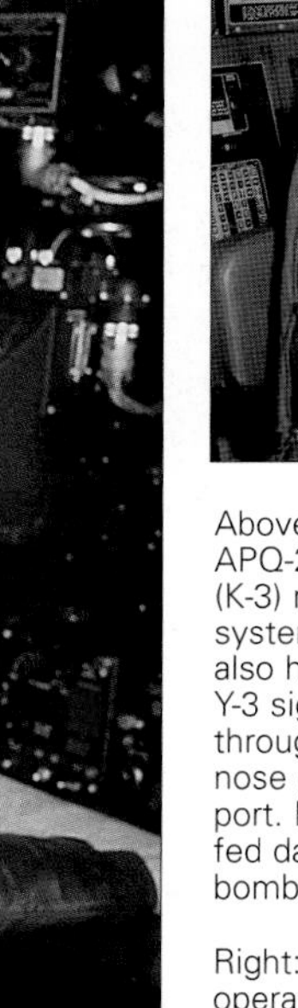

Above: As well as the APQ-24 (K-1) or APS-23 (K-3) radar bombing system, the bombardier also had an optical Y-1 or Y-3 sight which peered through a flat pane in the nose glazing, offset to port. Both radar and sight fed data to the Sperry A-1 bombing computer.

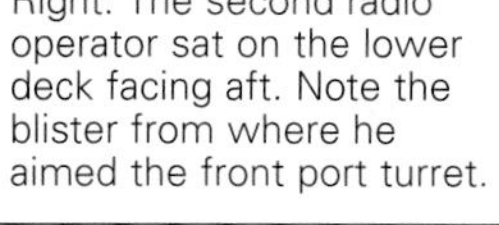

Right: The second radio operator sat on the lower deck facing aft. Note the blister from where he aimed the front port turret.

Powerplant

Known as the 'corn cob' on account of its four rows of seven cylinders each, the Wasp Major was essentially created by placing two Twin Wasps back-to-back and driving a common crankshaft. The engine suffered initially from cooling problems due to the difficulty of getting fresh air to the aft row of cylinders.

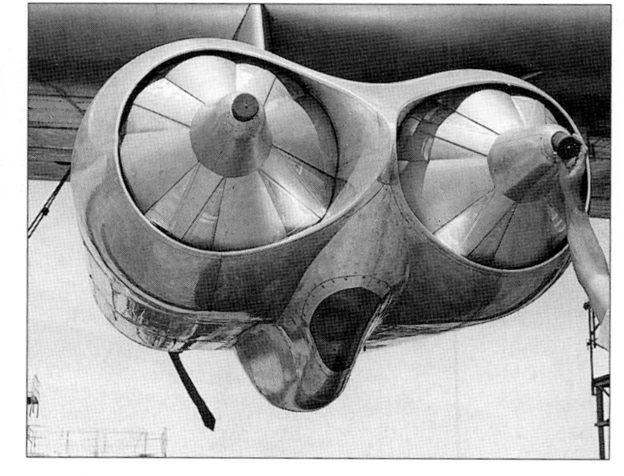

Although the 'prototype' B-36D flew with Allison J35s, the J47-GE-19 was fitted to all jet-augmented B-36s. The jets used the same fuel as the piston engines. For much of the flight they were shut down, incorporating iris-type shutters which deployed from the centrebody to blank off the intake to prevent the engines windmilling inflight.

Project Tom-Tom

Following FICON trials, the GRB-36F (49-2707, JRB-36F after 1955) was further modified for Project Tom-Tom, which involved the carriage of RF-84s in a wingtip-to-wingtip arrangement. This followed earlier trials with a C-47/Culver PQ-14 combination, and later with an ETB-29B and EF-84Bs, but which had ended after a crash on 24 April 1953 claimed the Superfortress, both of the Thunderjets, and their crews. For Tom-Tom, the GRB-36F had a nose air data boom added, and an APX-29 rendezvous beacon in a prominent fairing above the forward fuselage. The wingtip attachment installation (right) included an arm which swung out, on to which the RF-84 latched by means of a gripping fork arrangement. Tom-Tom was conducted from April 1956 until 23 September, when it was stopped following an incident in which an RF-84 was torn away from the wingtip mounting.

Two RF-84Fs (51-1848/9) were modified with the gripping fork wingtip system for Tom-Tom. Hooking up proved very difficult and ultimately led to the project's termination, but with the recce aircraft attached the B-36 suffered only minor performance penalties. The reduction in induced drag (imparted by the effective increase in the B-36 wing's aspect ratio with the Thunderflashes attached) helped offset the extra weight of the parasites.

As was the case with the XB-36, work on the XC-99 was delayed by more pressing wartime commitments, but in 1945 it began to take shape. The fuselage, with the largest volume ever built for a non-seaplane, was constructed at Lindbergh Field in San Diego, and the wings and tail – actually B-36 wings and tail – were brought in from Fort Worth. Like the early XB-36, the XC-99 was to be powered by six Pratt & Whitney R-4360-25 piston engines. Indeed the wing layout and 230-ft span were the same for both aircraft.

When Russell Rogers and Beryl Erickson took off in the XC-99 on 23 November 1947, they were flying the largest aircraft to ever roll down a runway. Though it had the same wing span as the B-36, it was 185 ft (56.39 m) long. The only XC-99 to be built was delivered to the Air Force in 1949, and it was used during the Korean War for weekly flights between the west coast of the United States and Japan. The US Air Force retired the XC-99 in 1957 and it was put on static display near Kelly AFB outside San Antonio, where it remained until moved to Dayton in 2004.

In the early 1950s, Pan American World Airways expressed an interest in a commercial version of the XC-99 that would have been built under the Convair Model 37 designation, but decided speed was more important than volume and consequently bought smaller jetliners instead.

Fighter conveyor aircraft

In the 1949 Capitol Hill showdown between the US Navy and the US Air Force, the US Navy lost an aircraft-carrier and the Air Force won a continued B-36 production line. At that time, there were already plans afoot to turn the B-36 into an aircraft-carrier – an airborne aircraft-carrier!

During World War II, it had become obvious that long-range bombers benefit greatly from having fighters escorting them. Early in the war, fighters did not have the range to accompany the bombers, but this was gradually extended. For example, by 1944, P-51D fighters based in England were flying with Eighth Air Force bombers deep into Germany.

After the war, the range of the large bombers was to grow further, and the extension of fighter range capability would not keep pace. The B-36 had a range that greatly exceeded that of any fighter, so the idea emerged to carry fighters with the bombers. As the enemy attacked, these 'parasite' fighters could be released to do battle, and would then be recovered and pulled back into the big bombers for the trip home.

The first tests of this concept were carried out in 1948, using the tiny McDonnell F-85 Goblin parasite jet fighter, which was released and recovered from a Boeing B-29 using a trapeze-like device slung beneath the bomber. The idea was that, after the tests, the F-85s would be carried by B-36s, which could accommodate up to three of the little Goblins entirely within modified bomb bays. The B-29 tests were not entirely successful, and the programme was cancelled in 1949 before it was tested with a B-36.

However, the US Air Force revived the concept in January 1951 under its Fighter Conveyor Aircraft (FICON) programme. This now involved using RB-36Fs as fighter conveyors, and Republic RF-84F Thunderflashes (originally designated as YF-96A) as the fighter aircraft. Actually, the idea under FICON was to use the parasites as fast reconnaissance aircraft, although they could have also been used as escort fighters.

The first FICON tests involving a GRB-36F conveyor and an F-84E fighter began in early 1952 out of Eglin AFB in Florida. In FICON operations, the GRB-36F recovered the fighter by extending a cradle-like apparatus from its modified bomb bay. The fighter pilot attached his aircraft to this device with a hook on its nose. When the fighter was thus attached, the cradle was lowered over the top of the F-84,

grasping it and pulling it up into the GRB-36F's bomb bay. The F-84 was released by slowly lowering the cradle and simply letting go. The first full test of this whole cycle occurred in April 1952, and tests continued throughout the year.

The FICON system involving the GRB-36F and the F-84 was far more successful than had been the B-29 and F-85 tests just four years earlier. Nearly 170 FICON test flights are said to have been made, include some at night. The operations were so successful that, in 1953, the Air Force ordered Convair to convert 10 RB-36Ds as GRB-36D conveyor aircraft. Republic, meanwhile, supplied a number of RF-84Ks. These operational FICON pairs were delivered to Strategic Air Command early in 1955. They served for less than a year and are not known to have been used operationally against foreign targets during that time.

NB-36H Crusader and the X-6 programme

Shortly after World War II, as the US Navy was working on plans for nuclear ships and submarines, the US Air Force began to lay plans for nuclear-powered aircraft. It was more than inter-service rivalry – a strategic bomber with unlimited range was the ultimate fulfilment of the doctrine of strategic air power. Accordingly, the Nuclear Energy for Propulsion of Aircraft (NEPA) programme was initiated in 1946.

In 1947, the Atomic Energy Commission, a United States government agency responsible for the development of nuclear reactors, was created. This organisation would work up the initial parameters for the design of shipboard nuclear reactors for the Navy, and it would absorb the activities of NEPA as well. In turn, the commission established the National Reactor Testing Station. It opened in 1949 in a remote location in the lightly populated high desert country of southern Idaho near the town of Arco.

In 1951, the commission decided that it was theoretically possible to produce a reactor for an aircraft. The problem, of course, would be the size and weight of the reactor, as well as its radiation shield and cooling system. Therefore it would have to be a very large aircraft.

Because no one had ever designed a nuclear-powered aircraft before, the development process would be methodical, moving step-by-step. Before a nuclear-powered strategic bomber could become an operational reality, a nuclear-powered experimental aircraft would have to be tested as a proof-of-concept demonstrator. This aircraft would be built under the designation X-6.

Before the X-6 could fly, however, it was necessary to build and fly a functioning nuclear reactor aboard a conventional aircraft in order to evaluate shielding techniques. As a reactor test bed for the eventual X-6, the US Air Force chose the largest aircraft in its fleet, the B-36H.

***Above:** The operational FICON pairing consisted of the GRB-36D, of which 10 were modified by Convair in 1953-55, and the RF-84K, 25 of which were modified from RF-84Fs by Republic. The GRB-36Ds served with the 348th SRS, 99th SRW at Fairchild, while the RF-84Ks flew with the 91st SRS at nearby Larson AFB, both in Washington state. FICON operations lasted for less than a year during 1956: inflight refuelling and other reconnaissance platforms, such as the U-2, had rendered the B-36/RF-84 combo obsolete, while the rendezvous process proved difficult for the fighter pilots, who suffered several collisions with the trapeze. The final straw was a collision in which the GRB-36's cradle was knocked off the trapeze by a fighter: General LeMay swiftly ordered the programme's termination.*

***Above left:** Mating the RF-84K with the GRB-36D on the ground required the B-36's main undercarriage to be taxied on to platforms to provide sufficient clearance for the RF-84K to be manoeuvred underneath (illustrated), or over a special pit into which the RF-84K could be placed before the B-36 was taxied into position. The GRB-36D had vestigial bomb bay 'fillet' doors which were contoured to the RF-84's fuselage when the parasite was hauled into the bomb bay. The fin of the fighter nestled into the former no. 4 bomb bay, which had a concave cover and a rubber-edged slit to enclose the fin and maintain an element of aerodynamic integrity.*

Wearing the badge of ARDC ahead of the 'U.S. Air Force' titles, this aircraft (49-2694) was the first to be converted to the operational GRB-36D standard, and is seen undergoing trials at Edwards AFB. The view highlights the limited ground clearance available when the RF-84K was carried, especially if was outfitted with drop tanks.

The reactor-carrying B-36 – 51-5712 – was initially designated as the XB-36H, and it carries that title on the forward fuselage on this early test flight (above). In late 1956 it was changed to the better known NB-36H designation, as seen in this view of the five-man crew (right). The aircraft was one of the many that had been damaged at Carswell/Fort Worth by a tornado on 1 September 1952. Its nose was badly damaged, so it made sense to use this airframe for the NEPA (Nuclear Energy for the Propulsion of Aircraft) project as it required an all-new forward fuselage.

The NB-36H undertook its test flights over sparsely populated areas of Texas and New Mexico, the reactor not being started until the aircraft was over 'safe' territory and at altitude. It was always followed by a chase aircraft, in this instance a Boeing B-50 bomber, which carried engineers and measuring equipment. If the NB-36H had gone down or had to land at another airfield, the chase aircraft could disgorge paratroops to immediately secure the top-secret test vehicle. Note that the outboard jet engines are shut down.

Early in 1951, as the Atomic Energy Commission undertook the construction of the first airborne nuclear reactor, designated as R-1, the Air Force ordered Convair to specially modify a B-36H under the test bed designation NB-36H, and the given name Crusader. Convair was also given the contract to modify an additional pair of B-36s that would become the first two nuclear-powered X-6 aircraft. The NB-36H would simply fly with a functioning reactor aboard. The X-6s would fly under nuclear power, with the R-1 reactor powering General Electric turbojet engines. The total propulsion system, incorporating the reactor and engines, was designated as P-1.

The design of the propulsion system would allow the engines to transition between nuclear power and jet fuel, permitting a conventionally fuelled take-off followed by an extremely long cruise under reactor power. Nuclear cruise would easily permit a non-stop flight around the world.

Meanwhile, work was under way to construct a test base for the X-6 programme. While the NB-36H Crusader flights were conducted from Carswell AFB near Fort Worth, the Air Force wanted a location for the X-6 project that was farther from the prying eyes of just anyone. The site chosen for what was to be designated as 'Test Area North', was near the small town of Monteview, in the Idaho desert, about 40 miles northeast of the Atomic Energy Commission's National Reactor Testing Station.

The huge hangar that was constructed at Test Area North had thick, lead-lined walls and was designed to contain radiation if necessary. General Electric moved its nuclear powerplant operations here, and installed robotic equipment so that work could be done on the engines and reactors without exposing humans.

The plan was that the first X-6 flight would come in 1957. The first operational nuclear bomber would, in turn, follow the X-6 in the early 1960s. The first operational wing of such aircraft was to be in service in 1964. Not only Convair, but Boeing, Douglas and Lockheed were invited to submit proposals for the development of these bombers.

X-6 cancelled, NB-36 continues

In 1953, however, the incoming Eisenhower Administration revisited the concept of nuclear-powered aircraft and red-lined the X-6 programme. As noted above, the conventional YB-60 programme was also axed at around the same time, as the US Air Force chose to acquire the Boeing B-52 instead. Once again, though, the NB-36H remained a live programme, as did numerous feasibility studies for nuclear-powered operational aircraft for the future. While the X-6 was not to be the first of this new breed, the concept of a nuclear-powered aircraft remained alive and well.

At Test Area North in Idaho, work on the 15,000-ft (4572-m) runway that was planned for the X-6 was not begun, but the other work continued. In 1955, the Atomic Energy Commission initiated a series of Heat Transfer Reactor Experiments at Test Area North's big nuclear-shielded hangar. These experiments were part of the ongoing development of airborne nuclear powerplants, but the engines tested were many orders of magnitude larger than the P-1 that had been earmarked for the X-6. The idea was to refine the concept, then scale down the reactor. The goal was a thermal output of at least 50 megawatts delivered by a reactor the size of the one-megawatt reactor that would be test flown in the NB-36H.

Also in 1955, the NB-36H was finally ready for its debut. Except for a completely redesigned nose section, it was similar in outward appearance to a conventional B-36H. Inside was a different story. The gross weight had been pushed up to 360,000 lb (163296 kg). The one-megawatt reactor itself weighed nearly 18 tons, and the crew shield – consisting of lead plates and water tanks – added another 24,000 lb (10886 kg).

The first flight of the NB-36H occurred on 17 September 1955, and the first top-secret flight tests of an airborne nuclear reactor were soon underway. These missions were

With its smart cheat lines and nuclear symbols, the 'Crusader' brought a splash of colour to the B-36 world. The crew sat in a lead-lined capsule with a windscreen of up to 11 in (28 cm) thickness and heavily shielded hatches. Behind them in the bomb bay was the Aircraft Shield Test Reactor (ASTR), which was installed in the bomb bay after the aircraft was taxied over a special loading pit. The reactor was rated at around 1 megawatt output – much lower than would be needed for propulsion – and together with its shielding weighed around 35,000 lb (15876 kg). TV cameras were mounted on either side of the rear fuselage so that the crew could monitor the engines – a job typically done in the regular B-36 by crew in the aft compartment.

considered so sensitive that, on every flight, the NB-36H was accompanied by a C-119 transport aircraft or B-50 carrying paratroopers. Should the NB-36H have crashed, or been forced to land at a civilian airport because of mechanical difficulties, it would be the job of the paratroopers to surround the aircraft and prevent any unauthorised individuals from reaching it.

During 18 months of flight testing, all phases of airborne nuclear reactor operations – from shielding to power output – were evaluated. When the NB-36H made its 47th and last flight on 28 March 1957, sufficient data now existed to move ahead to the next phase of nuclear aircraft development. That phase never reached its conclusion. After a series of promising reactor tests conducted through the end of 1960, the incoming Kennedy Administration ordered the airborne nuclear propulsion programme to be terminated the following year. Secretary of Defense Robert McNamara had little enthusiasm for manned bombers of any kind. Today, Test Area North remains an active site.

The end of the line for the B-36

After barely half a decade at anything resembling full strength, Strategic Air Command quietly began the phase-out of the operational B-36s early in 1956. The career of SAC's biggest bomber had been sorrowfully short, and the great aircraft had served against the backdrop of everyone knowing this. Ever since the Peacemaker fleet had reached critical mass with SAC in 1951, everyone had known that it was just filling in for its ultimate replacement, the Boeing B-52. By 1957, the B-52 would reach its own critical mass within SAC. SAC had ended 1956 with 247 B-36s, and just 97 B-52s. At the end of 1957, there would be 127 B-36s to 243 B-52s. In 1958, the last year that the B-36 was carried on the inventory, there would be just 22, all B-36Js, compared to 380 B-52s.

By this time, SAC's focus was on not only the B-52, but the process of integrating an entirely new type of weapon into its inventory. In 1959, after experimenting and deploying various short-range ballistic missiles, SAC began testing its first operational Intercontinental Ballistic Missile. This weapon, the B-65 (later SM-65) Atlas, was built by, of all the companies in the military industrial complex, Convair.

The great B-36 Peacemaker fleet was flown, one by one, to Davis-Monthan AFB, near Tucson, Arizona, where they were cut up for scrap. The last B-36J to be scrapped went under the torch on 12 February 1959, less than six years after the first B-36J had been delivered in October 1953.

Only four intact Peacemakers survive today, although large sections of disassembled B-36s still exist under private ownership around the United States. On 12 February 1959, the same day that the last scrapping occurred, the B-36J carrying tail number 52-2827 was flown back to where it all started, to the Convair factory at Fort Worth. Coincidentally, it was the last Peacemaker to come off the Convair assembly line and the last to be flown by a SAC crew. It remains at this location – now owned by the Lockheed-Martin Corporation – to this day.

In April 1959, the last two Peacemakers to be flown – both B-36Js – went to their final resting places. The one carrying tail number 52-2217 was flown to SAC headquarters at Offutt AFB outside Omaha, Nebraska, on 23 April. It was donated to the Strategic Air Command Museum, which is now known as the Strategic Air & Space Museum, in Bellevue, Nebraska.

On 30 April, 52-2220 arrived at the US Air Force Museum at Wright-Patterson AFB near Dayton, Ohio. It served as the centrepiece of the museum's first display hangar, which was literally built up around the B-36J. This Peacemaker is preserved in mint condition, as the museum makes it a practice to maintain most of its collection in theoretically flyable condition. With the building now completely enclosing it, however, it will never fly.

The only surviving Peacemaker that is not a J-Model is the RB-36H located at the Castle Air Museum, near the former Castle AFB, on the outskirts of Atwater, California. This aircraft, tail number 51-13730, was parked at Chanute AFB in Illinois for many years and was moved to California after the base was closed. It was reassembled and formally dedicated as a museum piece in 1994.

Bill Yenne

B-36 survivors

Right: 52-2220 is the B-36J now on display at the USAF Museum, Wright-Patterson AFB. Here it is seen parked out on Wright Field pending restoration and the construction of a new building, which opened in 1971. On 30 April 1959 it made the last flight by a B-36.

Left: RB-36H 51-13730 is now on display in its correct 28th SRW colours at the Castle Air Museum, in California. It had spent many years outside at Chanute AFB, initially with no markings other than that of the local 69th Student Squadron, and then for a period masquerading as 'B-36D' 44-92065 of the 92nd BW.

Right: Of the 385 B-36s built only five initially cheated the scrapmen. One of these was RB-36E (formerly YB-36) 42-13571, seen here on display at the USAF Museum. It was broken up in 1970/71 after 52-2220 arrived at the museum, although sections of this historic aircraft still survive in private hands.

B-36 operators

Including prototypes, a total of 383 B-36s was delivered to the US Air Force. The most that would be in service with the Strategic Air Command at the end of any calendar year was the 338 that were in the inventory at the end of 1954. The accompanying table charts the rise and fall of the B-36 in SAC service.

Strategic Air Command B-36 year-end inventory

Year	B-36	RB-36	Proportion of total SAC fleet
1948	35	0	4 percent
1949	36	0	4 percent
1950	38	20	6 percent
1951	98	65	14 percent
1952	154	114	16 percent
1953	185	137	18 percent
1954	209	133	13 percent
1955	205	133	11 percent
1956	247 total*		8 percent
1957	127 total*		5 percent
1958	22 total*		1 percent
1959	0		0 percent

** Beginning in 1956, SAC no longer considered RB-36s as distinct from B-36s*

5th Strategic Reconnaissance Wing (Heavy) (15th AF, circle X)

23rd Bombardment Squadron
31st Bombardment Squadron
72nd Bombardment Squadron

Established as the 5th Strategic Reconnaissance Wing on 1 July 1949, it was activated on 16 July 1949 and redesignated as the 5th Strategic Reconnaissance Wing (Heavy) on 14 November 1950, as the 5th Bombardment Wing (Heavy) on 1 October 1955, as the 5th Wing on 1 September 1991 and as the 5th Bomb Wing on 1 June 1992.

Based at Fairfield-Suisun AFB (later Travis AFB), near Fairfield, California, the wing operated B-36s from January 1951 through September 1958. It began maintaining proficiency in strategic bombardment in July 1953 but was not redesignated as a bombardment wing until October 1955.

6th Bombardment Wing (Heavy) (8th AF, triangle R)

24th Bombardment Squadron
39th Bombardment Squadron
40th Bombardment Squadron

Established as the 6th Bombardment Wing (Medium) on 20 December 1950, it was activated on 2 January 1951 and redesignated as the 6th Bombardment Wing (Heavy) on 16 June 1952, as the 6th Strategic Aerospace Wing on 1 May 1962, as the 6th Strategic Wing on 25 March 1967, and as the 6th Strategic Reconnaissance Wing on 1 April 1988. Inactivated on 1 September 1992, it was redesignated as the 6th Air Base Wing on 22 December 1993, activated on 4 January 1994 and redesignated as the 6th Air Refueling Wing on 1 October 1996.

Based at Walker AFB, near Roswell, New Mexico, the wing operated B-36s from August 1952 though August 1957.

7th Bombardment Wing (Heavy) (8th AF, triangle J)

9th Bombardment Squadron
436th Bombardment Squadron
492nd Bombardment Squadron

Established as the 7th Bombardment Wing (Very Heavy) on 3 November 1947, it was organised on 17 November 1947, redesignated as the 7th Bombardment Wing (Heavy) on 1 August 1948, as the 7th Wing on 1 September 1991, as the 7th Bomb Wing on 1 June 1992, and as the 7th Wing on 1 October 1993.

Based at Carswell AFB in Fort Worth, Texas, the wing was the first to be equipped with the B-36. It had trained with B-29s in global bombardment operations from November 1947 to December 1948 and began converting to B-36s in June 1948. The wing controlled two B-36 groups (the 7th and 11th) from December 1948 through February 1951, and three B-36 squadrons from February 1951 to May 1958. The wing also flight-tested the XC-99 in June 1949 and evaluated the RB-36 in 1950. It operated B-36s from June 1948 through May 1958.

A scene from a US air show in the 1950s depicts an RB-36H from the 5th SRW (badge of winged skull on the nose) at Travis. Initially the B-36 usually carried wing badges on both sides of the nose, but when SAC introduced the 'Milky Way' sash, as seen on here, the wing badge on the port side was replaced by SAC's famous badge of a mailed fist grasping an olive branch and lightning bolts.

Built as the first B-36B, 44-92026 of the 7th Bomb Wing is seen after it had been modified to B-36D standard. The 7th was the first wing to get the Peacemaker. Until 1953 the fleet wore geometric tailcodes to signify unit (letter) and Air Force (shape) assignments. This was a hangover from the wartime period when the system had been used primarily on the B-29 force.

11th Bombardment Wing (Heavy) (8th AF, triangle U)

26th Bombardment Squadron
42nd Bombardment Squadron
98th Bombardment Squadron

The 11th Bombardment Wing (Heavy) was established on 18 November 1948 and activated on 16 February 1951. It was consolidated with the 11th Bombardment Group. The 11th Bombardment Group was established as the 11th Observation Group on 1 October 1933, redesignated as the 11th Bombardment Group (Medium) on 1 January 1938, activated on 1 February 1940, redesignated as the 11th Bombardment Group (Heavy) on 1 December 1940, as the 11th Bombardment Group (Heavy) on 3 August 1944, and as the 11th Bombardment Group (Very Heavy) on 30 April 1946. Inactivated on 20 October 1948, it was redesignated as the 11th Bombardment Group (Heavy) and activated on 1 December 1948. The 11th Bombardment Wing (Heavy) was redesignated as the 11th Strategic Aerospace Wing on 1 April 1962, as the 11th Air Refueling Wing on 2 July 1968 and inactivated on 25 March 1969. The consolidated unit retained the designation as the 11th Strategic Group. Inactivated on 7 August 1990, it was redesignated as the 11th Support Wing on 2 June 1994 and activated on 15 July 1994. It was redesignated as the 11th Wing on 1 March 1995.

Based at Carswell AFB in Fort Worth, Texas, the wing was the second unit to be activated with the B-36. It operated B-36s from December 1948 through December 1957, initially as the 11th Bombardment Group.

28th Strategic Reconnaissance Wing (Heavy) (8th AF, triangle S)

72nd Bombardment Squadron
717th Bombardment Squadron
718th Bombardment Squadron

Established as the 28th Bombardment Wing (Very Heavy) on 28 July 1947, the wing was organised on 15 August 1947. It was redesignated as the 28th Bombardment Wing (Medium) on 12 July 1948, as the 28th Bombardment Wing (Heavy) on 16 May 1949, as the 28th Strategic Reconnaissance Wing on 1 April 1950, as the 28th Strategic Reconnaissance Wing (Heavy) on 16 July 1950, as the 28th Bombardment Wing (Heavy) on 1 October 1955, as the 28th Wing on 1 September 1991 and as the 28th Bomb Wing on 1 June 1992.

Based at Rapid City AFB (later Ellsworth AFB) near Rapid City, South Dakota, the wing operated B-36s from July 1949 through May 1957. The wing performed global strategic reconnaissance from 1950 to 1955, with bombardment as the a secondary mission in 1954 and 1955. The wing trained primarily as a bombardment wing from 1955, but retained a reconnaissance capability until September 1956.

42nd Bombardment Wing (Heavy) (no tailcode)

69th Bombardment Squadron
70th Bombardment Squadron
75th Bombardment Squadron

Established as the 42nd Bombardment Group (Medium) on 20 November 1940 and activated on 15 January 1941, the wing was redesignated as the 42nd Bombardment Group (Medium) on 6 September 1944. During World War II, it had served throughout the South Pacific. It ended its combat service attacking isolated Japanese units on Luzon in August 1945. Inactivated on 10 May 1946 it was consolidated with the 42nd Bombardment Wing (Heavy) which was established on 19 February 1953 and activated on 25 February 1953. Redesignated as the 42nd Wing on 1 September 1991, it became the 42nd Bomb Wing on 1 June 1992. Inactivated on 30 September 1994, it was redesignated as the 42nd Air Base Wing and activated on 1 October 1994.

Based at Limestone (later Loring) AFB in Maine, the wing operated B-36s from April 1953 through September 1956. In 1954 and 1955, portions of the wing deployed to RAF Upper Heyford and RAF Burtonwood in the United Kingdom, and the entire wing deployed to RAF Upper Heyford in October and November 1955.

Along with the 95th BW at Biggs, the 42nd BW was one of two wings to equip with the B-36 in 1953, the last units to get the type. This is a B-36D, wearing the wing badge of two bombs on a blue shield, with four red circles on a yellow band.

The nose of this B-36 proudly wears the pteranodon badge of the 92nd Bomb Wing, which was changed in 1957 to one featuring a sword, olive branch and lightning bolt. The small glazed dome on the nose covered the sight for the nose guns. Note the covered flat glazed panel for the optical bombsight.

Second unit to get the B-36 was the 11th Bomb Group (later Wing), which joined the 7th BG/BW in operating from Carswell. These are B-36Hs. Although not worn here, the squadron colours within the 11th BW were striped, as opposed to solid – standard practice when there were two wings co-located in one Air Division.

72nd Strategic Reconnaissance Wing (Heavy) (2nd AF, square F)

60th Bombardment Squadron
73rd Bombardment Squadron
301st Bombardment Squadron

Established as the 72nd Observation Group on 21 August 1941, it was activated on 26 September 1941, redesignated as the 72nd Reconnaissance Group (Special) on 25 June 1943 and disestablished on 1 November 1943 in the Canal Zone. It was activated again as the part of the reserve forces of Fourth Air Force between July 1947 and June 1949. Activated on paper on 16 June 1952, it was not operational until it absorbed residual resources of the 55th Strategic Reconnaissance Wing in October 1952. The 72nd Strategic Reconnaissance Wing (Heavy) was established on 4 June 1952, activated on 16 June 1952 and redesignated as the 72nd Bombardment Wing (Heavy) on 1 October 1955. Inactivated on 30 June 1971, it was redesignated as the 72nd Air Base Wing on 16 September 1994 and activated on 1 October 1994.

Based at Ramey AFB, near San Juan, Puerto Rico, the wing operated B-36s from October 1952 through January 1959.

92nd Bombardment Wing (Heavy) (15th AF, circle W)

325th Bombardment Squadron
326th Bombardment Squadron
327th Bombardment Squadron

Established as the 92nd Bombardment Wing (Very Heavy) on 17 November 1947, it was redesignated as the 92nd Bombardment Wing (Medium) on 12 July 1948, and as the 92nd Bombardment Wing (Heavy) on 16 June 1951. It became the 92nd Strategic Aerospace Wing on 15 February 1962, the 92nd Bombardment Wing (Heavy) on 31 March 1972, the 92nd Wing on 1 September 1991, the 92nd Bomb Wing on 1 June 1992 and the 92nd Air Refueling Wing on 1 July 1994.

Based at Fairchild AFB (formerly Spokane Army Air Field) near Spokane, Washington, the wing operated B-36s from July 1951 through March 1956. Before converting to the B-36, it had served as a double-sized B-29 bombardment wing for most of the period from November 1947 to April 1951.

95th Bombardment Wing (Heavy) (no tailcode)

334th Bombardment Squadron
335th Bombardment Squadron
336th Bombardment Squadron

Established as the 95th Bombardment Group (Heavy) on 28 January 1942, it was activated on 15 June 1942 and was redesignated as the 95th Bombardment Group (Heavy) on 20 August 1943. Inactivated on 28 August 1945, it was redesignated as the 95th Bombardment Group (Very Heavy) on 13 May 1947. Activated in the Reserve on 29 May 1947, it was inactivated on 27 June 1949. It was consolidated with the 95th Bombardment Wing (Medium) which was established on 4 June 1952, activated on 16 June 1952 and was redesignated as the 95th Bombardment Wing (Heavy) on 8 November 1952. Discontinued and inactivated on 25 June 1966, it was redesignated as the 95th Strategic Wing and activated on 8 August 1966. Inactivated on 30 September 1976, it was redesignated as the 95th Air Base Wing on 16 September 1994 and activated on 1 October 1994.

Based at Biggs AFB, near El Paso, Texas, the wing operated B-36s from August 1953 through February 1959. The wing deployed to Andersen AFB, Guam, and operated under control of the 3rd Air Division from July to November 1955.

The blue/yellow badge on the nose of this RB-36H identified its operating unit as the 28th SRW, which was based at Ellsworth AFB, South Dakota.

99th Strategic Reconnaissance Wing (Heavy) (15th AF, circle I)

346th Bombardment Squadron
347th Bombardment Squadron
348th Bombardment Squadron

Established as the 99th Bombardment Group (Heavy) on 28 January 1942 and activated on 1 June 1942, it was redesignated as the 99th Bombardment Group (Heavy) on 30 September 1944. Inactivated on 8 November 1945, it was redesignated as the 99th Bombardment Group (Very Heavy) on 13 May 1947. Activated in the Reserve on 29 May 1947, it was inactivated on 27 June 1949. It was consolidated with the 99th Strategic Reconnaissance Wing (Heavy), which was established and activated on 1 January 1953, and was redesignated as the 99th Bombardment Wing (Heavy) on 1 October 1955. Inactivated on 31 March 1974, it was redesignated as the 99th Strategic Weapons Wing on 22 June 1989 and activated on 10 August 1989. It was redesignated as the 99th Tactics & Training Wing on 1 September 1991, as the 99th Wing on 15 June 1993 and as the 99th Air Base Wing on 1 October 1995.

Based at Fairchild AFB, near Spokane, Washington, the wing operated B-36s from August 1951 through September 1956. From January 1955 to February 1956, the wing participated in Project FICON, in which one squadron's GRB-36D bombers were modified to carry RF-84K reconnaissance fighters on long-range flights. Strategic bombing became the wing's primary mission in late 1954, but it was not redesignated as a bombardment wing until October 1955. The wing deployed to Andersen AFB on Guam from January to April in 1956. The RB-36s were replaced by B-52s beginning in August 1956.

Second Air Force's only B-36 unit was the 72nd Strategic Reconnaissance Wing at Ramey AFB, Puerto Rico. Ramey had been a regular B-36 deployment base, from where attacks could be launched into the south of the Soviet Union via the Mediterranean, with forward deployment and recovery bases in French Morocco and Turkey. The 72nd was assigned as a permanent wing in late 1952. Here one of the wing's RB-36Fs is being worked on next to a 15th Air Force aircraft.

SAAB J 29 Tunnan

The Flying Barrel

Nicknamed 'The Flying Barrel', the SAAB J 29 was designed as a fighter, but it also found a niche in the reconnaissance and ground attack roles. This highly successful programme produced Europe's first post-war swept-wing fighter, which went on to capture important speed records. In time and performance it was to be compared with the American F-86 Sabre and the Russian MiG-15. It served the Swedish Air Force well and was sold to Austria, but it attracted the most justified international attention when it served as air cover for UN troops who were engaged in heavy fighting during the Congo crisis in the 1960s.

After World War II Sweden was determined not to rely on any other nation for its need for arms, partly because of the difficulties it had experienced buying them from abroad during the conflict, and partly because Sweden wanted to remain neutral in the rising conflict between NATO and the Warsaw Pact states. The SAAB (Svenska Aeroplan Aktiebolaget) aircraft factory at Linköping – close to the F3 Wing at Malmen and the most important Swedish aircraft manufacturer – had by the end of World War II produced an interceptor (the J 21), and two bombers (the B 17 and the B 18), and it was planning at that stage a new fighter, the project J 27, which was based upon the 2,500-hp (1865-kW) Rolls-Royce Griffon piston engine.

It was a well trained group of skilled craftsmen who in 1945 began to study a request from the Swedish Air Board for a jet fighter with a top speed of Mach 0.86 (about 1050 km/h; 650 mph), a weight of around 5000 kg (11000 lb), a wing area of 23 m² (247 sq ft), and armed with four 20-mm cannon. The main responsibility for the development of what would become the J 29 lay in the hands of engineer Lars Brising, who began to make sketches of the type in October 1945. He also coined the nickname of the type – Tunnan, or 'Barrel' – and called his staff of collaborators 'the cooper's guild'.

At first, SAAB had to abandon the development of a civil project, a replacement for the faithful old Douglas DC-3, called the SAAB 90 Scandia, since it took up too much space in many respects, while the aforementioned J 27 military project also fell into oblivion. One of the first problems to solve was the question of the powerplant for the JxR (as the J 29 was known at this stage) project. Initially it was planned to equip the new fighter with a Swedish-designed jet engine. Orders for two different types of powerplant were placed, one with Svensk Flygmotor AB (SFA) for a centrifugal-flow engine and one with STAL (Svenska Turbinfabriken AB Ljungström) for an axial-

Main picture: Born in the early days of the Cold War, the J 29 survived in the front line for a surprising length of time given the rate at which fighter technology was advancing. The last fighters were retired in 1967, this pair of Sidewinder-armed F3 J 29Fs being typical of the final front-line machines.

Four J 29Bs 'clean up' after take-off from Kamina in the Congo. The 'Barrel' was at the heart of providing air cover for United Nations forces who were assisting the Congolese government in quelling a rebellion by the province of Katanga.

flow compressor jet engine, but a solution was closer to hand than that.

In 1945 Sweden had ordered a number of Vampires from the de Havilland Aircraft Company in Great Britain, and that aircraft was powered by the Goblin centrifugal-flow engine. In connection with the deal, the Swedes obtained a clear insight into the development of different de Havilland jet engines, not only of the Goblin but also of the classified Ghost, which was intended for installation in the DH Comet airliner. The difference between the two engines was, among other things, that the Goblin (with 3,100 lb/13.79 kN thrust) had a divided air intake, while the Ghost (with 5,000 lb/22.25 kN thrust) could operate with either a divided or single central intake. To use the Goblin in the J 29 was not possible since de Havilland had no interest in manufacturing a Goblin with a central air intake.

However, SFA got permission to manufacture the Ghost under licence in March 1947, long before de Havilland itself started series production. It was very unusual to grant a customer a licence for manufacturing before the tests are concluded and a certain degree of series production has been achieved, but the gesture of benevolence de Havilland showed SFA in giving a permit to participate in the development of the Ghost also revealed from the British part a confidence in the ability of SFA. The fact that the Goblin engine was fitted into the SAAB J 21R fighter (Sweden's first jet aircraft) in 1945, 60 of which were delivered to the Swedish Air Force, permitted SAAB to explore the transonic speed range.

"The whole aircraft was built around the combination of the engine and central air intake," stated Lars Brising about the J 29, so the diameter (1.35 m/4 ft 5 in) of the Ghost engine was the deciding factor in establishing the barrel-shaped fuselage. This was not so popular from the aesthetic point of view, ugly on the ground as it was, but access to the classified British engine at an early stage of the development of the J 29 saved the Swedes at least three years of research for an engine of their own.

Another problem was the swept wing, unknown in those days to most aircraft constructors. Knowledge of the swept wing came to the Swedes in a fortuitous way. During a business trip to Switzerland in 1945, a SAAB employee happened upon some copies of German wartime reports, regarding wind tunnel trials with the swept wing. He brought these papers home to Sweden, where Lars Brising – after studying them – soon understood that he had found the wing for the J 29. The new 'Swedish' wing had a sweepback angle of 25° and was so thin that there was no space for the undercarriage, which was instead housed in the fuselage. In turn this meant it had to be very narrow to fit within the confines of the circular section fuselage.

Tests and improvements

Wind tunnel tests at the Royal University of Technology and the Aeronautical Research Institute showed that there would be problems with the swept wing at lower speeds and therefore to minimise these problems the angle was set at 25°. The Germans had experimented with up to 45° sweepback, but that incurred greater instability and a higher accident rate, especially during landing with inexperienced pilots. The wing of the J 29 was fitted with automatic leading-edge slots and, on prototype 29001 and 29003, connected ailerons and flaps. This last feature stirred up the Air Board, and orders were given that prototype 29002 and mass-

There goes the future – four J 29As from F13 taxi out past two Flygvapen 'dinosaurs' which dated from pre-war times, the B 3 (Junkers Ju 86) bomber/transport and the SK 12 (licence-built Focke-Wulf Fw 44).

Swept-wing aerodynamics were in their infancy in 1946, so Aircraft 201 was produced to test the wings intended for the J 29. This Safir was modified with half-scale wings to test their flying properties at low speeds. The Safir later tested the wing of the Lansen.

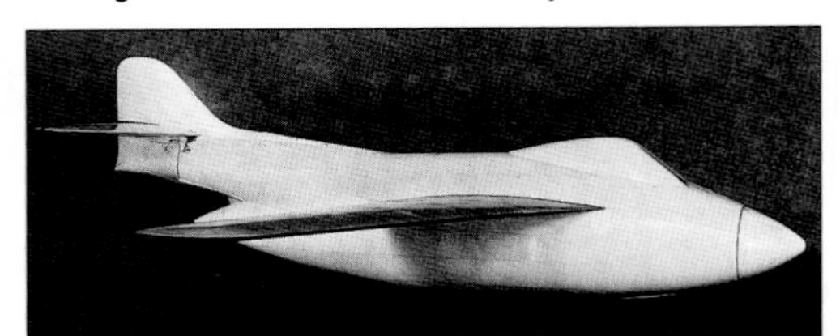

Before the J 29's shape was fixed, numerous tests were performed with models in the wind tunnel. Above is a very early incarnation, before swept wings were adopted, but the fuselage is unmistakably that of the 'Barrel'. Below is a later model with swept wings and looking much more like the eventual product.

produced J 29s should have separate ailerons and flaps. To test the wind-tunnel results, in 1946 one-half scale models of the swept wings were mounted on a piston-engined trainer, a SAAB Safir, which was called Aircraft 201, and flown successfully. For extra power, the Safir was equipped with a Lycoming engine with a variable-pitch propeller. To further improve low-speed qualities the wingtips of the J 29 were made thicker, probably under the influence of British investigations.

On 1 September 1948 it was time for the first test flight of J 29 prototype no. 1 (of four scheduled to be built). It had been allocated the number 29001. (In Sweden the two first figures in the serial number refer to the type while the last three are the number of each individual aircraft). Pilot on the first test flight was Squadron Leader Robert 'Bob' Moore, a British graduate from the Empire Test Pilots School. He had been in Sweden for a while, flying various SAAB aircraft, including the above mentioned Safir. He had also participated in the cockpit design, where his opinion was warmly appreciated. He was used to British fighters which, as he wrote: 'at this time had instruments and controls in the cockpit in a completely haphazard way'. However, it was test pilot Bengt R. Olow who decided the final cockpit lay-out.

Before the first test flight, in the summer of 1948, taxiing trials had started with 'Bob' Moore at the controls. They included a series of accelerations with increasing speeds until the first 'hop' was achieved. But on 1 September the final 'hop' was transformed into the first test flight of 30 minutes. On this occasion the flying was a little bit 'shaky', since the speed was kept to a minimum because the slots were locked in an extended position and the landing gear doors were accidentally only partially closed. But Robert Moore's first impressions of the J 29 afterwards were: "It was love at first flight. An ugly duckling on the ground but in the air a swift."

During the following test flights, problems occurred. There were unacceptable vibrations when the airbrakes in the wings were used which were transmitted to the fuselage. 'The Barrel' also had a tendency to 'snake' and Dutch Roll when reducing power, even at high speed. Adding a fillet between the engine cowling and the rear fuselage section – a spot where unstable, pulsating separation was found – solved this problem. Prototypes 29001, 29002 and 29003 did all the test flying during the years 1948 and 1949, while 29004 was aimed to be as close to mass-production examples as possible, with all tested and accepted modifications. 29004 flew for the first time in July 1950 and fulfilled the majority of SAAB's expectations. A speed of 950 km/h (590 mph/Mach 0.85) at 24,000 ft (7315 m) was attained without vibrations or similar problems, and the Tunnan was cleared for production.

Left: 29001 – the first prototype J 29 – sits on the airfield at Linköping. The aircraft was completed without gun armament (as was the second), and was skinned in hand-hammered aluminium, giving it a highly polished appearance. The two small pitot tubes behind the intake did not appear on production aircraft.

Squadron Leader 'Bob' Moore climbs aboard the prototype for its maiden flight on 1 September 1948 (above). At right Moore brings a gun-armed prototype in to land using the approved aerodynamic braking method.

Operational service

The first Swedish unit to receive the new fighter (designated J 29A) was, by tradition, F13 Wing at Norrköping-Bråvalla. In May 1951 the replacement began of its Vampire FB.Mk 1s, which it had operated since 1946. F13 kept its J 29s until 1962 when they were totally replaced by the J 35 Draken. From 1951 to 1958 SAAB delivered 661 J 29s in four main versions: the A-version (c/n 29101 - 29324), the B-version (c/n 29325 - 29685), the C-version (the photo-recce variant and numbered 29901 - 29976) and finally the F-version (210 rebuilt B- and E-models equipped with a 'saw-tooth' wing and an afterburner). A few J 29Bs were, in 1953, rebuilt as test aircraft for the new wing and called J 29E, and the sole J 29D (c/n 29325) was equipped in 1954 with an afterburner and increased fuel capacity. C/n 29325 never regained its normal behaviour in the air after the rebuild and was scrapped in 1961 in spite of considerable measures. "Not approved flying qualities," the record says.

The J 29 was a pilot's aircraft, but was considered difficult to fly under some conditions. When introduced into operational units in the early 1950s, a number of crashes occurred in the checking-out phase, mostly due to the new flap/aileron interaction during landing. SAAB was aware of the problem and had informed the air force. At the first units the message was received and training was altered accordingly. However, it was later lost for many reasons, with occasionally fatal results.

Even fresh paint influenced the flying qualities, so far as an individual aircraft did not have the same flying qualities after painting as it had before. When about 100 of the J 29As had been delivered a new problem emerged. During delivery tests certain aircraft had a tendency to 'dodge' right or left. No explanation could be found despite a rigorous investigation, but someone came up with a solution: a small plate was attached to the trailing edge of one of the flaps, close to the fuselage. It was deflected up or down, depending on the nature of the swerving. Nobody could explain how this worked, but the diversions ceased.

During the planning period the Air Board demanded that the J 29 should have a high combat readiness. That implied a rugged aircraft with easy maintenance. The largest strains during flying were turns and landings. The J 29 could take 7 *g* in turns and the undercarriage could withstand very hard landings. Easy access to different parts of the aircraft was also essential. Therefore, the J 29 was littered with 145 doors, hatches and covers to facilitate ground service, all designed and positioned in such way that surface finish and strength specification were not compromised. Even engine changes in the open did not take more than a couple of hours. The ability to service the J 29 in all kinds of weather and environments – from the bitter cold in northern Sweden to the humid heat of the Congo, from a warm and cosy hangar to a muddy war zone – was excellent, and was a key feature in the effectiveness of the Tunnan.

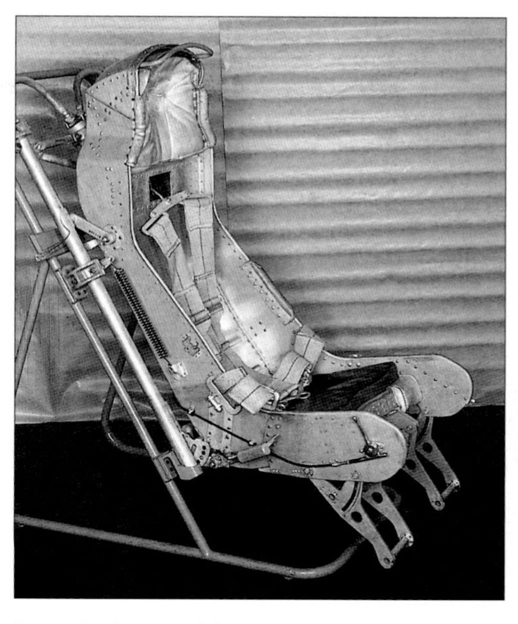

Eager to be as self-sufficient as funds and capability would allow, Sweden produced its own SAAB-designed ejection seat for the J 29 (this trend was only broken when a Martin-Baker seat was specified for the Gripen). The seat did not provide true 'zero-zero' capability, but was tested from a stationary aircraft.

Varying roles

In the fighter role the J 29 was delivered to the following Flygvapen fighter wings: F3 at Linköping-Malmen, F4 at Östersund-Frösön, F8 at Stockholm-Barkarby, F9 at Göteborg-Säve, F10 at Ängelholm, F12 at Kalmar, F13 at Norrköping-Bråvalla, F15 at Söderhamn and F16/20 at Uppsala. For an experienced pilot the J 29 was an excellent aircraft with good manoeuvrability even at low speed, and first-rate stall characteristics. The powerful engine made it a feared adversary in all speed ranges, especially following the introduction of the afterburner in the J 29F.

Second in importance was the recce role. Experience from World War II indicated the importance of aerial photo reconnaissance, and in the late 1940s Sweden obtained from Great Britain a number of Spitfire PR.Mk XIXs (Swedish S 31) that were built for this purpose. These aircraft were operated by F11 Wing at Nyköping-Skavsta until S 29Cs arrived in August 1953, and the 'Barrel' stayed in operation there for 12 years, before being replaced by the S 35E Draken. Twelve S 29Cs were also placed in service with F 21 Wing at Luleå.

Strap-on RATO rockets were tested by the J 29A/B for heavyweight take-offs, such as this aircraft carrying four rockets and two 900-litre (198-Imp gal) tanks.

J 29A – the first production 'Barrel'

The J 29A was similar to the fourth prototype (29004), which first flew in 1950 and was intended to be a production-representative aircraft. The first 32 aircraft, including 29106 (right), were built with airbrakes in the wings, but subsequent aircraft had brakes located in the fuselage sides, below the leading edge of the wing. Deliveries of J 29As began in May 1951 and all 224 were in service by the summer of 1953.

F13 at Norrköping (above) was the first recipient of the J 29A. Note the SK 16 (T-6) trainers in the background. Another five wings (F9, F12, F13, F16 and F20) converted to the A-model in 1952. The aircraft at left are from F16 at Uppsala, marked for participation in the 'Air Pentathlon'.

In essence the S 29C was a 29B with a modified fuselage to house cameras, and it satisfied all the demands that were required of a photo-recce aircraft: high speed (most were equipped with the new 'E-wing' which increased top speed from M 0.86 to M 0.89) and high altitude capability. The service ceiling was 12900 m (42,323 ft) but occasionally it could reach 15000 m (49,212 ft), and because of its speed, defensive armament was not needed. It was a reliable, popular and rugged recce aircraft, and a pioneer in the development of modern aerial reconnaissance systems.

To house all the photo equipment the front section of the S 29 had to be rebuilt. It was enlarged to accommodate two vertical and two forward oblique 92-cm (36.2-in) focal length, large-scale SKa 10 cameras, and one wide-angle 15-cm (5.9-in) camera for orientation of the high-altitude vertical pictures and for mapping. The forward-looking cameras peered through glass windows in the nose. The windows were protected by doors, flush with the fuselage. In addition to their normal function, the doors could be used as air brakes in evasive manoeuvres, especially in dives. The fighter J 29s, against which the S 29s regularly practised, did not have the same ability. A camera sight was installed to permit exact orientation above the target to reduce film consumption. Three SKa 16 continuous strip cameras were also added, introducing a low-level photo-recce capability. To permit long-distance photography against heavily defended targets, the AGA Company developed new 150-cm (59-in) lenses. Vibration and focusing created problems in the beginning, but they were solved and the system was a success. Rear Warning Radar and DME navigation system were also fitted, as well as an improved compass and altitude gyro.

The predecessor of the A 29 Tunnan attack version in the ground support role was the SAAB A 21, a twin-boom, piston-engined (Daimler-Benz DB 605) pusher type, which was later rebuilt and equipped with a Goblin jet engine and called the A 21R (1950). Later (in 1952) the A 28 Vampire appeared and in 1954 the two ground support wings, F6 at Karlsborg and F7 at Såtenäs, changed to the A 29. The A 29 was not really constructed for the ground attack role and its career as such was short-lived (until 1957) – but quite successful. However, its successor, the A 32 Lansen, was far better suited to the role.

Further improvements

During the first test flights the critical Mach (M_{crit}) number of the wing was verified as 0.86, but the effects of transonic flow appeared when the aircraft dived. To pull up too fiercely could cause a wing stall. It was necessary to raise the M_{crit} to widen the margins and improve handling. This was achieved by reducing the relative thickness of the wing. The leading edge profile of the outer wing section was lengthened and the 'saw tooth' wing was born. This meant that the leading-edge slots were deleted but it did not matter, because the new wing caused no impairment in low-speed handling. The M_{crit} was raised from 0.86 to 0.89, and the load factor of the wing was also raised. Manoeuvrability at high speed increased enor-

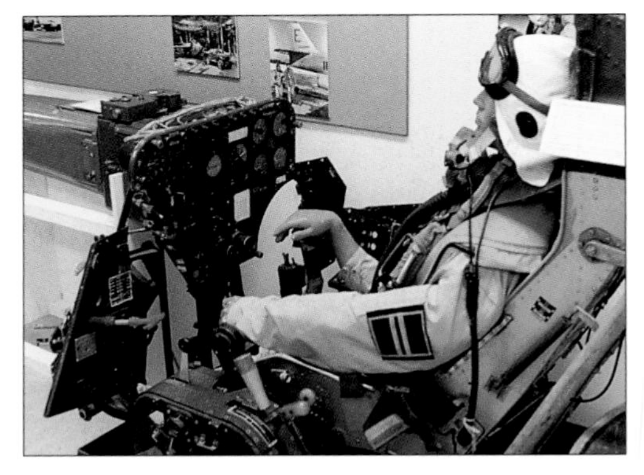

This display shows the J 29's conventional cockpit set up. The stick was inspired by that in the Bf 109.

The longest serving J 29As were those at Uppsala, where F16 (operational) and F20 (air force academy, illustrated) hung on to the type until 1959. The lack of wing tanks in the A-model meant that it was not modified for further use.

Considered the state-of-the-art in centrifugal-flow turbojets, the Ghost was built under licence by SFA. The bifurcated intakes identify this engine as an RM2A, for installation in the Venom.

J 29B and A 29B

First flying on 11 March 1953 and delivered between May 1953 and December 1955, the 29B was the most produced version, with 332 built. The primary difference was the adoption of wing tanks. Although there was a slight drop in fuselage tank capacity, the wing tanks greatly extended the Tunnan's range. Aircraft were supplied to fighter wings (such as F10 above) as J 29Bs, and briefly served with two attack wings (F6, right, and F7) as A 29Bs. In the attack role the principal weapons were unguided rockets.

mously, especially when the thrust of the new afterburner was added. The new wing was tested on a number of J 29Bs from December 1953, designated J 29E, and the successful results of these tests led to a rebuild programme. Together with fitment of the afterburner, this became the J 29F.

A considerable part of the J 29's success rested on SFA, which had a licence deal with the de Havilland Engine Company to series manufacture the Ghost engine. SFA not only manufactured the engine, it also participated in its development, which in Sweden with a central air intake was known as the RM2, and with a divided intake as the RM2A (intended for installation in the Swedish DH Venom night fighters).

As early as 1944, Östen Svantesson, an engineer at the Royal Swedish Air Board, had started the construction of a Swedish designed jet engine after having examined, among other things, the turbine compressor connected to the engine exhaust of a Lockheed F-5E Lightning, which had force-landed in Stockholm 1944. This Swedish jet engine, called the Dovern, was first run in 1945 and the tests continued until 1952 when the project was abandoned in favour of an offer from Rolls-Royce to manufacture the Avon under licence, intended for installation in a new SAAB ground support aircraft – the A 32 Lansen.

The Dovern was intended to have an afterburner and when the project was discontinued, Östen Svantesson was asked to work on an afterburner for the RM2. So he did, and in 1953 it was time for a test flight in the flying test bench, a Lancaster, with good results. Thrust was increased by 25 per cent, but the most radical feature with this new afterburner was that the pilot could regulate the revolutions of the engine when it was running. In 1954 SAAB received the first example of the new RM2B engine. One J 29B, known as the J 29D (c/n 29325), was equipped with the afterburner and test flew for a number of hours before it was decided to transform 210 J29Bs and Es into the J 29F version.

Tunnan proposals

In 1950 SAAB presented drawings for a two-seat trainer (the SK 29) and a radar version (the J 29R). There was always a need for a training version, especially in light of the number of crashes that occurred in the beginning of its

The A 29B was essentially similar to the J 29B, its different designation only reflecting its attack-dedicated tasking. The 'Barrel' was an effective attack platform, as demonstrated by F22 in the Congo. This group of A 29Bs is from F7 at Såtenäs.

Record-breakers

The development of the J 29 – Europe's first swept-wing jet (if one discounts wartime German aircraft) – was a national triumph for Sweden, and to underscore its achievement the type was used to establish two world speed records. At right Captain Anders Westerlund poses with his 500-km closed-circuit record-breaking J 29B – note the balsa wood filling in the gun apertures – while below Hans Neij and Birger Eriksson are congratulated after their successful S 29C assault on the 1000-km record.

S 29C – the recce 'Barrel'

29901 (right) was the prototype S 29C tactical reconnaissance platform, first flying on 3 June 1953. The airframe was essentially similar to that of the J 29B, with the exception of a reworked forward fuselage which had slab-sided bulges added to accommodate cameras. Early S 29Cs had the original wing, but most were built with the 'sawtooth' wing. Visible on the prototype is a rear-facing EW antenna at the base of the tail, later relocated inside the tailcone.

Left: A reconnaissance pilot poses alongside an early S 29C (note absence of wing fences) and some of the cameras it could carry. The standard operational fit comprised four SKa 10s (the larger cameras flanking the aircraft), arranged in split-vertical and forward-oblique pairs, and a fan of three SKa 16 cameras (in the front row) for low-level coverage.

career in the hands of inexperienced pilots. SAAB offered two solutions, one with the pupil/instructor in tandem and one with them seated side-by-side. In comparison with the original 29, the following changes were proposed: canopy hinged at the rear, dual gunsights and controls, increased oxygen supply, no armament, no armour plating, no lighting, no IFF and the M_{crit} reduced to 0.78. A batch of 20 SK 29s was suggested, but SAAB declared that under no circumstances could it undertake the manufacturing of these aircraft with the desired rapidity. Instead it recommended that the Air Board produce them under its own regime. The Air Board approved the drawings but could not accomplish the manufacturing on its own and since the problems seemed to be unsettled, the whole project was cancelled.

The contemporary day fighters of the J 29, the MiG-15 and the F-86 Sabre, were given an all-weather capability by means of radar, and there were proponents for such a version of the 29. Antenna installations proposed by SAAB included a radome on top of the upper part of the air intake, which both the MiG and the Sabre had, a pod in the intake or a plate lens in the intake orifice. It was intended to use a 3-cm wavelength transmitter with an output of 100 kW. In October 1950 the Air Board received from SAAB an inquiry concerning the radar version, but as the company could not pursue the matter further due to shortage of space, the Air Board decided to put the plans on ice. As a matter of fact, it was not until 1959 that the Swedish Air Force received its first radar-equipped aircraft in the shape of the J 32B Lansen.

Record-setter

In 1953 a group of J 29s, together with a group of F-86 Sabres, visited Waterbeach in Great Britain where they were invited to a sort of competition – a dogfight between the J 29, F-86 and British Gloster Meteor Mk VIII. Neither the Sabres nor the J 29s had any problem with the Meteors, and at an altitude of around 9000 m (29,500 ft) the 'match' between the Sabres and the J 29s finished even. The 'Barrels' out-turned the Sabres in the horizontal plane while the latter had better performance on top of loops, where it could use its better acceleration and adjustable slots.

During the 1950s it was also popular to let jet aircraft try to break speed records. Sweden chose the J 29 to tackle the 500-km (310.7-mile) closed-circuit speed record. A supervisor was placed at the turning point of the course to ensure that the aircraft passed this point. A flying start at full throttle was used and the time was measured twice, once on the way 'out' and again on the way 'home' and then the average speed was calculated. When the J 29B was supplied to the Swedish air force, it had an aircraft with enough fuel capacity to try to beat this record which had, not surprisingly, recently been set by the American F-86 Sabre.

Two aircraft from the F16 Wing at Uppsala were chosen for the attempt and they were polished, the engines were checked with the utmost care, and the muzzles of the guns were filled with streamlined balsa wood. A distance of 250 km (155.3 miles) was measured from Uppsala, northward along the Swedish coast. On 6 May 1954 the weather forecast was promising. Captain Anders Westerlund took off in one of the two J 29Bs, passed the starting line at a speed of more than 1000 km/h (621 mph), flew as fast as he could up to the turning point, passed the controller in his aircraft who by radio verified the pass, pulled hard and used as much power as he dared, since the engine was overheating, and crossed the goal line with an average speed of 977 km/h (607.05 mph) over the 500-km closed course. This speed record lasted more than a year and received considerable attention abroad before an American F-86 Sabre regained the record. The Sabre crashed on the line during that attempt, but the new mark was allowed to stand.

The following year two S 29Cs from F11 were prepared for an attempt on the world's speed record over a 1000-km (621.3-mile) closed circuit. This record was set in 1950 by Group Captain Cooksey, RAF, in a Gloster Meteor at a speed of 822 km/h (510.8 mph).

J 29 Tunnan production by year

	1951	1952	1953	1954	1955	1956	1957	1958	Total
J 29A	12	138	74						224
J 29B		107	207	47					361
S 29C				31	37	8			76
J 29F*					1	88	97	24	(210)
** Rebuilt -B and -E models*									**661**

As befitted the seriousness of the times and the perceived nature of the Soviet threat, J 29s were built at a prodigious rate. For a time in 1953 the Linköping works was producing the equivalent of one new aircraft per day.

For operations at low level most S 29Cs adopted two-tone camouflaged top surfaces. The light-coloured panel at the base of the rudder covered a rear-facing radar warner. A fleet-wide addition was a 'sugar-scoop' heat shield below the jetpipe.

This view of an F21 S 29C highlights the 'E' wing, which was fitted to most reconnaissance 'Barrels'. The extended leading edge on the outer panels and new wing fences improved high-speed manoeuvrabiliy and safety. Note the lack of national insignia on the wings.

The course for the Swedish attempt was similar to that of the previous year: from Nyköping northward, out over the sea and following the coast up to Örnsköldsvik where the turning point was situated. Captain Hans Neij (later General and Chief of the Air Staff) of F11 at Nyköping was commissioned to make the attempt, together with a short-commissioned pilot, Birger Eriksson.

After a few trial runs the real attempt was made on 23 March 1955, a cold and clear winter's day. Flight level was set at 5000 m (16,400 ft) and both aircraft were equipped with drop tanks, which had to be dropped by theoretical calculations since there were no gauges indicating the fuel level of the drop tanks. The turning point had to be passed at low level for visual identification by the controller. So far, so good. But on the home run the drop tanks were jettisoned into the water too early, which meant that towards the end of the run the fuel warning light came on and the speed had to be reduced to ensure completion of the course and a safe landing at F11. However, the S 29s had done enough and a new record of 900.6 km/h (559.6 mph) had been set.

Congo – the first real test

In June 1960 Belgium decided to grant independence to its colony in the Congo, following a spate of riots, looting and murder. The victims of these atrocities were mostly black, since fights had started between different tribes, and white Belgian civilians who were protected with varying degrees of success by Belgian troops stationed in the country. The new Congolese government, which had dismissed the European officers in the Congolese army, could not maintain law and order, even over its own soldiers, who were now commanded by unqualified Congolese former sub officers. The situation was chaotic, to say the least, and it grew worse, when the Province of Katanga, perhaps the richest province of them all, claimed autonomy under its leader Moise Tshombe, who asked Belgium for help to maintain his regime. The Congolese government requested in their turn help from the United Nations, which demanded that Belgium withdraw its troops from the Congo in general, and Katanga in particular.

At the same time, the UN decided to send troops to help the Congolese government maintain law and order. In February 1961 the UN Security Council passed a resolution to urge all foreign military and paramilitary personal to leave Katanga and ordered UN forces to occupy the radio stations, post and telegraph offices in the province. The Katangese tried to stop this action by using force and fighting broke out in September 1961.

Katanga had at its disposal a French-built jet trainer, the Fouga Magister, which could be armed with machine-guns, small bombs and rockets. This warplane, flown by white mercenaries, together with other aircraft (among them a number of North American Harvards), created havoc on the government side. Civil crews of the charter airlines hired by the UN, and many of the military crews in UN transports, refused to fly as long as the Fouga was a threat. But there were rumours which said that Katanga also had access to MiGs, French Mystère fighters, Canberras and P-51 Mustangs. As far as the Mustangs were concerned, the rumours were true. Moise Tshombe had bought 14 Mustangs from Israel, to be delivered in January 1963 by boat via Portugal to Angola, where they were to be assembled but the conflict ended before they were put into action.

The idea of the UN possessing combat aircraft was not new. The issue had already been discussed in the spring of 1961 within the UN, but was rejected, since it was considered that the UN should not use what was termed 'offensive violence'. However, the UN changed its mind and, after much hesitation, the UN

Two Flygvapen wings operated the S 29C in the recce role: F11 at Nyköping (below) in the south and F21 at Luleå (right) in the north. Before camouflage was widely adopted, the fins of S 29s were usually painted black to provide a quick and easy distinction between them and their fighter/attack brethren.

J 29F – the definitive fighter

A need to improve the performance of the Tunnan led to the adoption of an SFA-designed afterburner – in fact, the first such unit to be installed with a British engine. The J 29F also incorporated the 'sawtooth' wing from the J 29E. Fitment of the afterburner entailed a redesign of the engine nozzle area, which was more bulged and incorporated small airscoops. Conversion of J 29Bs and the small number of Es was conducted between February 1955 and May 1958.

inquired of Sweden, Ethiopia and India whether they could provide combat aircraft. The reason why these countries were chosen was that they were considered neutral.

The Swedish government approved and acted surprisingly fast in its – also surprisingly – positive spirit, probably in the light of the death of the Swedish UN Secretary General, Dag Hammarskjöld, who had died in an aircraft accident in Zaïre in September 1961. One of the rumours said that the aircraft was shot down. The Swedish government also feared that the respect for the UN was going to vanish if the operations in the Congo could not be concluded in a successful way. Also, the lives of Swedish soldiers were at stake, since Swedish troops fought on the UN side in the conflict and they needed protection from the air. Within a week the Swedish authorities had discussed, planned and acted in the matter.

Flygvapnet was ordered to form a task force of five aircraft with all the necessary personnel and equipment, including Swedish surveillance radar. The official designation of the unit was F22. The unit was to have a Swedish commander, but it was to receive orders from the United Nations headquarters.

Regardless that the more advanced J 29F was available, five J 29Bs (c/n 29374, 29393, 29398, 29440 and 29475) were chosen, due to the type's easier maintenance and longer range. The distance between Sweden and the Congo is 12000 km (7,460 miles), so there were a lot of countries that had to approve an overflight by the Swedish force. There were two alternative routes: either the west route over France, Spain, Morocco, Algeria and Nigeria, or the eastern route via Austria, Greece, Egypt, Sudan, Ethiopia and Uganda. The western alternative was, from the Swedes' point of view, the most favourable. But, since France objected, the other alternative was used. Obtaining a permit to fly over the "eastern route countries" was problematic, too. Austria, for example, emphasised that foreign military aircraft with loaded guns were prohibited to fly over the country in units, and the 29s were armed and ready for combat for the flight to the Congo.

After a few days, all problems were solved, and on 30 September 1961 the five F22 J 29s took off from Ängelholm in Sweden to Leopoldville in the Congo, where they landed on 4 October, ready for immediate action, if necessary. The long journey was carried out without any major mishaps, and the rapidity (six days from take off in Sweden to be ready for action in the Congo) caused international attention. Four days later the unit was transferred to Luluabourg in the Kasai Province. The intention was to co-operate with the Ethiopians in their F-86 Sabres, but they refused to leave Leopoldville for Luluabourg, which they considered to have a runway of insufficient length for their Sabres.

The main tasks for F22 were air combat, attacks against airport installations and equipment, and close co-operation with UN ground

Left: A row of zero-length launchers under the inner wings allowed the J 29 to carry a variety of rockets, for either air-to-ground work or against bombers in the interceptor role.

Below: An F15 J 29F patrols over the Gulf of Bothnia from its base at Söderhamn, on the coast around 220 km (140 miles) north of Stockholm.

Flygvapen operations – dispersal, protection and exercises

Some Swedish bases, notably F9 at Säve (above), had underground tunnels carved out of the rock in which aircraft were kept, safe from even nuclear attack. Various kinds of protection and camouflage were employed at the war bases (dispersal airfields), including full grass-covered shelters (below) and earth berms (below left).

The 'Barrel' was designed to be serviced anywhere (above), with all the necessary equipment being transported by truck. At the main bases, wooden dummies were used to fool attacking pilots (right). When operating in extreme cold, hot air was blown through the engine prior to start-up (below).

Above: As well as operations from established war bases – usually including highway strips – in winter J 29s also operated from frozen lakes.

For J 29 pilots the job consisted of training for war and periods on alert. By necessity the training was hard and uncompromising, with regular simulated air combat (above left) and attacks (above, attacking B 3/Ju 86s and B 18s at F1 Västerås). When on air defence duty, pilots often sat at cockpit alert, taking their meals where they sat (right).

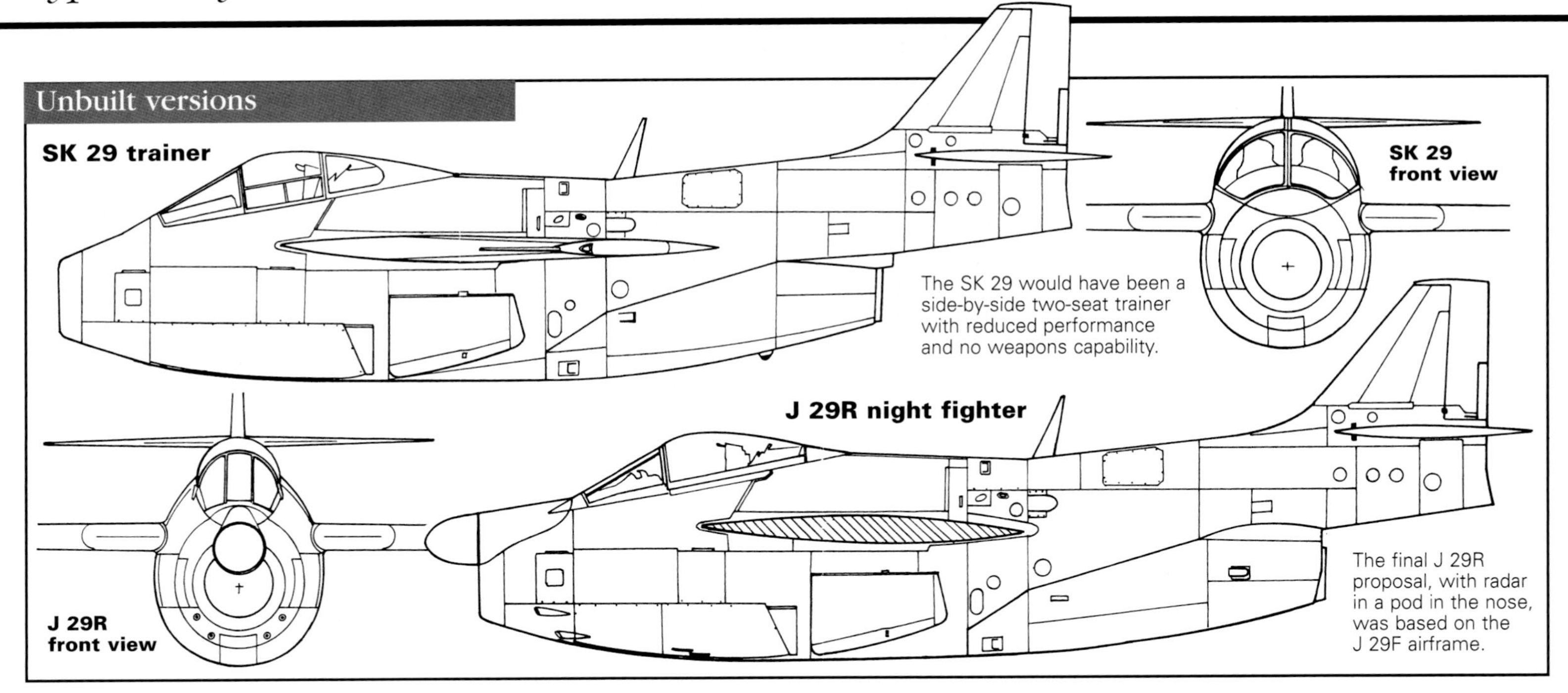

forces. At the end of October 1961, the Congolese government sent troops into Katanga in order to defeat Moise Tshombe. It tried time after time to persuade the UN to join the Congolese forces, but in vain. The offensive ended up in a total fiasco, mainly because the Katangese used their air force skillfully.

As a consequence, the UN agreed to give the Congolese government limited air support, not forgetting the armistice agreement the UN had with Katanga. The UN declared that all aircraft flying in the border area would be shot down. In reality, that meant that transport aircraft, patrolling the area and looking out for the Katangese enemy, could call for the J 29s, which were held at high readiness at the base, and lead them to the target. However, this was a hazardous operation since civilians, including aid organisations, often used aircraft as a means of transportation in the area. The mere thought of shooting down such an aircraft was something of a nightmare for the UN in general and for the Swedish fighter pilots in particular, so it was not so strange that this procedure did not lead to any encounter between the 29s and Katangese aircraft.

The declaration from the UN regarding air cover boosted the morale of the Congolese troops, while the Katangese aircraft were largely frightened off. Katangese pilots had a great respect for the 29. One of them wrote in his diary. "0800. Arrival at the airport and loading of aircraft. 0930. Departure for the objective. Drop 50 bombs then machine-gun but three UN jets dive on us, we peel off and go back to Kaniama". The way the 29s could master the rugged field conditions also drew international attention, as did the response of the pilots to scramble calls. On one occasion, two 29s were in the air four minutes after the call came, even though the pilots were in the dressing room at the time of the alert.

A connection with F8 Wing outside Stockholm contributed significantly to the high service rates achieved by the 29s. Whenever a spare part was needed, contact with F8 Wing was taken and the spare part was hurriedly sent on its way. This service worked around the clock, throughout the year. Pilots faced many difficulties in the air with hardly any ground navigation aids available, having to rely on the traditional compass and stopwatch. The weather was also very unpredictable, with sudden fierce rainstorms, thick clouds with low base and dense fog to contend with. To make things worse, the combination of heat and short runway forced the unit to load the 29s with lower quantities of fuel and ammunition than normal. The strength of the headwind for the day was often the deciding factor in what could be loaded.

Renewed campaign

In the beginning of December 1961 the Katangese forces started their long-planned offensive which was aimed to split the UN troops into isolated units and then defeat them. The first objective was the UN stronghold at Elisabethville. The UN was not by any means unprepared and had, in turn, plans for how to deprive the Katangese of their air support and remove the air threat against UN transports by letting the 29s and the Indian Canberras attack their air bases. In order to have as much time as possible over the battle area, the 29s moved base to Kamina, situated in the middle of Katanga.

One of the Katangese airbases was at Kolwezi. On the morning of 6 December it was to be attacked by a pair of 29s. The weather was inclement and the 'Barrels' could not find the air base due to the low cloud base, so, after

Above: Thanks to its lack of sophistication and numerous access panels, the J 29 was considered an easy aircraft to maintain, which was just as well considering that much of the work was to be done by conscripts. Note the mesh intake cover which was widely used for ground running after a fatal accident in which a conscript was ingested.

Right: Like the S 29Cs, some fighter 'Barrels' acquired top-surface camouflage in recognition of the fighter threat when operating at low level. These F3 examples are seen with standard-finish J 29Fs from F10.

To ensure the survivability of its fighter force in the event of an attack, Sweden used all manner of dispersal and protection techniques, and the J 29 had to be able to operate under the harsh conditions encountered in the field in the depths of the Arctic winter. Here a Tunnan is pushed back into its hide deep in the woods, between snow banks and under netting.

Until the Draken era, most Swedish pilots received basic training on F5 at Ljungbyhed, before being posted to their 'home' wings, where they then underwent advanced training. F20 at Uppsala was the air force academy, which flew J 29As from 1952 alongside F16 (an operational unit) until the J 29F (illustrated) was phased in during 1959.

some ground strafing in which they destroyed a locomotive, they returned to Kamina. The Indian unit, which participated in the attack with the Canberra, had better luck. The weather situation improved when they reached the airbase at Kolwezi and they managed to destroy two transport aircraft. In the afternoon two 29s made a second attempt to attack Kolwezi in which they destroyed a helicopter and fired at the buildings on the air base. The air base was heavily defended by anti-aircraft fire and the 29s were hit several times. Both aircraft returned to Kamina safely, however one was so badly hit that it had to be taken out of service for many days.

On the following day the air base at Jadotville was the objective, and there F22 strafed two hangars, which apparently contained fuel and/or an aircraft, since a fierce fire broke out. With this attack the mission to annihilate the Katangese air force was considered accomplished. But that did not mean that F22 could rest. On 8 December a Katangese force, three times stronger than before, attacked an Indian UN ground unit of 200 soldiers. F22 was called upon to protect the UN reinforcement troops, flown in by transport aircraft, and to strafe the Katangese forces. A number of lorries and an armoured vehicle were destroyed, and so effective was the operation and so great was the respect for the 29s that, on one occasion when a lorry was attacked by a 29, the soldiers on the lorry jumped off and began waving with a white flag in order to surrender – to a jet fighter.

After these incidents, between 9 and 21 October F22 was busy protecting UN troops, which on many occasions were involved in heavy fighting with an enemy often superior in number. Other military objectives that were attacked were locomotives, oil plants, anti-aircraft batteries, training facilities for military personnel and a factory, which was presumed to be military but which later turned out to be a brewery. Many later said that this last mentioned attack contributed greatly to the decline of the fighting spirit of the Katangese forces. By now the level of fighting had declined into sporadic skirmishes and the lull in major actions lasted almost a year. Neither side was defeated and it was time for the UN to evaluate the autumn actions of F22 and investigate the problems.

One of the main problems was the transport of fuel to the various UN air bases. It worked at Elisabethville but not with satisfaction at the others. Due to the unsafe roads, every drop had to be transported by air. At the bases there were no fuel bowsers, so the aircraft had to be refuelled by hand from a jury-rigged tank on a lorry. Fuel availability dictated UN flight activity, particularly during November. Another problem which needed attention was the lack of communications and protective services for the aircraft. During the first months, the almost complete lack of navigation aids did not meet the demands of jet fighter operations. During the summer of 1962 a badly needed improvement arrived, which also included radio communications and the provision of weather and rescue services at the air bases.

Higher efficiency

The need for recce aircraft in Congo grew stronger, as the target had vanished by the time of the attack on a growing number of missions. There were also accusations from the Katangese of 29s attacking civilian targets and, not least, the UN wanted to investigate the degree of truth in these allegations. The question of sending recce aircraft had originated as early as the autumn of 1961, and in April 1962 it was discussed in Stockholm and at the UN HQ in Leopoldville.

After negotiations between the UN and the Swedish government, which laid bare certain disagreements concerning the use of the Swedish fighters in the Congo, it was decided to send two disassembled S 29Cs by air in a transport provided by the USA. The problem with the efficiency of the S 29 in the Congo was that it was equipped with a Swedish navigation system called PN50/A, which was useless in the Congo since it relied on radio beacons and such a sophisticated aid did not exist in Africa. It was therefore suggested that a radio compass should be adapted. Unfortunately, this was not practical which meant that, during operations, it needed an escort of J 29Bs with radio compasses fitted. On 10 November both S 29s (c/n 29944 and 29906) were ready for duty and now a symbiotic co-operation started between J 29s and S 29s that came to be commonplace in the coming actions in the Congo, where the J 29 guided the S 29 to the desired objective.

In order to cut down the cost of the UN activities in the Congo, Assistant Secretary-General Ralph Bunche scrutinised the efficiency of the Swedish 29s, the Indian Canberras and the Ethiopian Sabres and stated in his report to the UN HQ in New York as follows:

'(a) Swedish unit: excellent logistic back-up,

From the end of 1963 the J 29Fs remaining in front-line service were equipped to carry two Rb 24 (AIM-9B) Sidewinders, greatly expanding their fighter capabilities. F3 at Linköping-Malmslätt (Malmen) had been the first unit to receive the J 29F, and flew it until J 35D Drakens arrived in 1965.

Ground crew refuel a J 29F. Like all the Tunnan's support equipment, the fuel tanks and pump were designed to be transported rapidly out into the field, and were compatible with either road or rail. The aircraft is fitted with Sidewinder launch rails.

high rate of serviceability, excellent fire power, best short field performance of three available fighter types.

(b) Indian unit: unsatisfactory logistic back up because of having to obtain spares from the UK, inspections have to be done in India. Low serviceability rate and reduction of aircraft available. India is unable to guarantee spares.

(c) Ethiopian unit: logistic back up is complex because of obtaining spares through various channels. Aircraft performance requires longer runways than other types. Keenness and willingness of personnel do not make up for lack of experience.'

He probably looked at these figures, which shows combat readiness percentage during three months in 1962 when the fighting restarted:

Combat readiness (percentage)

Type	October	November	December
J 29	100	97.5	82.6
Canberra	98	78	82.7
F-86	72.3	75	86.8

So when a general reinforcement of UN air power in the Congo was discussed, contact again was taken up with the Swedish authorities. On 4 July the UN requested the augmentation of F22 with more fighters i.e. J 29Fs. The UN also suggested sending the Indian and the Ethiopian aircraft home. The Swedish government, however, took an unsympathetic attitude towards the request, mainly because of the above mentioned dispute but also because of the cost to the Swedes of the Congo engagement. The answer was accordingly "no".

In November 1962 the situation in the Congo, from the UN point of view, had deteriorated and a new request was conveyed. The Ethiopian Sabres had left for home without warning. Only one Indian Canberra was operational and the Katangese air force had received reinforcements. After many hard and long discussions, the Swedish government decided to accede to the UN request, particularly since Italy, the Philippines and Iran also promised to send fighters.

When fighting began again in December the 29s found themselves on their own as the Italians wanted to celebrate Christmas at home and the fighters from the other two countries did not show up until the action was over. The UN wanted the Swedes to send the F version of the 29 since its afterburner meant shorter take-offs and heavier loads, but as it had a shorter endurance the Swedes decided to use J 29Bs again, and during 7-16 December four more J 29Bs (c/n 29364, 29445, 29371 and 29365)

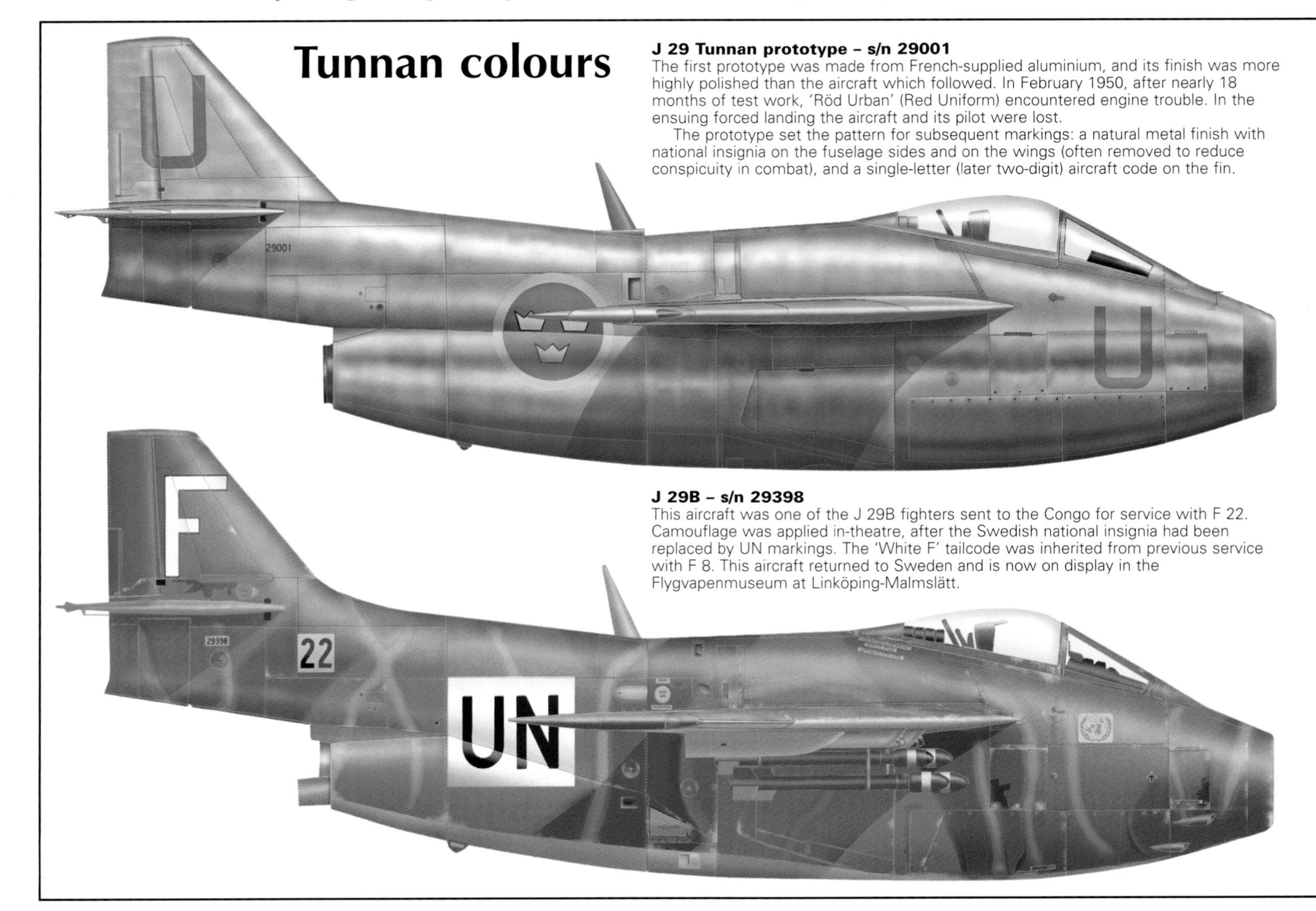

Tunnan colours

J 29 Tunnan prototype – s/n 29001
The first prototype was made from French-supplied aluminium, and its finish was more highly polished than the aircraft which followed. In February 1950, after nearly 18 months of test work, 'Röd Urban' (Red Uniform) encountered engine trouble. In the ensuing forced landing the aircraft and its pilot were lost.

The prototype set the pattern for subsequent markings: a natural metal finish with national insignia on the fuselage sides and on the wings (often removed to reduce conspicuity in combat), and a single-letter (later two-digit) aircraft code on the fin.

J 29B – s/n 29398
This aircraft was one of the J 29B fighters sent to the Congo for service with F 22. Camouflage was applied in-theatre, after the Swedish national insignia had been replaced by UN markings. The 'White F' tailcode was inherited from previous service with F 8. This aircraft returned to Sweden and is now on display in the Flygvapenmuseum at Linköping-Malmslätt.

A J 29F sits with ground power on during a Flygvapen exercise. The black bar markings on forward and rear fuselage were temporarily applied to identify the aircraft as the 'enemy' in air defence exercises. The Tunnan also carries radar reflector pods.

with spare engines were transferred to the Congo, just in time to participate in the action around New Year.

Final action

In December 1962 the informal truce turned more towards renewed military activity, and after a reorganisation and a reinforcement of the air force, Katangese troops attacked the UN on 24 December. Some days later the UN troops responded. As in the previous year, the main task for F22 was initially to destroy the Katangese air force and it did so in two days. During 29 and 30 December the air bases at Kolwezi and Jadotville were under constant and fierce attack by the 29s. 5,000 cannon rounds and 90 rockets were expended in 30 missions. Photos taken by the S 29s showed that the Katangese air force was completely wiped out, which meant that the UN ground forces could carry out operations unmolested from the air, finally leading to the collapse of the resistance in Katanga.

During this phase of combat the 29s again performed well under the rugged field conditions and showed that they could take a lot of punishment. The air base attacks were always conducted under heavy fire from the ground, and on one occasion a 29 returned with 12 bullet holes in the fuselage and the wings, while during another attack a 29 had its canopy damaged by bullets. The whole confrontation between the UN and the Katanga ended up with a big ground assault on Kolwezi on 21 January 1963, with the 29s acting as a protective umbrella.

After that, the general situation in the Congo improved and in April four 29s (c/n 29398, 29371, 29944, and 29906) flew back to Sweden, while the remaining five were destroyed *in situ*. Thereby F22 Wing ceased to exist. Its actions had borne the stamp of great efficiency with no losses to its skillful personnel, who had at their disposal an outstanding aircraft in the shape of the 'Flying Barrel'. Together, man and machine had been thrown into an alien environment, so different from that which they were used to in Sweden. The Swedish air contribution can be summed up by a foreign air attaché in Leopoldville, who wrote: "I consider that the operations the Swedish air unit has executed is the peak of what can be done by a modern air force. The skill of the pilots and the ability of the mechanics to keep the aircraft in shape are completely unprecedented and should not be surpassed. I don't have the knowledge of any example in modern time when such a small air force has achieved such a decisive result."

The last days of front-line service

After participation in the Congo campaign the Tunnan completed another four years of active duty in the Swedish Air Force. On

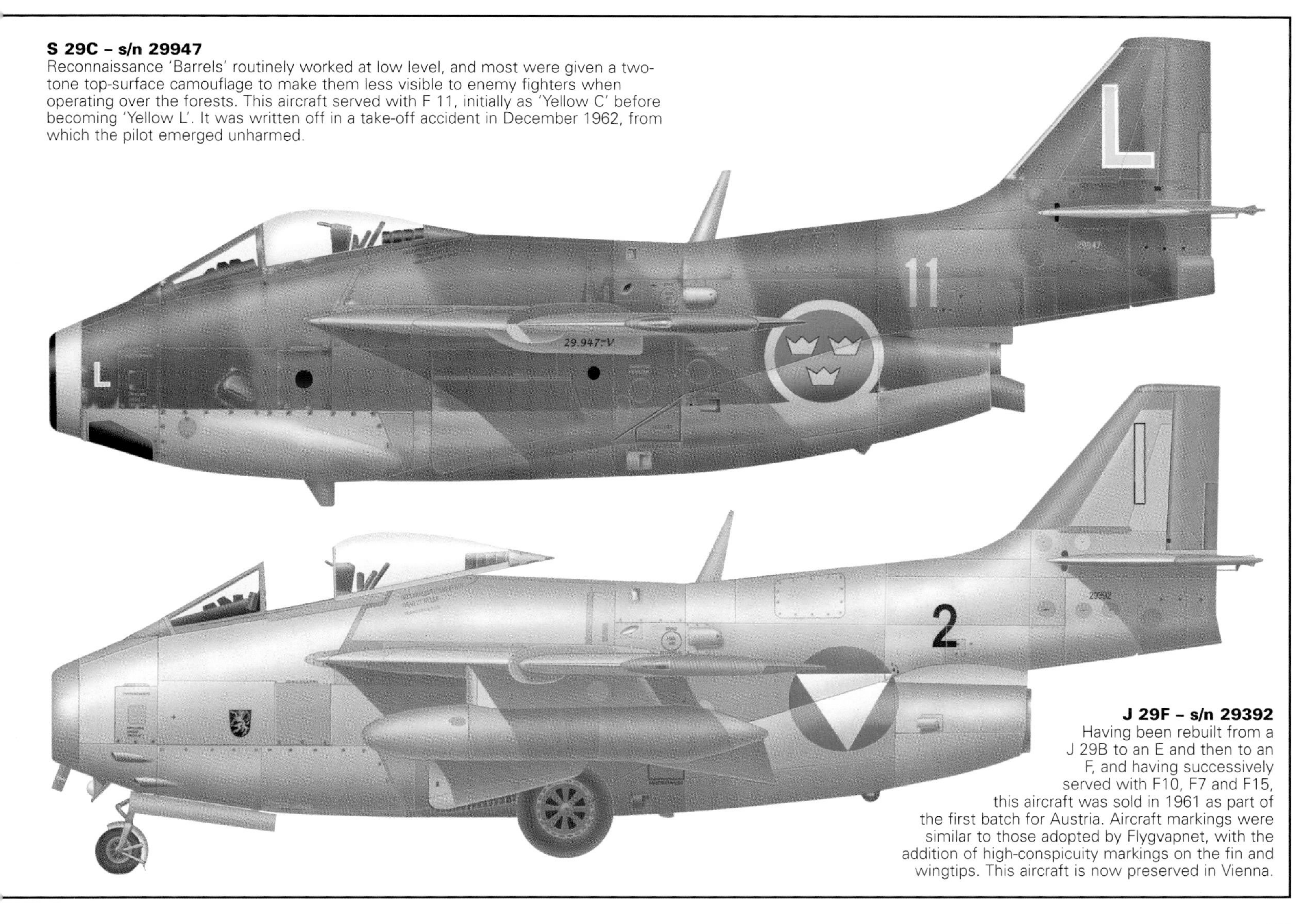

S 29C – s/n 29947
Reconnaissance 'Barrels' routinely worked at low level, and most were given a two-tone top-surface camouflage to make them less visible to enemy fighters when operating over the forests. This aircraft served with F 11, initially as 'Yellow C' before becoming 'Yellow L'. It was written off in a take-off accident in December 1962, from which the pilot emerged unharmed.

J 29F – s/n 29392
Having been rebuilt from a J 29B to an E and then to an F, and having successively served with F10, F7 and F15, this aircraft was sold in 1961 as part of the first batch for Austria. Aircraft markings were similar to those adopted by Flygvapnet, with the addition of high-conspicuity markings on the fin and wingtips. This aircraft is now preserved in Vienna.

F22 was established in the Congo with the J 29B, this variant being chosen because of its greater range. The first base was at Luluabourg, but the 'Barrels' moved much closer to the action, settling at Kamina inside Katanga itself. The first priority was to establish air superiority over the Katangese air force, in turn greatly boosting morale on the UN/Congolese side.

12 May 1967 the last eight J 29Fs undertook their final duty as front-line fighters when they, representing F4 Wing at Östersund in northern Sweden, flew over the province of Jämtland in a lap of honour. But it would be another nine years before the final Flygvapen flight took place.

High-speed jet fighters need, particularly in peacetime, shooting practice on targets flying as fast as they do themselves. Therefore it was natural to use the J 29 as a target tug at the end of its flying career. Six J 29s were chosen for the assignment: c/n 29333, 29441, 29507, 29575, 29578 and 26624. The home base was F3 Wing at Malmen. The area which was used for practising was the Baltic Sea between the islands of Öland and Gotland, or over the bay of Hanö.

The Bülow Co. and SAAB constructed tow targets in realistic sizes with representative heat and radar signatures. The target also included recording devices for automatic scoring, winches, cordite wire-cutters, reels and tracer ignition system. There were two types of targets: the dart target was used for attacks with the help of radar, and the wing target when practising pursuit curves. Since both the towing J 29 and the target had the same colour – fluorescent red – the pilots suggested a change of colours for the 29s to yellow. This was not immediately implemented, despite the obvious safety issues involved. During the years the J 29 worked as a target-tug (1967 to 1975), 500 dart targets and 4,000 wing targets were towed.

When target-towing ceased, one example (c/n 29507) was put aside in order to be used at the Swedish air force's 50th anniversary celebrations in 1976. During that year it flew three times at air shows and performed the final landing at the air base of Malmen at 3 p.m. on 29 August 1976, in front of a big crowd and television cameras. After taxiing in and the engine had stopped, the J 29 was crowned with a laurel wreath in commemoration of a long and faithful service. Thank you and good-bye after nearly 30 years.

'Barrels' for Austria

Between the years 1945 and 1955 Austria was, as a defeated part of Hitler's Third Reich, occupied by the four victorious Allied nations, but in 1955 it regained its autonomy. The constitution of the country stated that it should be neutral, free of military alliances and not allow foreign military bases inside its boundaries. Already in 1947, at a Moscow conference, it was decided that a future air force in Austria should not contain more than 90 aircraft, and in the beginning the new Austrian air force got some training aircraft from the occupying countries.

The first military jet aircraft was the de Havilland Vampire T.Mk 55, which was delivered in 1957 but it was considered too advanced to start with so a purchase of 12 French Fouga Magisters was accomplished. There were plans at an early stage to buy a real fighter, the Italian Fiat G 91, but because of political reasons (the South Tyrol incident), the deal was not concluded. Among other offers received were several modern American and Russian types, but the aircraft chosen was the J 29F, partly because it originated from another neutral country but also because the purchase sum was fairly low – 15.7 million Kronor for 30 J29Fs, including training.

Evaluation of the 29, pilot training and education of mechanics was in progress during 1960 and in July 1961 the first 15 overhauled J 29s landed at Vienna's international airport. These aircraft were used to form the Jagd-Bomb-Staffel (1. JaBoStaffel with yellow tail-codes). Soon afterwards, in the autumn of 1961, Austria wished to increase its J 29 inventory and to establish a second JaBoStaffel. A request for an additional 15 J 29Fs was forwarded to the Swedish authorities, but this was initially rejected as deliveries of the new J35 Draken to Flygvapnet were delayed. Fifteen J 29As were offered on loan until January 1963, by which time the J 29Fs could be delivered. In the event, the loan offer was not taken up, and the delivery of the second batch of J 29Fs started in January 1963, ending in June 1964.

Before delivery the second batch of aircraft was overhauled before entering a modification process. At the Swedish air service company Svenska Flygverkstäderna in Malmö both the Hispano-Suiza cannon on the port side of the fuselage were removed and replaced by a Swedish-designed capsule containing three Vinten 70-mm cameras, which could be controlled by the pilot in the air. After a couple of years there were discussions in Austria to buy another eight to 10 J 29s as a replacement for the aircraft that had been lost in crashes, but the purchase was never concluded.

The attributes which commended the J 29 for operations from off-base facilities in the bitter Swedish cold also allowed the type to post extraordinarily high in-service rates in the primitive and humid conditions to be found in the Congo.

Congo missions were dangerous for F22, as evidenced by this hit on a canopy (left). Nevertheless, between the major bouts of fighting there was time for less serious flying, such as this spectacular low-level beat-up over UN 'blue hats' on the runway at Kamina (above).

Austrian training began with a group of 15 pilots and 40 mechanics, located at F15 at Söderhamn. Pilot training began with 10-15 hours in a Vampire trainer, combined with theory training about the 29, covering safety equipment, navigation aids, and the controls and regulators of the cockpit. Before taking off for the first time, taxiing and rapid accelerations were practised. Forty hours in the J 29 followed. No tactical exercises were carried out and the only shooting practice was with cannon against ground targets. German was spoken in all instructions, except in radio communications, where English was used. After the 40 hours in Sweden, the pupil could continue his flight training in Austria in the Link trainer, which the Austrians had constructed together with the English company Redifon.

Austrian operations

Schwechat, the international airport of Vienna, came to be the home for five of the newly delivered J 29s for nearly a year, while they waited for their new base at Linz to be made ready in May 1962. From the beginning it was intended that they be based in the southern part of Austria, at Klagenfurt, but the tourist industry and environmental activists complained about the noise the J 29s made, which was said to keep out the valuable dollar- and Deutsch Mark-tourists. But in spite of the complaints, 10 of the 29s were based at Klagenfurt for a year. The 15 aircraft in the second batch formed 2. JaBo Staffel (with red codes), with its home base at Graz-Thalerhof.

In Austrian service the principal role of the J 29 was close air support, but neither rockets nor bombs were used, only the cannon. The secondary role was surveillance of Austrian airspace. An Egyptian military transport aircraft, an Antonov An-12, one day passed without permission through Austrian air territory on its route from Belgrade to Dresden. Two J 29s were practising in the vicinity and they were ordered to identify and turn away the intruder. In spite of the fact that international signs were given to persuade the Egyptian aircraft to leave Austrian airspace, it just flew on until it reached the Czech border, its crew apparently not impressed at all by the warlike 'Barrels'.

During its 11 years of service in the Österreichische Luftstreitkräfte the J 29 was much appreciated by the people who worked with it and it was also a popular sight at air shows around Austria. But in 1969 it was time for the first J 29 (Yellow B, ex-Swedish c/n 29447) to retire, but not so far away since it was placed as a gate guardian at its home base at Linz. As the replacement, the SAAB 105, arrived one by one, so *die fliegende Tonne* was withdrawn, and in July 1972 the last J 29 made a round trip to all the military air bases. In Austrian service the J 29 racked up 13,205 hours 8 minutes in 18,301 flights.

Export plans

Both the American F-86 Sabre and the Russian MiG-15 were sold in large numbers to other countries. Their equal competitor, the SAAB J 29, had no such luck, but that was not because of any lack of ambition from SAAB's point of view. On the contrary, SAAB tried very hard to sell the 29 to other nations, not least inspired by the purchase by the Austrians. As early as 1953 SAAB put the 29 up for sale and among the countries that responded was Israel. It wanted to buy 12 J 29s but the negotiations came to an abrupt end in the spring of 1954 for reasons never revealed. Since it has always been Swedish policy not to sell weapons to countries that were involved in war, or close to it, or had an uncertain domestic situation, like civil war for example, many countries were excluded from buying the 29.

Two nations that were interested in manufacturing the 29 under licence were Finland and Yugoslavia. Finland, which as an old part of Sweden had since its independence in 1917 maintained good relations with Sweden, was also in a delicate position, balanced between the east (the Soviet Union) and NATO. Consequently it was necessary either to manufacture its own weapons or to buy them from

Photo-recce in the Congo

These four photos show examples of the imagery gathered by F22's S 29Cs. Above is a high-altitude view of the Kamina air base from where they operated. The other photos show a camouflaged Vampire trainer at Kolwezi (above right), a destroyed bridge (right) and the damage inflicted on Kolwezi airfield by a J 29 strike (below).

F22's first batch of five fighter 'Barrels' acquired locally applied camouflage for their low-level operations. Behind the aircraft above is an unpainted aircraft from the second batch. They had just returned from the Congo. Both schemes are seen flying together at left. The aircraft below is 29906, one of two S 29Cs sent to provide reconnaissance, missions being flown with fighters for escort and navigation lead.

another 'neutral' country. As early as 1952 Finland licensed or purchased a number of types, but since SAAB did not have the capacity to manufacture more aircraft, the only possibility for the Finns to get the 29 was by building it themselves.

Under any circumstances the Finns had to rely on the Swedes when it came to the question of powerplant, since they had the licence from de Havilland to manufacture the Ghost engine, so consequently their approval to let the 29 be equipped with this engine was necessary. The Swedish Ministry for Foreign Affairs had doubts about the deal. It stated in a report: "It would be detrimental [to Sweden] if Russia in an attack on Finland got in its hands aircraft whose fighting qualities could be evaluated in test flying. From a military security point of view the deal raises certain hesitations." All these difficulties put an end to the whole affair.

Also in 1952, Yugoslavia wanted to purchase a licence to manufacture the 29, but the Swedish authorities had the same reservations with regard to Yugoslavia as they had concerning Finland, and there was no purchase. Yugoslavia later bought aircraft from the Soviet Union. Chile was a serious prospective buyer in the mid-1960s. Unlike its neighbours, Chile had not acquired the F-86 Sabre, so great efforts were made to persuade the Chileans to buy the 29. They were also offered the SAAB 105, but neither this aircraft nor the 29 were sold to Chile. Argentina, Mexico and Peru were other nations that for a short while were interested in the J 29 but that interest faded away due to different reasons. Tunisia came so far in the purchase negotiations that four of its pilots were checked out on the J 29, but since no export licence was granted the Tunisian purchase was reduced to a number of SAAB 91D Safir trainers.

'Barrel' crashes

At first glance, 190 crashes with 99 fatalities are frightful figures, particularly since they occurred during peacetime. But the figures must be considered from the correct perspective. Many thought that after the Korean conflict, World War III would break out in Europe and, with that in mind, the Swedish air force felt obliged to practise its operations in a 'warlike' scenario. That meant regular ultra low-level flying (10 to 20 m/30 to 65 ft above ground and/or sea level) at high speed, which undoubtedly led to accidents. It also meant realistic dogfights where squadrons fought each other in a very restricted area.

Nine J 29s are seen on the Kamina ramp (left) towards the end of the Congo adventure. Only four aircraft returned to Sweden, including the two S 29Cs. The remainder had any valuable items removed and were then destroyed (below).

Furthermore, the 29 was a difficult aircraft to handle. It was a wide gap to bridge for a flying pupil to convert from the benign Vampire trainer to the swept-wing 29. The Dutch roll was a problematic feature of the 29 that caused a lot of fatal crashes. The roll was the result of 'unclean' flying i.e. the aircraft is not moving straight forward but yawing to the right or to the left, resulting in a wingtip drop. A normal reaction by an inexperienced pilot was to correct the drop by using the ailerons, but that move on a swept-wing aircraft can be disastrous, particularly on approach when the flaps are out and the speed is low. When the aileron is lowered on the drooped wing, drag increases, compounding the problem. The aircraft then enters a screw-like manoeuvre which in many cases at low level was fatal. SAAB was aware of this weakness and informed the air force about it, but for many reasons the information did not reach all the operational units in a reasonable time. The right measure to correct the yaw-roll disturbance is to raise the flaps and use the rudder to correct the yaw diversions.

Many aircraft also exhibited their own individual peculiarities. For example, S 29 (29962) was well-known as what the Swedish called a 'dodger' due to its behaviour in the air. To correct the swerving, a taller fin was mounted on that particular aircraft. Another problem had to do with the noise. Due to the new afterburner the noise of the J 29F increased enormously, and that caused vibrations and created metal fatigue in the rear part of the 29. The tail could all but break off, throwing the aircraft into a violent bunt manoeuvre. After the problem was understood the rear end was reinforced. The phenomenon of vibrations causing metal fatigue was discovered during investigations into the DH Comet accidents in the 1950s.

Technical defects at the beginning of the career of the 29 also contributed to the unfavourable statistics of crashes. Particularly, the cathode ray display was the cause of death

Above: After replacement in the reconnaissance role by the S 35E Draken, several S 29Cs were reassigned to target facilities duties with fighter units, including F3 and F12. This quartet from F3 carries radar reflector pods under the wings to enhance the aircraft's signature for intercept training.

The last active 'Barrels' were the six J 29F target-tugs used by F3 at Malmen as successors to S 29Cs used earlier in the target facilities role. Belatedly the aircraft received yellow markings to distinguish them from the red targets. Below is one of the 4,000 wing targets towed by the J 29Fs up to their retirement in 1975.

This aircraft was the first J 29F for Austria, seen on air test prior to delivery. The 'Barrels' were thoroughly overhauled in Sweden before entering Austrian service in mid-1961, initially at Vienna-Schwechat. The last one was retired in 1972.

for many pilots. Sometimes it 'capsized' and locked itself. The worst accident of this sort occurred in 1955, when four 29s plunged into Lake Glottern. None of the pilots survived. After that all 29s were equipped with double cathode ray displays.

In 1964 a ban on flying was imposed on the Austrian 29s due to a couple of accidents where the exchange of Swedish altimeters and certain navigation equipment played a role. The airworthiness of the 29 was questioned due to its age, but after praise for the basic aircraft design from the minister of defence, different military personnel and SAAB, the ban was lifted. Considering the 26 years the 29 was in active Swedish service, with 700,000 hours of flying by more than 1,000 pilots, the crash statistics are not so conspicuous. There are contemporary aircraft around the world that exhibited worse figures.

Gul Rudolf – a surviving 'Barrel'

In 1985 the dream of having a J 29 take to the air again excited, among others, the former F10 C in C, Bertil Bjäre. Together with two other enthusiasts he found in July 1991 an example in the collection of the Air Force Museum at Linköping that was worth trying to restore. It had the Royal Swedish Air Force c/n 29670, was delivered as a J 29E to F9 at Göteborg/Säve in April 1955 and was modified to F-standard in 1956. On 25 March 1957 it arrived at F10 at Ängelholm, where it got its present colour/name: *Gul* (yellow) *Rudolf.* In 1964 it was transferred again, at first to F3 at Malmen and then to F16/F20 at Uppsala. (The two wings were co-located, but F16 was the fighting wing while F20 was the Air Force Academy). It performed its last flight before the restoration on 30 July 1968. J 29 c/n 29670 had

Left: Aircraft from the first Austrian batch, operated by 1 JaBo Staffel (with yellow codes), were pure fighter/attack machines, lacking the reconnaissance capability devised for the second batch.

Below: The 'red' squadron was 2 JaBo Staffel at Graz-Thalerhof. This aircraft from the second batch is fitted with the camera capsule in place of the port guns, with windows just forward of the partially deployed airbrake (itself just ahead of the mainwheel door). The cameras could be steered from the cockpit. The process of changing guns for cameras could be performed in as little as 30 minutes.

Die fliegende Tonne ***holds a special place in Austrian air force history as it was the reborn nation's first true combat aircraft. The laurels on the nose recorded flight hours milestones, the nearest aircraft racking up a creditable 700 in Austrian service.***

a total flight time of 1,195 hours, of which 479 hours were since overhaul. It had been preserved in a dry environment since 1970.

In May 1992 *Gul Rudolf* had its wings removed, and when the fuel pipes were loosened it was discovered that the rubber tanks still contained 800-900 litres (176 to 198 Imp gal) of fuel, which meant that they were still tight. *Rudolf* was transported to F10 at Ängelholm again for an inspection before it was moved to premises where the restoration work began. The engine was removed and sent to the Aerotech company at Arboga, which performed and sponsored an overhaul. All detachable cylinders and valves were attended to by the workshop at F10. The Air Force Museum contributed with new brakes, new tyres, new lines for the rudders and directions for the overhaul of the fuselage and the wings. The restoration continued with a labour force of around 15 retired air technicians and pilots. The aim was to have *Rudolf* ready for the first flight on 5 August 1995, which was the date of the 50th anniversary of F10.

So *Rudolf* was. After some delays the engine was delivered and tested in July 1995. Lieutenant Colonel Rolf Grimsby was entrusted with the first test flight, which occurred on 11 July. A SAAB 105 accompanied it for safety reasons and to provide a camera platform. "As it was picked out of the display window at SAAB", was the praise from Colonel Grimsby when he climbed out of the cockpit afterwards. Unfortunately, due to a bicycle accident he was not able to be the pilot at the anniversary show, whose main attraction was, of course, the immaculately restored 'Flying Barrel'.

Lennart Berns and Robin Lindholm

Saab J 29 Tunnan specifications

Span: 11.00 m (36 ft 1 in): **Length**: 10.227 m (33 ft 6.5 in): **Height**: 3.75 m (12 ft 3.5 in)

	J 29A	J 29B	J 29F
Weights			
Empty	4580 kg (10,097 lb)	4640 kg (10,229 lb)	4845 kg (10,681 lb)
Loaded	6880 kg (15,167 lb)	7520 kg (16,578 lb)	7720 kg (17,019 lb)
Max	7530 kg (16,600 lb)	8170 kg (18,011 lb)	8375 kg (18,463 lb)
Fuel capacity			
Fuselage tanks	1430 litres (314.5 Imp gal)	1400 litres (308 Imp gal)	1400 litres (308 Imp gal)
Wing tanks	–	750 litres (165 Imp gal)	750 litres (165 Imp gal)
Drop tanks	900 litres (198 Imp gal)	900 litres (198 Imp gal)	900 litres (198 Imp gal)
Performance			
Max speed	1035 km/h (643 mph)	1035 km/h (643 mph)	1060 km/h (659 mph)
Cruising speed	800 km/h (497 mph)	800 km/h (497 mph)	800 km/h (497 mph)
Climb to 10000 m (32,810 ft)	7.3 minutes	8.5 minutes	5.2 minutes
Service ceiling	13700 m (44,947 ft)	13700 m (44,947 ft)	15500 m (50,853 ft)
Armament	4 x 20 mm cannon and 14 x 105-mm rockets or 4 x heavy rockets	4 x 20 mm cannon and 14 x 105-mm rockets or 4 x heavy rockets	4 x 20-mm cannon and 24 x 75-mm rockets or 8 x 140-mm rockets and 2 x firebombs
Missiles	–	–	2 x AIM-9B

Gul Rudolf, ***the last flying Tunnan, wheels over Ängelholm where its restoration was undertaken by F10 volunteers. F10 has since closed and the J 29 was moved to F14 Halmstad, where its engine was removed for an overhaul by Volvo Aero at Trollhättan. The aircraft was due back in the air again in time for the Flygvapen show to be held in August at F17 Ronneby.***

Operators – Flygvapnet

Kungliga Östgöta Flygflottilj (F3)

Situated to the west of the city of Linköping, the meadows at Malmen have since the 16th century been a 'playground' for Swedish military infantry units, but in 1926 the Swedish Air Force took over and formed a reconnaissance unit there. It remained in this role until 1948, when it was tranformed into a fighter unit and was initially equipped with a Swedish product – the FFVS J 22. This was followed by the Vampire, and in 1953 the J 29A arrived. That same year the J 29B also appeared. The first rebuilt B-version, called the J 29E, landed in 1956 and the E-version with the new afterburner, called the J 29F, came in 1957. The Tunnan was replaced in the late 1960s by the J 35F Draken. The unit was disbanded in 1974 and today houses several old military aircraft. Some of them are on display at the nearby Swedish Air Force Museum. Malmen is also home to FMV:Prov, Sweden's main aircraft test establishment.

Kungliga Jämtlands Flygflottilj (F4)

Outside Östersund lies an island, Frösön, where the wing was established in 1926. In the beginning it was a bomber wing, but from 1947 it has been a fighter wing, first equipped with the P-51 Mustang, in Sweden called the J 26, and later with the Vampire, called the J 28. The unit received its first J 29s in 1956, in the form of J 29Es. The following year the J 29F arrived, and the 'Barrel' era lasted until 1966 when the J 32B Lansen replaced them.

Kungliga Västgöta Flygflottilj (F6)

Situated at Karlsborg, near Lake Vättern, the wing was formed as a ground support wing in 1939 and during World War II it was equipped with, among others, the Douglas/Northrop Div. 8A-1 Helldiver, which was replaced by the twin boom SAAB A 21A. The jet age came in 1954 with the attack-orientated A 29B, which stayed for only three years until the A 32 Lansen arrived. In 1956 the J 29E was received, followed by the J 29F in the following year. The wing was disbanded in 1994, having flown AJ 37s in its later years.

Kungliga Skaraborgs Flygflottilj (F7)

This wing was activated in 1936 as a light bomb wing, initially connected with the F6 Wing at Karlsborg but moving in 1938 to its present location at Såtenäs. The SAAB A 21R was the predecessor to the 'Barrel', which in the shape of the A 29B was only operated for two years (1954-1956), succeeded by the A 32 Lansen.

Kungliga Svea Flygflottilj (F8)

Outside Stockholm lies Barkarby, the home of the first Swedish fighter wing, which began its operations in 1940 with Gloster Gladiators (J 8) and American Seversky Republic EP-1s (J 9). In 1955 the J 29B entered the stage, followed in 1956 by the E-version and in 1957 by the F-version. The 'Barrel' flew with the wing until 1973, but was gradually replaced by the Hawker Hunter. In 1974 the wing was disbanded.

Kungliga Göta Flygflottilj (F9)

This wing was inaugurated in 1940 at Säve, Gothenburg, and after World War II was equipped with the SAAB J 21 and the DH Vampire. In 1952 the J 29A arrived, succeeded in 1956 by the E-version and in 1957 by the F-version. The Hawker Hunter replaced the J 29 in 1962 and the wing was disbanded in 1969.

Kungliga Skånska Flygflottiljen (F10)

At first allocated to Bulltofta, Malmö, and then to Ängelholm-Barkåkra, F10 wing was founded in 1940. After World War II the wing was equipped with the SAAB J 21R. In 1953 the J 29B entered service with the wing. The J 29E came in 1955 and in 1957 the J 29F followed. In the mid-1960s the J 29 gave way to the Hawker Hunter. The wing disbanded at the end of 2003.

In the later years of its career the Tunnan was used widely as a 'faker', acting as a radar target for other fighters. Operating in this role were this S 29C of F3 (above, with radar reflector pods) and J 29F of F4 (below).

Above: Ground crew work on an A 29B of F6 at Karlsborg. This wing, and F7, were the only operators of the attack-dedicated Tunnan.

Above: Airbrakes out, an F7 A 29B slows to hold formation on the camera platform.

Right: F8 personnel service a J 29B in the workshop at Barkarby.

F9 operated the 'Barrel' (this is a J 29A) from Säve, making use of the airfield's tunnel hangars.

From its base at Ängelholm, F10 defended the southern part of Sweden, especially the Malmö region.

F11 at Nyköping was one of the two specialist reconnaissance wings flying the S 29C. This particular aircraft, 29962, was the infamous 'dodger' and had a taller fin fitted to correct its directional problems.

F3 – Malmslätt

F4 – Östersund-Frösön

F6 – Karlsborg

F7 – Såtenäs

F8 – Barkarby

F9 – Göteborg-Säve

F10 – Ängelholm

F11 – Nyköping

F12 – Kalmar

F13 – Norrköping

F15 – Söderhamn

F16 – Uppsala

F20 – Uppsala

F21 – Luleå-Kallax

This S 29C was used for target facilities by fighter wing F12 at Kalmar.

F13 at Norrköping was the traditional first recipient of new fighter types. This is a J 29A.

Kungliga Södermansland Flygflottilj (F11)

This long-range reconnaissance wing was set up in 1941 at Nyköping-Skavsta, and was equipped after World War II with the Spitfire PR.Mk XIX. The jet age began in 1954 with the arrival of the S 29 (serial numbers 901 - 976), at first with the C-version, but from 1955 with the improved E-version. For 11 years (until 1965) the S 29s made good service, until replaced by the S 35 Draken. The unit was disbanded in 1980.

Kungliga Kalmar Flygflottilj (F12)

This wing was established as a light bomb wing at Kalmar in 1941. In 1948 it was rearmed with fighters. It is the only unit which during its whole career has been equipped with only Swedish-built aircraft. In 1952 its J 21s were replaced by the J 29A, which it operated until 1958, when the J 32 Lansen arrived. For some curious reason, the unit had a single J 29F (29347) at its disposal as late as 1963. The wing was disbanded in 1980.

Kungliga Bråvalla Flygflottilj (F13)

In 1943 this unit started its activities as a fighter wing at Bråvalla outside Norrköping. It was the first wing to receive the Vampire in 1946, the J 29A in 1951 and the J 35 Draken in 1959. The wing ceased to exist in 1994.

Kungliga Hålsinge Flygflottilj (F15)

This fighter wing was established at Söderhamn in 1945 and was equipped with the Vampire before the J 29E came in 1956. The F-model of the 'Barrel' arrived in 1957. In 1961 the wing was reorganised as an attack wing and converted to the A32 Lansen. In 1997 the wing was disbanded.

Kungliga Upplands Flygflottilj (F16)

Since being established as a fighter wing in 1943 until disbandment at the end of 2003, this wing was based at Uppsala. After World War II the wing was equipped with the P-51 Mustang, but in 1952 it was time for the J 29A to make its debut. The first F-models arrived in 1959. The wing kept its J 29s until 1968, when the last ones were either scrapped or sent to other units. The J 29's replacement was the J 35 Draken.

Kungliga Flygkadettskolan (F20)

The increase in pilot training needs during WWII created the need for a wing to take care of those who wanted to become an air force officer. Basic flying training took place at Ljungbyhed in southern Sweden (F5 Wing) but the new school for advanced training was allocated in 1944 to Uppsala, from where F16 also operated. The jet age started with the Vampire but in 1952 the J 29A made its entrance, followed by the J 29F in 1959. The wing kept its 'Barrels' until the end of the 1960s, when they were either scrapped, given as a gift to foreign air forces during some of the wing's annual visits abroad, or sent to other Swedish wings, principally F3. The successor was the SAAB SK 60. F 20 closed as an Air Force Academy in 1998.

Kungliga Norrbottens Flygflottilj (F21)

Based at Luleå-Kallax, this unit was from the start in 1941 the home for a few liaison aircraft and an air ambulance, but in 1949 the wing got its first reconnaissance squadron, equipped with the S 26 Mustang and SAAB S 18. In 1954 the S 29C Tunnan arrived, followed in 1955 by the E-model with the improved wing. From 1961 the unit became a 'multi-purpose' wing, consisting of a fighter squadron, a ground-support squadron, and a recce squadron. In the latter half of the 1960s the S 29s were phased out in favour of the S 35 Draken.

F15 at Söderhamn was responsible for the defence of the central part of Sweden.

F16 at Uppsala was close to the capital, Stockholm. This is one of the unit's J 29As.

F20, also at Uppsala, was the air force academy and used the Tunnan (J 29F illustrated) as a trainer.

As well as its primary reconnaissance role, the S 29C was also used by F21 at Luleå for target duties.

INDEX

Page numbers in **bold** refer to illustration captions.

INDEX

Picture acknowledgments

Front cover: Jamie Hunter, Sergey Popsuevich, Lockheed Martin. **4:** Roberto Yañez, David Donald. **5:** Derek Bower, Eurofighter, via Tom Kaminski. **6:** Boeing, Aermacchi. **7:** Shlomo Aloni (three). **8:** Lockheed Martin, via Tom Kaminski, Shlomo Aloni. **9:** Bob Archer, David Donald. **10:** Peter R. Foster, AgustaWestland. **11:** via Tom Kaminski, Richard Collens. **12:** Chris Knott/API (three), Peter R. Foster. **13:** Roberto Yañez, via Tom Kaminski. **14:** Keith Riddle, Mike Crutch. **15:** Richard Collens, David Donald, Peter Liander. **16:** Tom Kaminski (three). **17:** Tom Kaminski, Sikorsky. **18:** Tom Kaminski (two), Rolls-Royce, Boeing. **19:** Tom Kaminski (two), Boeing. **20:** Boeing, Tom Kaminski, Rolls-Royce. **21:** Boeing (two). **22-23:** Robert Hewson. **24:** Sergey Popsuevich, Rostvertol via Alexander Mladenov. **25:** Rostvertol via David Donald, Mikhail Kuznetsov. **26:** Alexander Mladenov (two). **27-28:** Piotr Butowski. **29:** David Donald (two), Piotr Butowski (two), David Willis. **30:** Rostvertol via Alexander Mladenov, Piotr Butowski, Alexander Mladenov. **31:** Piotr Butowski (two), Alexander Mladenov. **32:** David Willis (three), Piotr Butowski. **33:** Daniel J. March (three), Alexander Mladenov, Piotr Butowski. **34:** Alexander Mladenov (two). **35:** David Donald (three), IAI via Alexander Mladenov, Alexander Mladenov. **36:** David Willis (two), Rostvertol via Alexander Mladenov. **37:** Piotr Butowski (four). **38:** BAE Systems (two), Bronco Aviation (two). **39:** Alexander Mladenov, Sergey Popsuevich. **40-45:** Emiel Sloot and Luc Hornstra. **46-47:** Jamie Hunter. **48:** Luigino Caliaro, EADS. **49:** Peter R. March, BAE Systems via Peter R. March. **50:** UK MoD (two), Luigino Caliaro. **51:** via Peter R. March, Dylan Eklund, UK MoD. **52:** BAE Systems (two), David Willis (two). **53:** Jamie Hunter, via Peter R. March. **54:** Jamie Hunter, MBDA, Peter R. March. **55:** BAE Systems via Peter R. March (two). **56:** Jamie Hunter, David Willis, UK MoD via Terry Panopalis. **57:** BAE Systems via Peter R. March, Peter R. March, David Willis. **58-59:** A.B. Ward. **60:** Jamie Hunter, BAE Systems via Peter R. March, A.B. Ward. **61:** BAE Systems via Peter R. March, A.B. Ward. **62:** Jamie Hunter, UK Mod (three). **63:** UK MoD (three). **64:** David Willis, Dylan Eklund (ten). **67:** UK MoD (two), BAE Systems, BAE Systems via Peter R. March, MBDA. **68:** UK MoD, EADS (two), BAE Systems, Daniel J. March (two). **69:** BAE Systems via Peter R. March (two), Jamie Hunter, David Willis (two). **70:** Luigino Caliaro, EADS (three). **71:** EADS, Daniel J. March (two), Emiel Sloot, David Donald. **72:** Peter R. March, Daniel J. March, EADS. **73:** Jim Winchester, Luigino Caliaro (two). **74:** Jim Winchester, Riccardo Niccoli, Daniel J. March. **75:** via Riccardo Niccoli (two), Luigino Caliaro. **76:** Riccardo Niccoli, Luigino Caliaro, via Riccardo Niccoli. **77:** Luigino Caliaro, via Riccardo Niccoli, UK MoD. **78:** Jamie Hunter (two), David Willis (six), Peter R. March, UK MoD. **79:** David Willis (four), Daniel J. March (two), Luigino Caliaro, BAE Systems. **80:** Piotr Butowski. **81:** Piotr Butowski (two), Hugo Mambour. **82-83:** Piotr Butowski. **84:** David Willis, Hugo Mambour. **85:** Piotr Butowski. **86:** Piotr Butowski, Hugo Mambour (two). **87-89:** Piotr Butowski. **90-93:** Chris Knott and Tim Spearman/API. **94:** via Peter R. March. **95:** Peter R. March. **96:** TRH, Bristol. **98:** TRH. **99:** via Peter R. March, TRH (two). **100:** Aerospace (two). **102:** Aerospace. **103:** TRH (two). **104:** TRH. **105:** Aerospace. **106:** Bristol. **107:** Aerospace. **108-115:** Larry Davis Collection. **116:** Lockheed Martin via Terry Panopalis (TP), Bill Yenne. **117:** Lockheed Martin via TP, USAF via Dave Menard (DM) via TP. **118:** Robert F. Dorr (two). **119:** Bill Yenne (two), Convair via TP. **120:** USAF, Convair (two), Aerospace. **121:** USAF, Convair (two). **122:** Convair, USAF. **123:** C. Graham via DM via TP, Lockheed Martin via TP, USAF. **124:** Lockheed Martin via TP, Bill Yenne (two). **125:** TRH, USAF, USAF via TP. **126:** Convair, USAF via David Donald, TRH, Dave Menard via TP. **127:** Robert F. Dorr, Lt Col C. Toynbee via DM via TP. **128:** Lockheed Martin via TP, Terry Panopalis. **129:** Robert F. Dorr (two). **130:** USAF, R. Williams via DM via TP, Skitt via DM via TP. **131:** Conair, USAF, Len Dotson via Bob Gerrard via DM via TP. **132:** Robert F. Dorr, Major D. Mikler via DM via TP, V. Hudder via DM via TP. **134:** C. Hinton via DM via TP, P. Paulsen via DM via TP, Bernard J. Schulte via DM via TP, Lockheed Martin via TP. **135:** Lockheed Martin via TP, Bill Yenne, C. Nelson via DM via TP. **136:** Lockheed Martin via TP, Bill Yenne (two). **137:** Bill Yenne, via TP (two). **138:** Bill Yenne (two). **139:** Aerospace, Convair (two). **140:** Larkins via DM via TP (two), USAF, National Atomic Museum, Terry Panopalis, Lockheed Martin via TP. **145:** Bill Yenne (two), USAF (three), Lockheed Martin via TP (two), Convair (two). **146:** Lockheed Martin via TP, Aerospace (two). **147:** Robert F. Dorr (three). **148:** Convair (two), Robert F. Dorr. **149:** Lockheed Martin via TP, via TP, Bill Yenne, T. Brewer via DM via TP. **150:** Col C.W. King via DM via TP, E.W. Quandt via DM via TP, A.I. Reveley via DM via TP. **151:** USAF, Len Dotson via Bob Gerrard via DM via TP, P. Paulsen via DM via TP, Robert F. Dorr. **152:** Saab, David Donald. **153:** via Swedish Aviation Historical Society (SAHS), Saab. **154:** via SAHS (six). **155:** via SAHS (two), Aerospace, Saab. **156-157:** via SAHS. **158:** Saab, via SAHS (two). **159:** via SAHS (three), Flygvapnet. **160-168:** via SAHS. **169:** Saab, via SAHS (two). **170:** Aerospace, via SAHS (two). **171:** TRH, via SAHS (two). **172-173:** via SAHS.